Springer Geography

The Springer Geography series seeks to publish a broad portfolio of scientific books, aiming at researchers, students, and everyone interested in geographical research.

The series includes peer-reviewed monographs, edited volumes, textbooks, and conference proceedings. It covers the major topics in geography and geographical sciences including, but not limited to; Economic Geography, Landscape and Urban Planning, Urban Geography, Physical Geography and Environmental Geography.

Springer Geography —now indexed in Scopus

More information about this series at http://www.springer.com/series/10180

Gouri Sankar Bhunia • Pravat Kumar Shit

GeoComputation and Public Health

A Spatial Approach

 Springer

Gouri Sankar Bhunia
Randstad India Private Limited
Gurgaon, Haryana, India

Department of Geography
Seacom Skills University
Santiniketan, West Bengal, India

Pravat Kumar Shit
PG Department of Geography
Raja N. L. Khan Women's College
Midnapore, West Bengal, India

ISSN 2194-315X   ISSN 2194-3168   (electronic)
Springer Geography
ISBN 978-3-030-71200-6   ISBN 978-3-030-71198-6   (eBook)
https://doi.org/10.1007/978-3-030-71198-6

This Springer imprint is published by the registered company Springer Nature Switzerland AG
The registered company address is: Gewerbestrasse 11, 6330 Cham, Switzerland

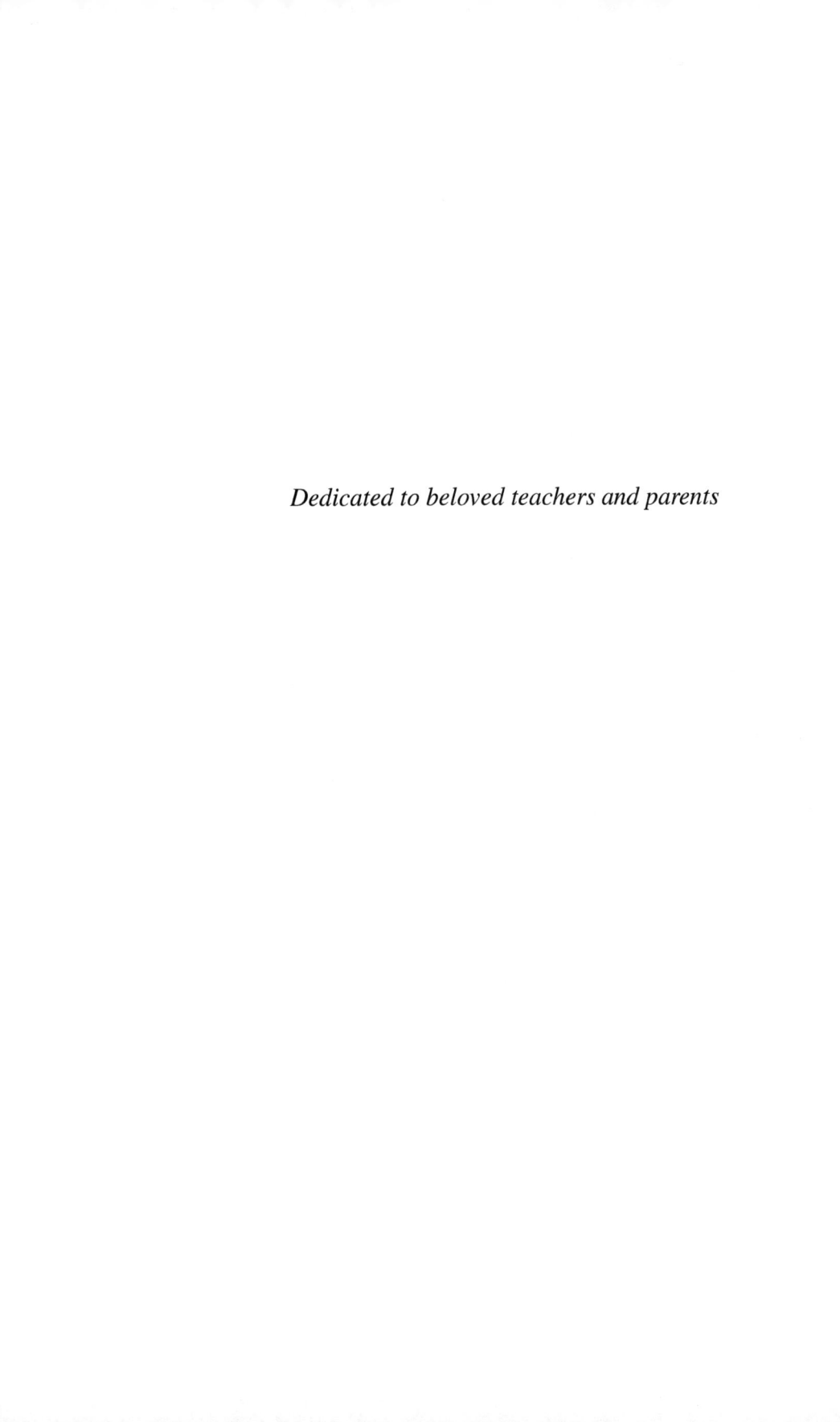

Dedicated to beloved teachers and parents

Foreword

The contemporary world has been witnessing unprecedented transformations both in the natural and human environments which are becoming more volatile, uncertain, complex and ambiguous. Often it is very difficult for planners and policy makers to take appropriate decisions for managing the different untoward situations, for example, the ongoing Covid-19 pandemic. Rapid growth of human population, increasing urbanization especially in the Global South, mass exploitation of natural resources and dumping of wastes have been worsening the livability and livelihood options, particularly of the marginalized people. Given the current trends of various hazards and disasters throughout the world, it will be difficult to achieve the targets of the Sustainable Development Goals within the stipulated time frame.

However, digital technology has given the opportunity to capture voluminous data at different spatio-temporal scales and analyse them almost at real time for taking more reliable decisions to overcome the crises for the benefit of humankind. Using artificial intelligence, machine learning and deep learning are the new vistas of man-machine interface to understand the complexities of the various events occurring every now and then. Reliable new algorithms and apps are constantly being developed for use in day-to-day affairs, making life more comfortable and relaxed.

Geographical sciences are mainly concerned with the interactions among the different components of man and environment to understand the typical patterns developed at various time-space dimensions. Geo-statistics or spatial statistics used to describe such patterns for decision-making and effective interventions for landscape planning and development. Now geo-computation has overtaken the traditional methods of geo-spatial analysis, as ICT-based platforms are now more powerful tools for big data analysis, being faster and at comparatively low costs. Real-time decision support can be easily derived out of the computational models which help human beings overcome the difficulties of handling the enormous amount of data being generated at every point of time.

The present volume, entitled *GeoComputation and Public Health*, authored by Dr. Gouri Sankar Bhunia and Dr. Pravat Kumar Shit is an important addition to the contemporary knowledge on this emerging field of study, particularly when the world is going through the nCovid-19 pandemic. The book has rightly incorporated the background information of the discipline with detailed account of the necessary tools and techniques along with their practical applications with suitable case studies. I congratulate the authors for this unique contribution, and I hope that the volume will be well accepted by its serious readers, intended to be the research scholars, faculty members of higher education institutions and scientists of different organizations across countries.

Professor of Geography &, Director, UGC-HRDC,   [L. N. Satpati]
University of Calcutta, India

Preface

GeoComputation is concerned with new-fangled computational practice along with artificial intelligence (AI) and computational intelligence (CI), algorithms and architypes that are reliant upon and can yield benefit of high-performance computing. It is an innovative computational-intensive archetype that progressively demonstrates its prospective to drastically modify up-to-date research exercise in public health analysis. The concept arose in the mid-1990s to reflect an emerging intersection of research between advance computational methods and geographical analysis and modelling. The acknowledgement of the geographical aspect in social science research occasionally vintages various non-deterministic problems and is motivating increasing curiosity in spatial investigation from outside the geographical arts. The diseases are geographically and temporally limited by variations in environmental variables such as temperature, humidity, rainfall, vegetation and land use pattern. Due to the absence of any suitable 'epidemic prediction tool', it is very difficult to forewarn epidemic outbreaks or to make any strategic plan and combat this menace. Hence, the tools of GeoComputation are typically concerned with customized software uses established to contrivance multifaceted techniques and often for addressing specific complications. Customary GIS tools or the open source GIS software have extended their presentations to operate both core functionality and inflate upon these utilities through software extensions. This book includes a practical coverage of the use of open source software solution in public health problems, particularly in vector-borne disease control programme through GeoComputation.

GeoComputation and Public Health is fundamentally a multi-product task, deals an overview, approaches and case studies to exemplify numerous methods and solicitations in addressing vectors borne diseases (e.g., visceral leishmaniasis, malaria, filaria). This book will also include a practical coverage of the use of spatial analysis techniques in vector-borne disease through open source software solutions. In this book, environmental factors (relief characters, climatology, ecology, vegetation, water bodies) and socio-economic issues (housing type and pattern, education level, economic status, income level, domestics' animals, census data) will also be investigated at micro and local level (village or cadastral) in addressing the various vector-borne diseases. The present book will also generate a framework for interdisciplinary discussion, latest innovations and discoveries on public health. One part of the book highlights the doles and precincts of new computational practices. Another part of the book contains geo-simulation; agent-based modelling; spatio-temporal modelling; geospatial data mining; various GeoComputational applications; accuracy and uncertainty of GeoComputational models; applications in environmental, ecological and biological modelling; and analysis in public health research. Therefore, this book aims to provide examples of GeoComputation by considering practical presentations and subjects that may ascend through use in public health programme.

This book advances the scientific understanding, development and application of geospatial technologies related to public health. Geostatistics and geospatial techniques for public health science assemble the most up-to-date techniques in GIS and geostatistics as they relate to public health, one of the most important human resources. Therefore, this book will help readers to find the recent advancement of the geospatial techniques and its application in public health and sustainable management.

Midnapore, West Bengal, India   Gouri Sankar Bhunia
                                                                        Pravat Kumar Shit

Acknowledgements

We are obliged to a group of experts for their kind support and valuable time to evaluate the chapters succumbed for insertion in this book. We are very much thankful to our respected teachers: Prof. L.N. Satpati, Prof. Malay Mukhopadhya, Prof. Sunando Bandyopadhyay, Prof. Ashis Kumar Paul, Prof. Ramkrishna Maiti, Prof. Soumendu Chatterjee, Prof. Nilanjana Das Chatterjee, Prof. Dilip Kr. Pal, Dr. Nandini Chatterjee, Prof. Sanat Kumar Guchait, Prof. Narayan Chandra Jana, Dr. Jatishankar Bandopadhay and Dr. Ratan Kumar Samanta for sharing their experiences, useful suggestions, continuous encouragement and immense support throughout the work.

We would like to thank the anonymous reviewers, who acted as independent referees. Their input was consistently constructive and has substantially improved the quality of the final product.

We would also like to thank Ranita and Debjani, whose love, encouragement and support kept us motivated up to the final stage of the book. Finally, the book has been several years in the making and we therefore want to thank parents, family and friends for their continuous support.

Dr. Pravat Kumar Shit would like to acknowledge the Department of Geography, Raja N.L. Khan Women's College (Autonomous), for providing logistical support and infrastructure facilities.

This work would not have been possible without constant inspiration from students, lessons from teachers, enthusiasm from colleagues and collaborators, and support from families.

Midnapore, West Bengal, India
Gouri Sankar Bhunia
Pravat Kumar Shit

Disclaimer

The authors of individual chapters are solely responsible for ideas, views, data, figures and geographical boundaries presented in their respective chapters in this book, and these have not been endorsed, in any form, by the publisher, the editors and the authors of forewords, preambles or other chapters.

Contents

About the Authors

Gouri Sankar Bhunia received his Ph.D. from the University of Calcutta, India. His Ph.D. dissertation focussed on environmental control of infectious disease using geospatial technology. His research interests include environmental modelling, risk assessment, natural resources mapping and modelling, data mining and information retrieval using geospatial technology. Dr. Bhunia has published more than 80 peer-reviewed articles in various leading journals and 5 books with Springer.

Pravat Kumar Shit is an assistant professor at the PG Department of Geography, Raja N. L. Khan Women's College (Autonomous), West Bengal, India. He received his M.Sc. and Ph.D. degrees in geography from Vidyasagar University and PG Diploma in remote sensing and GIS from Sambalpur University. His research interests include applied geomorphology, soil erosion, groundwater, forest resources, wetland ecosystem, environmental contaminants and pollution, and natural resources mapping and modelling. He has published eight books (five books with Springer) and more than 60 papers in peer-reviewed journals. He is currently the editor of the GIScience and Geo-environmental Modelling (GGM) Book Series, Springer Nature.

Abbreviation

ABM	Agent Based Model
ADS	Asian Dust Storms
AES	Acute Encephalitis Syndromes
API	Annual Parasite Index
ALI	Advanced Land Imager
ALOS	Advanced Land Observation Satellite
ANN	Artificial Neural Network
ANN	Average Nearest Neighbour
AML	ARC Macro-Language
AOI	Area of Interest
AI	Artificial Intelligence
ASTER	Advanced Spaceborne Thermal Emission and Reflection Radiometer
AVHRR	Advanced Very High-Resolution Radiometer
BI	Blend-then-Index
BoP	Bag of Pattern
CAR	Conditional Autoregressive
CC	Cloud Computing
CCA	Canonical-Correlation
CDC	Centers for Disease Control
CHRIS	Compact High-Resolution Imaging Spectrometer
CI	Computational Intelligence
CAM	Complementary or Alternative Medicines
CNN	Convolutional Neural Networks
CPC	Climate Prediction Center
CRS	Coordinate Reference System
CSR	Complete Spatial Randomness
CSV	Comma-Separate Values
CUSUM	Cumulative Sum
DBSCAN	Density-Based Spatial Clustering of Applications with Noise
DEM	Digital Elevation Models

DLG	Digital Line Graph
DMAP	Disease Mapping Analysis Programme
DTED	Digital Terrain Elevation Data
DWSI	Disease Water Stress Index
HER	Electronic Health Records
HPS	Hantavirus Pulmonary Syndrome
ETM	Enhanced Thematic Mapper
EO	Earth Observation
EOS	Earth Observation System
ESTAT	Exploratory Spatio-Temporal Analysis Toolkit
EXIF	Exchangeable Image File Formats
EVI	Enhanced Vegetation Index
FPM	Frequent Pattern Mining
FKNN	Fuzzy K-Nearest Neighbour Classifier
FTHSI	Fourier Transform Hyperspectral Imager
GBF-DIME	Geographic Base File-Dual Independent Mask Encoding
GC	GeoComputation
GDF	Geographic Data Files
GDEM	Global Digital Elevation Model
GIF	Graphic Interchange Format
GeoJSON	Geographic JavaScript Object Notation
GeoPKDD	Geographic Privacy-Aware-Knowledge Discovery and Delivery
GMA	Geographic Machine Analysis
GML	Geography Markup Language
GIS	Geographical Information System
GNSS	Global Navigational Satellite System
GPCP	Global Precipitation Climatology Project
GPU	Graphics Processing Unit
GWR	Geographically Weighted Models
GKDA	Gaussian Kernel Density Analysis
GPS	Global Positioning System
HIS	Hyperspectral Imaging Instrument
HTML	Hypertext Markup Language
HIV	Human Immunodeficiency Virus
HPS	Hantavirus Pulmonary Syndrome
IB	Index-then-Blend
IP	Internet Protocol
InSAR	Interferometric Synthetic Aperture Radar
IMD	Indian Meteorological Department
IoT	Internet of Things
JE	Japanese Encephalitis
JERS	Japanese Earth Management Satellite
JEV	Japanese Encephalitis Virus
JPEG	Joint Photograph Experts Group
JSON	JavaScript Object Notation

KDD	Knowledge Discovery in Databases
KML	Keyhole Markup Language
LiDAR	Light Detection and Ranging
LULC	Land Use\Land Cover
LISA	Local Indicators of Spatial Association
LISS	Linear Imaging Self Scanning
LST	Land-Surface Temperature
MAUP	Modifiable Areal Unit Problem
MBD	Mosquito-Borne Diseases
MDCOP	Mixed Drove Spatio-temporal Co-occurrence Patterns
MEDMI	Medical and Environmental Data Mashup Infrastructure
MODIS	Moderate Resolution Imaging Spectroradiometer
MOPITT	Measurements of Pollution in the Troposphere
mNDWI	Modified Normalized Differential Water Index
NDVI	Normalized Vegetation Difference Index
NDWI	Normalized Water Difference Index
NGA	National Geospatial-Intelligence Agency
NOAA	National Oceanic and Atmospheric Administration
NMHS	National Meteorological and Hydrological Service
OBIA	Object-Based Image Analysis
OGC	Open Geospatial Consortium
OSM	Open Street Map
OLI	Operational Land Imager
QBO	Quasi-Biennial Oscillation
RCMC	Restricted and Controlled Monte Carlo
RF	Radio Frequency
SAM	Spectral Angle Mapper
SDE	Standard Deviation of Ellipse
SPA	Statistical Pattern Analyser
SPMA	Sequential Pattern Miners Algorithm
SPOT	System Pour l'Observation de la Terre
SRTM	Shuttle Radar Topographic Mission
ST	Spatio-temporal
STAR	Spatio-temporal Autoregressive Regression
STDM	Spatio-temporal Data Mining
SMM	Spatial Microsimulation Model
SVM	Support Vector Machine
SVTT	Spatial Variations in Temporal Trends
TCW	Tasseled Cap Wetness
TM	Thematic Mapper
TRMM	Tropical Rainfall Calculation Mission
TIFF	Tagged Image File Formats
TIROS-1	Television and Infrared Observational Satellite-1
TIGER	Topologically Integrated Geographic Encoding and Referencing
TIN	Triangulated Irregular Network

TPM	Temporal Mining Pattern
TRMM	Tropical Rainfall Measuring Mission
UCMC	Unrestricted and Controlled Monte Carlo
USGS	United States Geological Survey
VGI	Volunteered Geographic Information
VL	Visceral Leishmaniasis
VRML	Virtual Reality Modelling Language
WCS	Web Coverage Service
WCT	Wireless Communication Technology
WFS	Web Feature Service
WHO	World Health Organization
WMS	Web Map Service
WMTS	Web Map Tile Service
WPS	Web Processing Service
WWW	World Wide Web
WSN	Wireless Sensor Network

Chapter 1
Introduction to GeoComputation

Abstract This chapter includes an introduction to the techniques of GeoComputing and GeoComputation. GeoComputation originally was supposed to be a straightforward overhaul of computational geography and trendy iterators. Its aim is to enrich geography with a toolbox of methods for modelling and analysing a set of highly dynamic, often non-deterministic problems. It's about neither compromising the geography, nor making use of unhelpful or simplistic representations more effective. As such, the term GeoComputation considers the key methods used and some of the most fundamental concepts and issues. Simply, GeoComputation is the use of computer processing to solve geographic problems. It is intended to suggest the adoption of a large-scale computationally intensive science paradigm as a method for all kinds of geographical research. For geographical problems, GeoComputation is this spatial or geographical component which differs from other types of calculations.

Keywords GeoComputation technique · Geographic information system · Problem and issues · Future trends

1.1 Concept of GeoComputation

GeoComputation (GC) is a revolution after the introduction of the geographic information system (GIS). It has been introduced and well-defined several times in the past. The series of declarations were examined by Stan Openshaw, Helen Couclelis, Bill Macmillan, Paul Longley, Mark Gahegan and several geoscientists (Longley et al. 2001). In 1996 the School of Geography at the University of Leeds was the first to host the GC International Conference (Openshaw and Abrahart 1996). This poses a perceptual difficulty when relying on spatial modelling and analysis, whether or not GIS is in place. The aim is to develop geography by using the toolbox of archetypal solutions and to look at a set of highly complex, often non-deterministic issues. GC allows the ongoing expansion of logical statistics to measure the association over space and time. Figure 1.1 illustrates the basic concept of GeoComputation.

G. S. Bhunia, P. K. Shit, *GeoComputation and Public Health*,
Springer Geography, https://doi.org/10.1007/978-3-030-71198-6_1

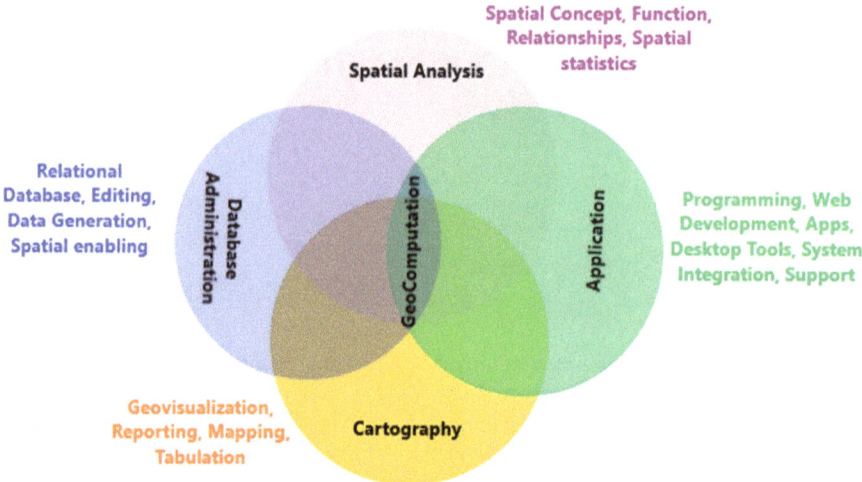

Fig. 1.1 Basic concept of GeoComputation

GeoComputation is intuitive in recognizing a wide range of algorithm-based approaches and techniques from the artificial intelligence (AI) and computational intelligence (CI) sectors. In general, GC is 'the art and science of solving complex spatial problems with computers'. Rees and Turton (1998) 'Geocomputations' are described as the method of applying IT to geographical problems. Couclelis (1998) states that 'geocomputation only applies to the field of spatial computing techniques'.

GC can be useful to solve the wide range of problems based on the computer science paradigm in geography and earth science context. According to GC 98 'spatial computation refers to the conjunction of punitive measures for informatics, geography, geomatics, computer science, mathematics and statistics. There are three aspects which make GC special:

- The information on geography or space.
- Finding new or higher results on the computer route for existing problems.
- Based on relief as a supernumerary of information lacking or supplementary, vast amounts of calculation. Computer experiments are used as a practical abstract method for learning how multifaceted systems operate through their subtleties and behaviour analysis and simulation.

GeoComputational research will embrace environmental data, modern computer technologies and high-performance computing hardware. There is an ample evidence to validate the observation from other computational sciences in geographical studies. Possibly this range of content and instantaneous recognition of submission of application will be its eventual undoing. There is a real challenge to discover how best to use the increasing and existing data treasures via supercomputer hardware with teraflop speeds and terabyte memories. The significant absent constituent is the computational paraphernalia that are required to feat the new prospects that now occur for undertaking original and existing snags in the context of numerous subjects and diverse problematic aspects.

Hence, GC is not a great clandestine. It is basically a computationally concentrated science-built archetype used to investigate an extensive series of physical and human geographical structures. It searches to expand the innovative opportunities for modelling, replication and examination of human and physical schemes through high-performance computing. According to Openshaw (2000), GC may be summarized as

- Not another name for GIS
- Not the nature of quantity
- Not excessively inductive
- Principle not to be evaded
- Profusion of philosophy
- Not the set of tools in a grab bag

1.2   GeoComputation vs Geography

Long before the invention of the computer, geography has played a crucial role in elucidating and informing mankind's relationship with the natural biosphere. Not only is GeoComputation (GC) concerned with the spatial use of computers. The aim is to use a large-scale computerized systemic architectural design as an instrument for all customs of geographic research. It is concerned with the geographical data and information derived through mathematical and statistical modelling. The quantitative revolution in geography took place in the late 1960s and early 1970s. Geographers started to look for more details and create more complex models. The quantitative revolution was first and foremost the product of problems with geographical analysis. In geographical and associated circles, there has been a new gush of interest in GC. However, geo-aspects have been either totally misplaced or underdeveloped and undervalued in quantitative geography research. Geographic Data pursues to involve with modern methods to data capture, transformation, processing, and analysis while enduring true to the censoriously aware traditions of Geography. Several geographers were deliberate to comprehend the inimitable and distinct features of geographical data that inadequate or condensed unacceptable any of these initial quantitative and science-based tools. Alternatively, GC is searching for a better output to current complications through computational method. It also discerning about attempting an innovative programmes of problems for applying science in a geographical context. Macmillan (1998) perceives that GC and quantitative geography share the same scientific philosophy. Today's quantitative geography can be observed as a fountain for numerous bequest statistical and mathematical knowledges that imitate a time of leisure computer applications. However, GC is more than quantitative geography ever aspired to.

Geographers have been carrying out and developing models for the last couple of decades and are well placed to contribute notability to interdisciplinary teaching and research projects in the computer sciences (Armstrong 2000). Earth observation (EO) from satellites and aerial data is decisive for understanding geographic system and to observe the progressions at the Earth's surface. The first EO images were collected in the 1970s, so we have captivating timeseries of almost 50 years accessible to investigate the changing aspects of the Earth. Alternatively, the computer replicas are indispensable to feign natural earth surface characteristics and to run set-ups of climate change and human influence. These models are very beneficial tools to evaluate the consequence of human decision-making and to comprehend the intricacy and responses of systems.

1.2.1   Geoposition

Geoposition represents a geodetic position with latitude and longitude referring to a specified data. It is common using Global Positioning System (GPS) and very useful for estimation of the real-world geographic location of an object based on radio frequency. The coordinates are indicated on a map with an entire address, which typically includes a country, city, town/colony, building name and street address. In addition to GPS, it is possible to identify geoposition by means of Internet Protocol (IP) address, MAC address, radio frequency (RF) system, Exchangeable Image File Formats (EXIF) data and any other wireless positioning system. Geolocation applications may use cell tower information to triangulate the approximate position on the earth when satellite navigation signals are not available.

1.2.2   Geoprojection

Geoprojection adds a supplementary level of notion between world coordinates and geographic coordinates. It acts as a manager for one or more GeoCanvas objects. It will switch from spherical coordinates (degrees) to Cartesian coordinates (in pixels). It can also be specified rotation (i.e., sets the projection's three-axis rotation to the specified angles λ, φ and γ (yaw, pitch and roll)), specified center (i.e., two-element array of longitude and latitude in degrees), specified translation (i.e. specified two-element array [x, y]), defined size (i.e. sets the scale factor of the projection to a given value), defined angle (i.e. sets the width at the specified angle to the specified angle), specified degree (i.e. sets the reach of the projection view to the specified pixels limits) and specified precision (i.e. sets the adaptive resampling threshold for the projection to the pixels value specified).

1.2.3 Geodisplacement

It refers to a geodetic dislocation of length or distance and initial bearing α from a geolocation. In addition to distance, natural barricades to movement can limit displacement. If spatial or temporal dislocation occurs, it is prospective to swing an entity to place and times very analogous to the places and times affected by the preclusion. Such changes entail less exertion, erudition and menace for reprobates than fluctuating to very dissimilar places and times. The temporal dislocation may be easier for crooks than spatial displacement because it entails less determination.

1.2.4 Geodistance

Geodistance computes remoteness in miles between a set of observations and a location. This allows to pursuit locations within a radius using latitude and longitude values based on eloquent models (i.e. query of data from the tables as well as pull out new records into the table). Geodistance can be measured through two aspects:

- *Latitude Movement*: Defining the distance from this point in north/south direction. The north direction is positive and south direction is negative.
- *Longitude Movement*: Defining the distance from this point in east/west direction. The direction to the east is positive, and the direction to the west is negative.

1.2.5 Geonearest

It bargains the shortest curve length between two points along the surface of an earth's mathematical model. It practices a gap and surmount tactic to decrease the number of distances. This method involves excruciating the initial set of positions into progressively smaller geographic regions. Usually, it will certainly find the adjoining neighbours in a linear time.

1.2.6 Geoidentify

It identifies the geographic entities in which the current geolocation is contained. The coordinates can be given by two vectors containing geographic locations. Geoidentify plots the labels if the points are identified. Using this technique the point is identified relative to locator position.

Most of the contemporary human geography is now in a mega-mess and is undefended. It accentuates statistical and mathematical applications that are instantaneously too multifaceted and too deterring. It is believed that the reworking of a GC archetype may support many extensions inside and outside of geography. Nowadays this is a very exciting time for geographers with computational intent, and expectantly GC continues to provide an opportunity to respond in the years to come.

1.3   GeoComputation vs GIS

The GIS revolution has twisted an enormous affluence of geographical evidence in a huge figure of various application parts (Fig. 1.2). The expansion of GIS can be mainly observed as the automation of pre-existing labour-intensive events and recognized machineries that were previously equitably established in research terms. The increase of accessible parallel hardware intensely raises the prospects within a GIS for

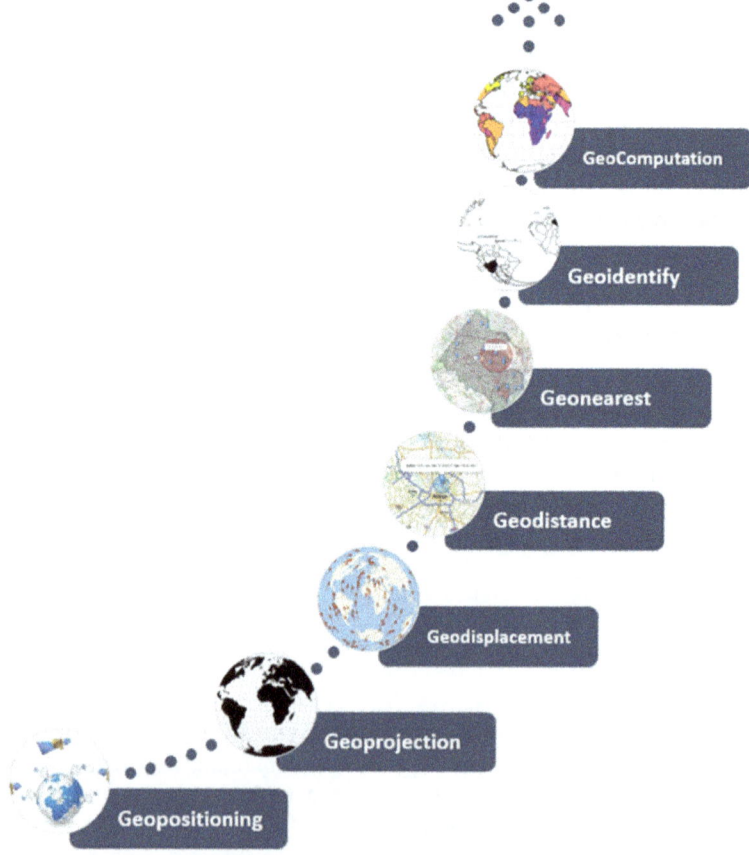

Fig. 1.2 Elements of geography in GeoComputational analysis

large-scale spatial analysis. GIS is merely a database infrastructure, which is useful but lacks research or philosophy outside of the calculation science on which it is based. It contends that large-scale computation can be employed as an archetype for explaining several problems with spatial analysis that is pertinent to GIS. GeoComputation, from a geographical viewpoint, is what comes after GIS in that it tries to use the data abundance provided by GIS and other advances in information technology. In the GISc domain, Goodchild (1991) identified five discrete themes as research plans: spatial analysis and spatial statistics; spatial connection theories; artificial and expert systems; visualization; and social, institutional and economic issues.

GeoComputation (GC) is an emerging field of research that promotes the practice of computational-intensive systems. GC as a follow-on revolution to GIS. GC is a paradigm that is clearly relevant to GIS, but also goes far beyond it. The relationship between GC and GIS is unclear. There are several aspects that are not covered by GC and others that are not included, which have been well described in GIS. According to Longley et al. (1998), GC is seeking to connect technology and the environment, advanced data models, geometry and application, local structure research and practice science. In contrast with GIS and geographical analysis, GC is perhaps an assemblage of practices without a uniting theory. GIS is extensively putative due to its compatibility with the suitable cartographic custom in geography. GIS is a connection but also with computer science or computational methods or statistics. GC has many interactions. If GIS is concerned primarily with digital map information, GC uses it in many distinct areas of application, where the focus is no longer on the original GIS elements.

The 'geodata revolution' drives demand for high-performance computer hardware and scalable software for managing and extracting noise signals, to understand and possibly change our world. GIS is essentially a database that lacks empirical or theoretical information other than measurement theory. Spatial bases permit the storage and generation of manageable subsets from extensive geographical data stores, which make interfaces essential tools in the future to gain knowledge from them. GC aids to analyse data with advance analysis, modelling and visualization capabilities.

1.4 Advantages of GeoComputation

In addition to having a space dimension, GeoComputation has the potential for impacting other space sciences, disciplines and application areas but also beyond universities and research centres. Continuous drastic growth of computing capacity and the ubiquity of user-friendly, portable computers are obviously in line with the electronic revolution. The opportunity to appeal to the interest of the coming generation of investigators is better than stubborn quantitative geography. There are several advantages of GeoComputation as follows:

- High-performance computing technologies represent the transition to the problem-solving study and modelling through a computational paradigm.

- The need to create new ways to handle and to take advantage of the increasingly large number of world information that is addressed spatially.
- The greater availability of existing and readily available artificial intelligence and computer intelligence methods for many geographical and earth sciences.
- Accelerates computer binding activities to allow for more in-depth experiments.
- Boosts result quality using computer-based methods to reduce the number of assumptions and eliminate shortcuts and simplifications that are no longer important in computational terms.
- Enables the study of larger databases and/or the obtaining of better results through the processing of finer data resolution.
- Creates new approaches and new methods based on computer technology, especially artificial intelligence and machine intelligence, developed by other disciplines.

1.5   Future of GeoComputation

GeoComputation is about using a measurement model to analyse geo-phénomènes of all kinds. GC deals with the use of very large equations to address big technical challenges. Morais (2012) indicated that 80% of data is geo-referenced and that much of it indicates the importance of the handling of big data. Traditionally, the geographic data have been collected through ground survey, remote sensing, photogrammetry, laser scanning, mobile mapping, geo-located sensor, geotagged web content, global navigational satellite system (GNSS), volunteer geographic information and so on. Raper et al. (2007) indicated that GC is a place-based operation, promotes navigation and routes and takes a whole new audience into the geospatial environment. Technology and developments are a significant factor in the modern period. Today's problems are mainly technological and allow us to decide how we communicate with our geography. New markets and applications have been developed for highly mobile computing devices. Pervasive computing has been integrated into regular objects and operations (Sui 2014). For a long time, the Internet has been altered from a *library* model containing HTML text-based content. The software's discrepancy in the desktop computing model has distorted, as server architectures, data format standards and services have become web enabled and accessible. The Geographic tools and Geographic browser interfaces of application programming have been overlapped and the provision of computational resources on demand via the World Wide Web (WWW). Such systems have become very widely distributed, and cloud computing has grown. Users permit to generate and contribute their peculiar information based on crowdsourcing (Crooks et al. 2015).

1.5.1 GeoQuantum Computation

Quantum computers are the nanoscale technologies that the dynamic state and sin space exploit to perform data operations by means of quantum mechanical phénomens such as overlay and interposition. Quantum computing requires a representation on the basis of q-bits or qubits of Boolean logic. A quantum computer has the same number of bits exponentially more complicated than a classical computer, since it takes 2n complex coefficients to describe the state of n qubits. GeoQuantum Computation would be a world in which all objects could compute. This technology helps increase computing power and storage; hence, even in difficult problems, the computation would be extremely fast. Computer technology can reinvent computing itself by linking to nanomachines, such as actuators and motes. Even any molecule may possess memory and network information and complete processes with a millimetre accuracy positioning within microseconds. In addition, a paper map may be a digital registry itself, and nanorobots would examine the human circulatory system from inside, detecting a health concern (Clarke 2014).

1.5.2 Data Science

Data science is a significant constituent of the next 50 years of the geographical analysis community. Consequently, spatial data science advances within geography have an ancient context of interdisciplinary research. Geographic data science uses new analytical techniques in order to solve location problems, as well as methodological innovations that make up some of the critical competitions to create models with geographic data. Since the geography of data science has a lot to gain, mainly with regard to the operational and technical aspects of 'big data'.

1.5.3 Data Science and GeoComputation

Data science has been developed as an interdisciplinary tactic that transforms 'big data'. Such information is obtained via numerous sources involving well-known technologies for earth observation; shrunk and long-lasting mobile phones (Batty 2013); broader sensor networks through the Internet of Things (Wilson 2015); and demonstration of public and private user interaction (Miller 2015). Nowadays, the unique increase in the convenience of the environmental spatio-temporal data is a key driver for the enhanced step of upgrading GeoComputational geography.

1.5.4   Internet of Things (IOT) and GeoComputation

The Internet of Things (IoT) is a substance to connect things, sensors, drives and other intelligent technologies, enabling communication between individuals and objects (Uckelmann et al. 2011). Today, the Internet and WWW are the information vehicles by providing file transfer, email, search engine, etc. Ubiquitous GPS-enabled wireless applications have created a virtual organization of users. The virtualization of the digital Earth will function as a mirror world that allows us to control the actuality and calculation of our reality. The web captures the house of the single, humanoid memory easily with Wikipedia and Google. Such a vision system permits us to regulate how much of our authenticity is genuine and how much is computed. In future, visualization and visual analytics information can be the functions of philosophy or mathematics today.

1.5.5   Develop Project on Practical Interest

The positive reaction between innovation, importance and the private sector's interest in it is well-known. We must identify appropriate niches where GeoComputation techniques can demonstrate a competitive advantage and develop the entrepreneurial skills required to make the private sector aware of it.

1.5.6   Must Develop Data Scientific Standards

Right now, reliability management is left to individual researchers' interests and abilities, many of whom may possess a clear perception of methodology expectations and limitations. The GeoComputation toolbox needs to be organized, recorded and cleaned, the warning labels need to be set up, and user guides need to be provided.

1.5.7   Grab Bag of Problem-Solving Techniques of Varying Degrees of Practical Utility

All computational theory and epistemology are well-defined. The former is a demanding field of mathematics; the latter reflects world views as a complicated process and raises deep and original questions regarding determinism, predictability, the character of thought and the limits of what can be understood.

1.5.8 Develop a Coherent Perspective on Geographical Space

Computational technology is commonly used in a number of fields nowadays, including psycho-computing and bio-computing, which has given birth to new areas, which have been highly original and intellectually exciting. In contrast, How is space changing in terms of formal and interpreted properties of computational approaches? What does GeoComputation say about geographical space?

1.5.9 Must Swap in Commands That Most Perceptively Analogus and Mutually Inspiring Computational Growths of the Information Age

The GeoComputation system instigates society and ends. It concerned the curiosity of geographers who do not normally make the use of data-oriented tools or numerical processes as a purpose of reproduction and evaluation (Pickles 1995). A certain extent of the GIS rebellion is the supremacy of this technology in the cause of thinking and disputing beyond techniques (Goodchild 1993).

In summary, there is a breach in acquaintance between the nonconcrete functioning of these paraphernalia and their fruitful positioning to the multifaceted applications and data sets that are common in GeoComputation.

1.6 Problems for GeoComputation

Some issues need to be addressed, especially in public health, before this new technology is efficiently utilized. Public health information is not regularly collected and processed in time and space without planning. Moreover, the health aspects are not always related to the location where they live. Subsequently, the person is visiting the healthcare centre may not correspond with the time of the incident. In several cases, people avoid stigmatic investigations. Many of them are methodological; advanced tools require sophisticated installation and operation. Although the use of these methods in earth science literature has many recorded examples, their use often involves significant investment in the design, set-up, testing and evaluation of them before results are obtained. Data science platform has to find creative organizational methods to support geography and the geographical research community in a data-driven and digital environment (Ash et al. 2018). In geography, data science concerns two important aspects: (i) What/where are the primary data representatives? and (ii) How deviating is the withdrawal of information within this framework from more extensively putative epistemologies? Moreover, 'Big data' are rarely raw (Boyd and Crawford 2012) and their geographic ontology is mainly ambiguous (Goodchild and Li 2012). Alternatively, Gorman (2013) deliberates the attainment of geographic data congregated through mobile, social and site-specific applications that have befell peripheral to GIS. In this context, software tools were not erected to accomplish such huge volumes of outwardly created data. This causes the development of the fragmented GIS ecosystem which turns into numerous disseminated but associated components that mandate a broader set of skills than may conventionally have been assimilated.

GeoComputing therefore needs to resolve certain big challenges, like

- The integration of 'domain knowledge' into the instruments to improve performance and accuracy.
- Techniques of appropriate data mining and know-how identification geographical operators.
- The variability of geospatial large data that indicates incompatibility is due both to the difference between different data channels and to their non-reliability.
- The veracity of geospatial big data extends irregular sampling rates (both spatial and temporal), entry errors, redundancy, corruption, lack of synchronization or a variety of collection purposes.
- The expansion of robust clustering algorithms to run across a succession of spatio-temporal scales,
- Procurement computability on spatial analysis multifaceted,
- Visualization and computer-generated reality paradigms that sustain a visual tactic to reconnoitering, understanding and collaborating environmental phenomena.

1.7 Conclusion

GeoComputation represents a conscious attempt to move the research agenda back to geographical analysis and modelling, with or without GIS in tow. Its concern is to enrich geography with a toolbox of methods to model and analyse a range of highly complex, often non-deterministic problems. It is about neither compromising the geography, nor enforcing the use of unhelpful or simplistic representations. It is a conscious effort to explore the middle ground from the doubly informed perspective of geography and computer science. It is a true enabling technology for the quantitative geographer and a rich source of computational and representational challenges for the computer scientist. However, GeoComputation is not another name for GIS, not quantitative geography, not extreme intuitionism, not devoid of theory, not lacking philosophy and not a grab bag of tool sets.

The challenge for GeoComputation lies in developing the ideas, the methods, the models and the paradigms able to use the increasing computer speeds to do 'useful', 'worthwhile', 'innovative' and 'new' science in a variety of geo-context. Geographers should know how to program, how to operate the web and how to assist data across the Internet and about the flukes of mathematical procedures. Correspondingly, the formulaic computer science scholar wants to boost his or her head up from a computer canopy and yield in the fruitfulness of the world around him or her. Possibly most of all, GC wants to preserve an eye on the future, to better know how to mount it.

References

Armstrong MP (2000) Geography and computational science. Ann Assoc Am Geogr 90(1):146–156
Ash J, Kitchin R, Leszczynski A (2018) Digital turn, digital geographies? Prog Hum Geogr 42(1):25–43
Batty M (2013) Big data, smart cities and city planning. Dialogue Human Geograph 3(3):274–279
Boyd D, Crawford K (2012) Critical questions for big data: provocations for a cultural, technological, and scholarly phenomenon. Inf Commun Soc 15(5):662–679
Clarke KC (2014, June) GeoComputation in 2061. In: GeoComputation. CRC Press, Boca Raton, pp 448–465
Couclelis H (1998) Geocomputation in context. In: Longley PA et al (eds) Geocomputation: a primer. Wiley, Chichester, pp 17–30
Crooks A, Pfoser D, Jenkins A, Croitoru A, Stefanidis A, Smith D, Karagiorgou S, Efentakis A, Lamprianidis G (2015) Crowdsourcing urban form and function. Int J Geogr Inf Sci 29(5):720–741
Goodchild MF (1991) Geographic information systems. Prog Hum Geogr 15(2):194–200
Goodchild MF (1993) Ten years ahead: Dobson's automated geography in 1993. Prof Geogr 45(4):444–446
Goodchild MF, Li L (2012) Formalizing space and place. In: Beckouche P, Grasland C, Gue´rin-Pace F, Moisseron J-Y (eds) Fonder les Sciences du Territoire. Editions Karthala, Paris, pp 83–94
Gorman SP (2013) The danger of a big data episteme and the need to evolve geographic information systems. Dialogues in Human Geography 3(3):285–291
Longley PA, Brooks S, Macmillan W, McDonnell RA (1998) Geocomputation: a primer. Wiley, Chichester
Longley PA, Goodchild MF, Maguire DJ, Rhind DW (2001) Geographic information systems and science. John Wiley and Sons, Chichester
Macmillan B (1998) Epilogue. In: Longley PA et al (eds) Geocomputation: a primer. Wiley, Chichester, pp 257–264
Miller HJ (2015) Spatio-temporal knowledge discovery. Geocomputation: A Practical Primer. SAGE Publications Ltd, Thousand Oaks, pp 97–109
Morais CD (2012) Where is the Phrase "80% of Data is Geographic" from http://www.gislounge.com/80-percent-data-is-geographic/
Openshaw S, Abrahart R (1996) Geocomputation. Proceeding 1st International Conference on Geocomputation
Openshaw S, Abrahart RJ (2000) GeoComputation (No. BOOK). Taylor & Francis, London
Pickles J (1995) Representations in an electronic age. In: Pickles J (ed) Ground truth: the social implications of geographic information systems. The Guilford Press, New York
Raper J, Gartner G, Karimi H, Rizos C (2007) A critical evaluation of location based services and their potential. J Locat Base Service 1(1):5–45
Rees PH, Turton I (1998) Geocomputation: solving geographical problems with new computing power. Environ Plan A 30(10):1835–1838
Sui D (2014, June) 16 ubiquitous computing, spatial big data and open GeoComputation. In: GeoComputation. CRC Press, Boca Raton, p 377
Uckelmann D, Harrison M, Michahelles F (eds) (2011) Architecting the internet of things. Springer Science & Business Media, Berlin, Heidelberg
Wilson MW (2015) Flashing lights in the quantified self-city-nation. Reg Stud Reg Sci 2(1):39–42

Chapter 2
GeoComputation and Spatial Data Operation

Abstract Spatial operations constitute a vital part of GeoComputing. In several ways spatial operations are distinct from non-spatial operations. This chapter shows how to modify spatial objects in a multitude of ways, based on their location and shape. We proposed a new GeoComputing method to represent the spatial data. The primary motivation of this chapter is to draw the public health community's attention to the new analytical possibilities offered by GeoComputing techniques.

Keywords Data operation · Topology · Spatial joining · Data aggregation · Spatial analysis · GIS data format

2.1 Spatial Data Operation

This chapter explains how spatial objects are constructed according to their geographical position and shape in a multitude of ways. It implies the computational manipulation of spatial objects that deepen our understanding of spatial phenomenon. The spatial operation functions from important components of a spatial data model, and it represents various features from the same real-world operation. Hence, the spatial operations have space-invariant properties (Karimipour et al. 2009). Spatial operation focuses on the issue 'What could have been the genesis of the observed spatial pattern?' It is a probing method to enumerate the pattern generated by observed model.

The integration of data into GeoComputational analysis requires the inclusion in certain data sets of different types of information collected from a variety of sources that require an efficient combination of similar entities with a consistency of information between data sets. A variety of sources, including maps, field surveys, photogrammetry and remote sensing, can be given with space data systems in different ways. In a geographical framework, with an increasing number of health-related data collected, GeoComputational techniques are increasingly prospective for research into health data. At present, spatial data analysis computer-based techniques have become a major field of knowledge.

G. S. Bhunia, P. K. Shit, *GeoComputation and Public Health*, Springer Geography, https://doi.org/10.1007/978-3-030-71198-6_2

2.1.1   Spatial Subsetting

Subsetting means the process of extracting part of an objective function of a spatial object that is used by hiding external data to extract information on a field of interest (FOI). The subsetting is analogous. Intersect is the default operator for spatial subsetting. Spatial joining and aggregation also have non-spatial counterparts. However, it is a form of spatial clipping that involves changes to the geometry columns of the affected features.

When an analyst wants to work with a definite focus area, one can dispose of the redundant spatial information without any loss of core data. It basically subsets one layer of the map using another layer's boundaries (Fig. 2.1). The clip is used to combine two input layers into one output layer, using point-to-polygon/line-to-polygon/polygon-to-polygon overlay. The cookie dough is matched with the basic input layer, and the clip layer is equal to the cookie (e.g. second input layer).

Several research projects involve integrating data from remote sensing images, scanned image and thematic maps; and their extent is larger than the area of interest. The raster subsettings are useful for unifying the spatial extent of input data. This operation reduces object memory exercise and is coupled with computational resources for succeeding investigation.

2.1.2   Topological Relations

The main constituent of geocomputation is the unification and assimilation of various data sets and building them accessible for logical manipulation analysis. According to the **Tobler law in 1970**, 'everything is related to everything else, but near things are more related than distant things' suggesting an overall structure of any geographic system. This legislation includes a network underlying it that is seen as a map in which the 'events' are vertices and the relationships between events are

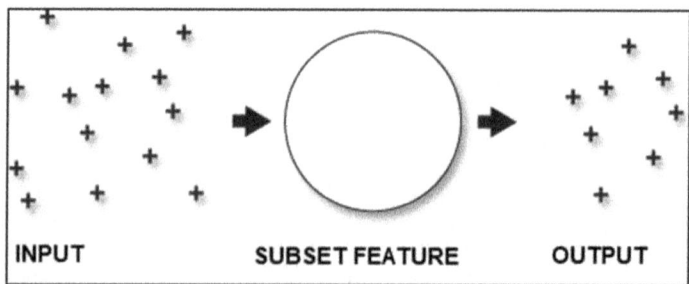

Fig. 2.1 The process of subsetting a smaller area from an original area is portrayed. The first frame depicts the original features (INPUT), followed by similar image with the area to be subset chosen by circle. Finally, the derived output illustrates the new layer (OUTPUT), a subset from original

edges. Topology deals with the spatial and structural properties of the geometric objects.

According to Saaty (1980), all topological features are invariant to any incessant deformation of space. In order to evaluate spatial topology, the relationships of all pairs of objects from one plane to another must be analysed based on the potential intersections of spatially extended internal, external and boundaries. Egenhofer (1993) and Worboys et al. (1994) classified the topological relationship into a nine-intersection scheme based on three basic elements – (i) boundary (e.g. consists of points/lines, separating the interior from the exterior; end points demarcate the edge of a line, and the polygon boundary defines the perimeter), (ii) an interior (e.g. consists of points, lines or areas which are in the object but do not belong to the boundary) and (iii) an exterior (e.g. exterior or complement indicated points, line or areas are not in the object).

Two features X and Y can both be represented by their interior (i), boundary (b) and exterior (e)

$$ObjectX^b \cap ObjectY^b \qquad ObjectX^b \cap ObjectY^i ObjectX^b \cap ObjectY^e$$
$$ObjectX^i \cap ObjectY^b \quad ObjectX^i \cap ObjectY^i ObjectX^i \cap ObjectY^e$$
$$ObjectX^e \cap ObjectY^b ObjectX^e \cap ObjectY^i ObjectX^e \cap ObjectY^e$$

The most important topological relations between objects used in GeoComputation are listed below.

Disjoint – No intersection area between object X and object Y (Fig. 2.2).

Meet – Two geometry objects (X, Y) meet at the boundary, but not the interior. If a geometry object has to intersect another geometry object, the geometry needs to be part of the dimension of the bigger object (Fig. 2.3).

Overlap – The boundaries and interior of both objects intersect (Fig. 2.4).

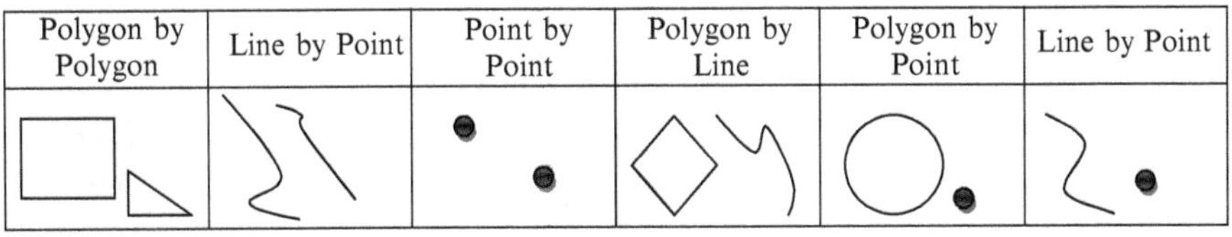

Polygon by Polygon	Line by Point	Point by Point	Polygon by Line	Polygon by Point	Line by Point

Fig. 2.2 Disjoint operation analysis in GIS

Polygon by Polygon	Line by Point	Point by Point	Polygon by Line	Polygon by Point	Line by Point

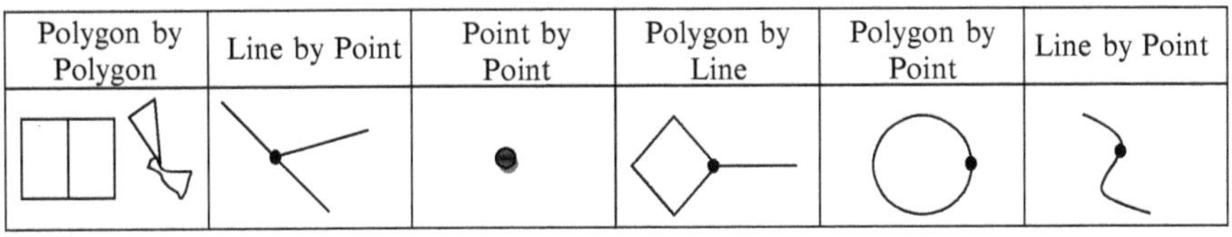

Fig. 2.3 Meet operation analysis in GIS

Contains – The interior and boundary of an object are completely inside the other object. A geometry of the object of higher order e.g., points cannot contain lines or polygons or lines cannot contain polygons (Fig. 2.5).

Inside – It is reverse of 'contain', such as if 'X' is inside 'Y', then 'Y' contains 'X' (Fig. 2.6).

Covers – The interior of an object is completely inside the other object and the boundaries intersect. A geometry object cannot include a geometry object of higher dimension (Fig. 2.7).

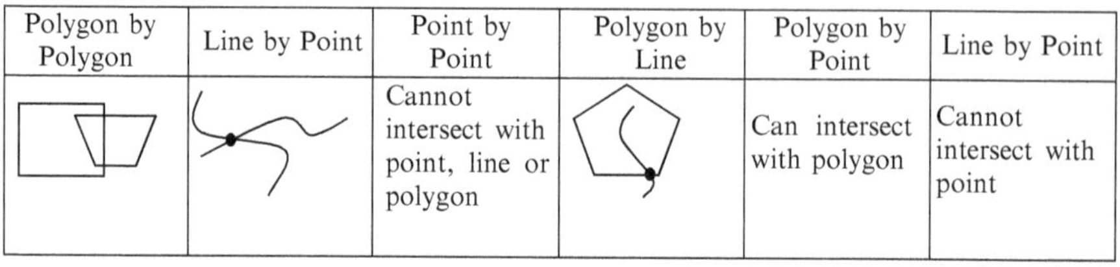

Polygon by Polygon	Line by Point	Point by Point	Polygon by Line	Polygon by Point	Line by Point
		Cannot intersect with point, line or polygon		Can intersect with polygon	Cannot intersect with point

Fig. 2.4 Overlay operation in GIS analysis

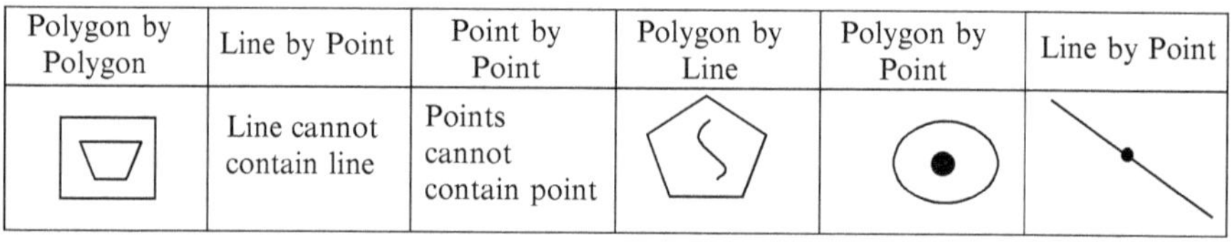

Polygon by Polygon	Line by Point	Point by Point	Polygon by Line	Polygon by Point	Line by Point
	Line cannot contain line	Points cannot contain point			

Fig. 2.5 Contains operation in GIS

Fig. 2.6 Inside operation in GIS

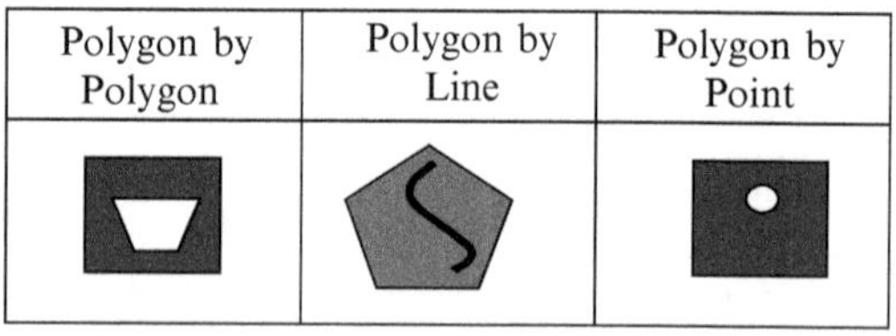

Polygon by Polygon	Polygon by Line	Polygon by Point

Fig. 2.7 Covers operation in GIS

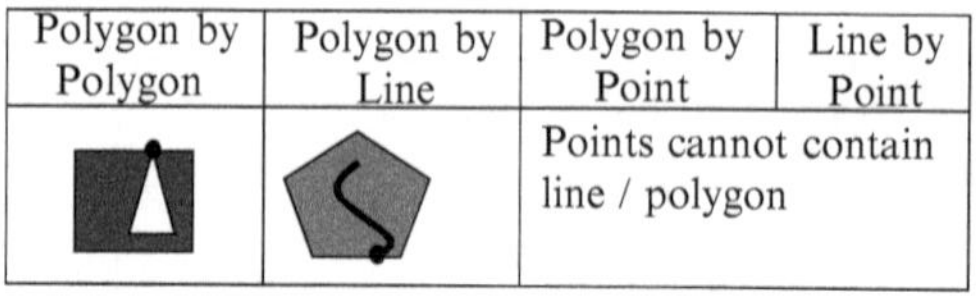

Polygon by Polygon	Polygon by Line	Polygon by Point	Line by Point
			Points cannot contain line / polygon

Equal – The interior and boundary of an object are lying on the boundary of the other object and vice versa. The compared geometry objects must be equal. This happens when a line falls exactly on the boundary of a polygon (Fig. 2.8).

Spatial operation uses the geometry of spatial functions to take spatial data as input and analyse the data. Topological operators are the components of spatial analysis functions of a GIS (Table 2.1). Each system has its own formulation of topological queries. Some of the topological operators used in GeoComputational analysis are discussed below.

2.1.3 Spatial Joining

Spatial linkage/joining is a way to easily add data from a class to another class of function. A common 'key' variable is used in two non-spatial data sets. Sometimes, two geographical data sets do not touch, but are still strongly geographically related to each other. In GeoComputation, the geographic proximity of spatial joining tools combines features, transferring the columns to the target features. Like the attribute data, connection adds a new column from the source object to the target object. The most common spatial join types are:

- Intersect – Two features touch to be nowhere.
- Within a distance – Two characteristics are within a fixed distance.
- Completely within – The connection feature is inside the target function.
- Identical – Both features correspond to one another.
- Closest – The link function is closest to the objective function.

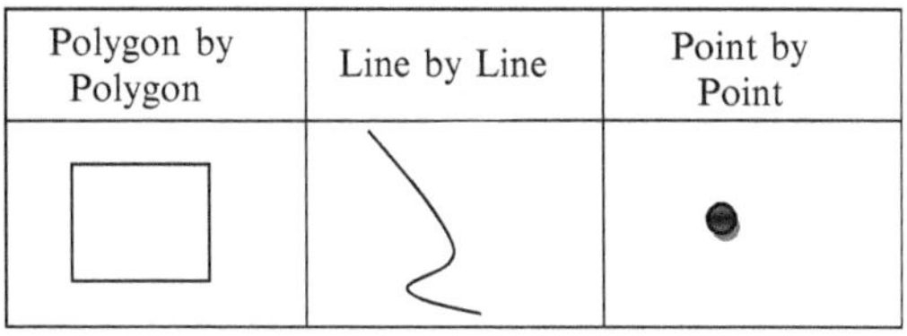

Fig. 2.8 Equal operations in GIS

Polygon by Polygon	Line by Line	Point by Point

Table 2.1 Topological operators for spatial analysis function

Topological relation	Oracle	Geomedia	ArcView
Disjoint	Disjoint	–	Are within a distance of
Meet	Touch	Meet	–
Overlap	Overlap by intersect	Overlap	Intersect
Contains	Contains	Entirely contains	Completely contains
Inside	Covers	Are entirely contained with	Contains the centre of
Covers	Inside	Contain	Have their centre in
Covered by	Covered by	Are contained by	Are completely within
Equal	Equal	Are spatially equal	–

Spatial join uses two types of relationships:

(i) One-to-one relationship – Upon completion of the spatial join, the goal of feature table should have the same number of rows as the final table. When more
than one function connected to the target function, numerical information will
be extracted from the connecting functions to the target function and applied of
the target function.

(ii) One-to-many relationship – The data from the join feature table can no longer
be clustered. The final table allows multiple entries from the joined-
characteristics table that may correspond to a single row in the chosen field in
the target characteristics table (Fig. 2.9).

2.1.4   Spatial Data Aggregation

A spatial data processing may be a means of condensing or sorting the spatial data
at an appropriate or more complex level of precision than at the data collection
level. A few component statistics are seen for the grouping factor in the aggregated
results. In GeoComputation, it not only simplifies accumulation through a diversity
of practices; this can also be used to identify issues with the use of consolidated
results. The integration of spatial information ensures quiet generation of individual
data, upgrading of data, simplifying the maps and distribution of spatial space in
various spatial tools in accordance with the estimated progression endorsed.

GIS software packages employ three simple data aggregation methods.

(i)   Dissolve: It works on the basis that they share the same attribute's value
(Fig. 2.10a).
(ii) Merge: It is a technique that enables the analyst to group artefacts in an editing
process (Fig. 2.10b).
(iii) Spatial join: It works on two layers of objects related to their locations. Spatial
data from one layer, which is often called the target layer, can be aggregated

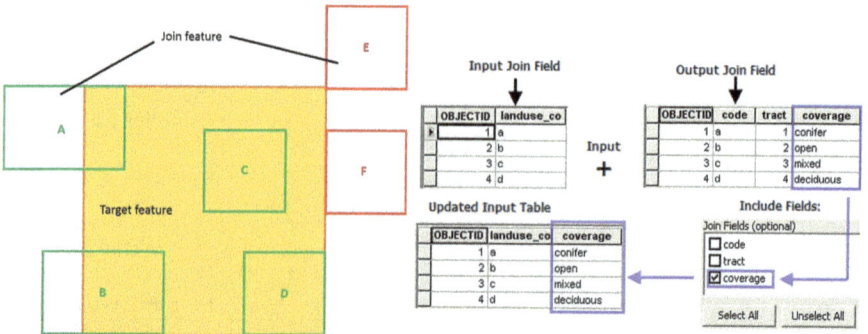

Fig. 2.9 Schematic diagram of spatial join

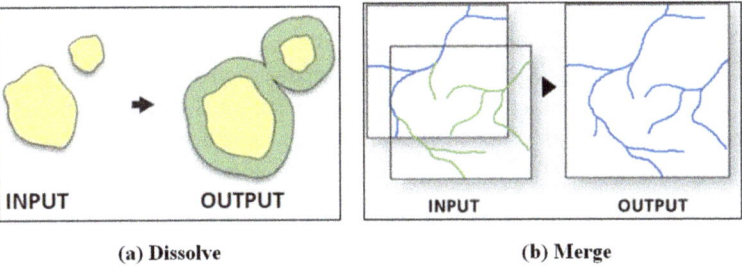

Fig. 2.10 Dissolve and merge function analysis in GIS

and added to objects from another layer. The aggregation is performed based on the distance criterion, both based on the target layer objects.

Aggregating raster data involves a decrease in resolution, e.g. increasing cell size. It is achieved by the cell size of the input raster being multiplied by a cell factor. A cell factor of 4 means that the cell number of the output raster is 4 times larger than the input raster or 16 cells. The value of an output cell is the sum, mean, average, minimum or limit of the output cell. Input cell values are determined by the frequency.

2.1.5   Issues and Problems

The collection of spatial data involves some issues, such as spatial congruence, MAUP, environmental degradation and cross-area aggregations. Space congruence is a complementary entity, because it meets similar boundaries, and is consistent with the target object. Subsequently, the incongruent object does not share the common boundaries with the target. The MAUP takes effect because the zoning method used for gathering aggregated spatial data is arbitrary and is not structured to avoid the mechanism underlying the data. Two factors can be differentiated between MAUP properties: size and zoning effects. The MAUP is systematically related to the ecological loss occurring when an individual is assigned to a statistical relationship identified by quantitative spatial data. Eventually, cross region consolidation refers to combined geographic data transfer from zoning schemes to zones that can be overcome by comparing regions in which data is uniformly dispersed through regions.

2.2   Spatial Query in GeoComputation

Spatial search is a *question-response system* derived from a data set which permits the cross-examination and exhibit of attribute values based on spatial criteria and vice versa. A spatial query language (SQL) was used to meet user demand in relation to GeoComputational data heterogeneity. Using familiar SELECT-FROM-WHERE statements in preference to practical commands like macro language in some GIS packages, users can articulate their enquiries more simply. Such extended SQLs (SESQLs) allow users to investigate the fundamental spatial characteristics of a spatial relation or metric constraints, including spatial information (e.g. point, line, polygon and image), as well as spatial operators (e.g. distancing, distance, overlap and contains). Such macro extensions allow users to tailor their space queries environment to their operating scenes and data display properties with space operators. Various GIS product packages are a challenge for technology professionals to their widespread use. The Geo-SAL survey language, distinct from SQL and QUEL, was developed to support spatial analysis. Two significant attempts to standardize spatial data processing and administration have been made recently: the Multimedia system (SQL3) for and the Open GIS Simple Function Specs for spatial data (OpenGIS SQL) (Jitkajornwanich and Elmasri 2017). OpenGIS SQL is being developed in line with the existing standard SQL92.

Spatial query can be executed in proportion to several approaches: thematic, geometric and topological. Geometry is represented by spatial reference that may be allocated to all substances. The topological possessions are represented by the relation of region, containment, overlapping, etc. In a direct search, there is a compartment derived from the database and the base data are not adapted from this process. In a manipulation, new space allied information elements are generated, which can be employed in supplementary analysis functions. There are many methods in GIS, such as layers and scale-based solutions, which reflect geographical details. The spatial data is seen in a sequence of thematic maps in the layer-based method (Fig. 2.11). It is a very useful method for fast database processing and spatial interpretation, as well as effective data division transfer management.

The size of the globe and eyes can instead vary with the world vision. For example, if a map is seen at a distance from us, several tiny objects in the map become indistinguishable; when a map is seen in a close field of sight, the vast scale of the spatial object becomes invisible. Nevertheless, the map is constructed at a given scale with expansive artefacts in a scale-based approach. In the latter organization, it is observed that the map defined by the scale method can easily be expanded. We

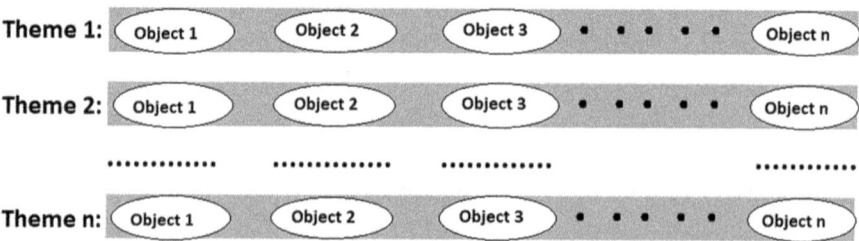

Fig. 2.11 The organization of map for spatial query for layer-based approach in GeoComputation

will expand the map in one block by expanding it by inserting the new space elements in the current block by connecting it to the original map (Fig. 2.12).

2.3 Spatial Overlay

Space operations are an essential aspect of GeoComputing. Overlay displays multiple layers of spatial information in a dynamic mapping environment. An example of an overlay option illustrated between point and polygon features which provide an output having all attributes of the original polygon (Fig. 2.13).

In GeoComputation, it also allows us to query and analyse large amounts of data in relation to each other across space and time. Spatial overlays contain at least two layers of input and produce at least one additional output layer. Clipping, intersect

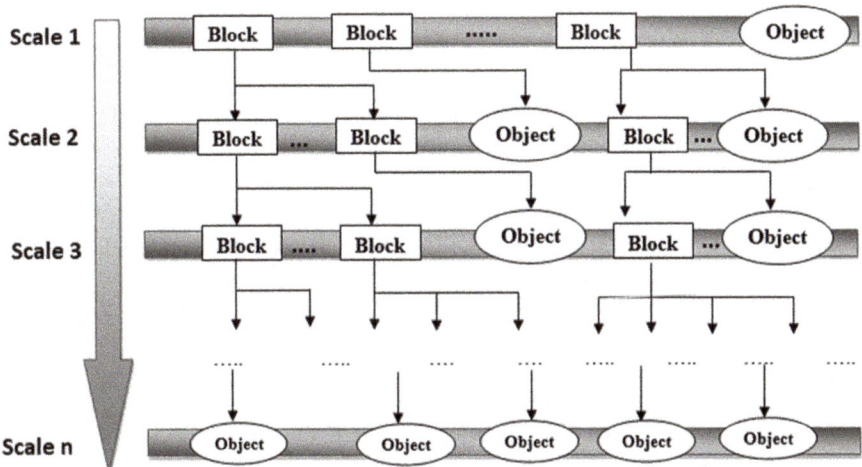

Fig. 2.12 The organization of map for spatial query for scale-based approach in GeoComputation

Fig. 2.13 Spatial overlay in GeoComputational analysis

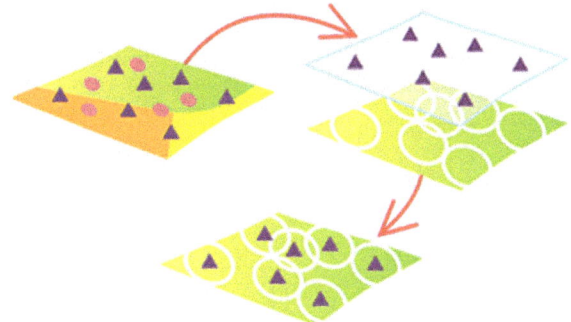

and union form a basic set of overlays. The layers can use the following Boolean algebra operators to combine the conditions of set algebra operations:

- Or (can accomplish two criteria)
- And (must fulfil two variables)
- No (not to fulfil the condition)

2.3.1 Types of Spatial Overlay

Spatial overlay can be classified as feature-based overlay (e.g. overlaying points, lines and polygons) and raster-based overlay.

2.3.1.1 Spatial Overlay Operators

- *Append* – The appending serves to combine several data sets which are the same, but contiguous, thematic data sets.
- *Union* – The two layers of data are merged. New features were created to overlap and overlay components, and a layer with all components from inputs and overlays has been created (Fig. 2.14). The union layer's attribute table comes with the attribute values for non-overlapping characteristics of the respective original layer. The attribute values for both overlapping features are also provided.
- *Identity* – The input-level identity function keeps all functions but retrieves features from the input layer that is superimposed by the input level (Fig. 2.15). The union function is very much like this.
- *Intersection* – This takes all layers into account as inputs and the output characteristics of both layers, which occupy the same field (Fig. 2.16). The properties of the overlapping functions are delegated to both the input and overlay layers of the intersection layer.
- *Update* – Update substitutes overlapping parts of the input layer with update layer features (Fig. 2.17). In the output layer, the attributes in the input layer are present.

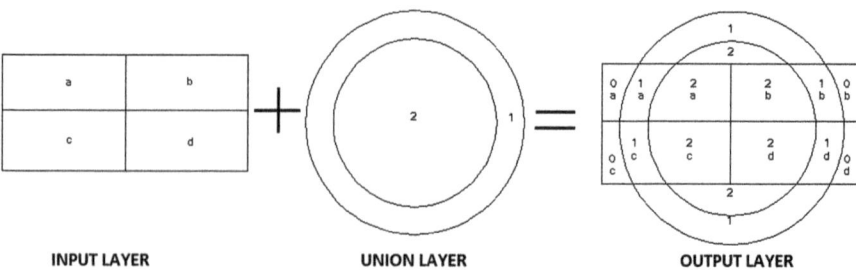

INPUT LAYER   UNION LAYER   OUTPUT LAYER

Fig. 2.14 Union operation

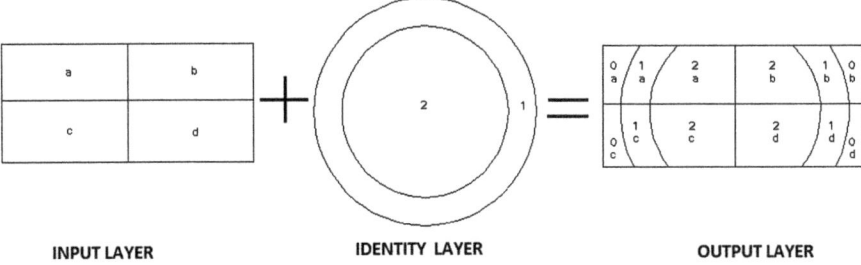

Fig. 2.15 Illustration of identity function

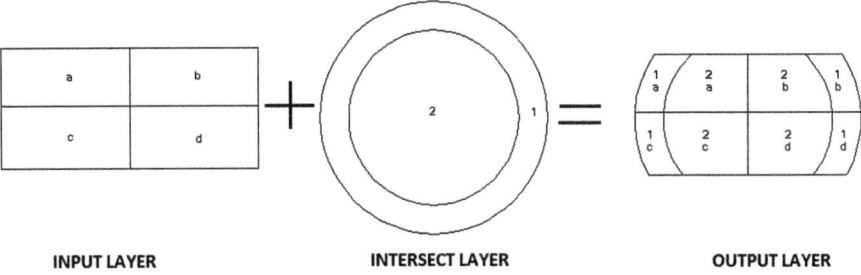

Fig. 2.16 Illustration of intersection function

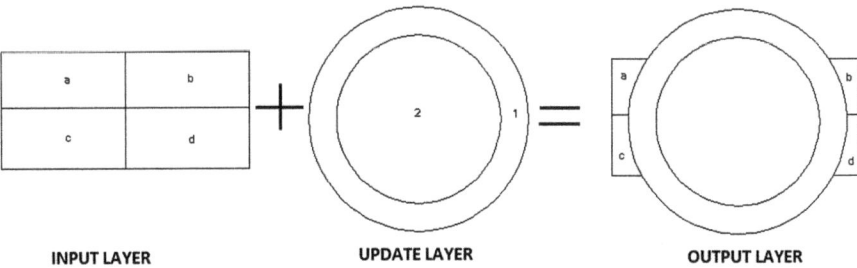

Fig. 2.17 Illustration of union function

- *Dissolve* – It unifies borders based on shared principles of attributes. Alternatively, the neighbours merge if they have the same attribute (Fig. 2.18).
- *Symmetrical Difference* – This operator provides a layer consisting of inputs and overlay layers, but removes the overlaps between the two layers. The symmetrical discrepancy layer attribute table includes all layer input and layer overlay attributes and fields (Fig. 2.19).

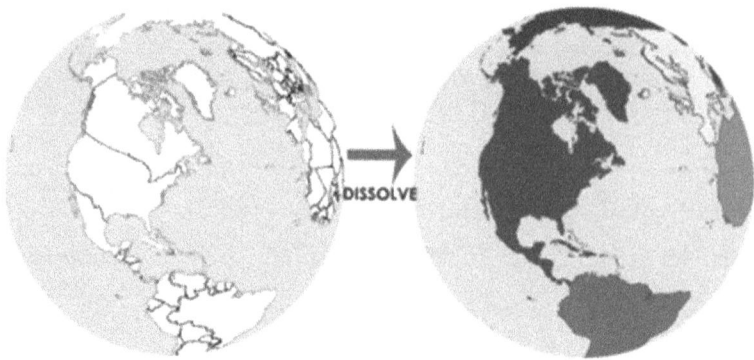

Fig. 2.18 Illustration of dissolve analysis

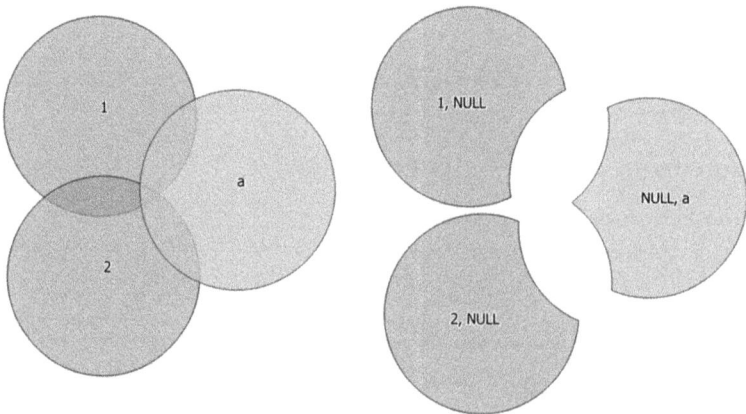

Fig. 2.19 Illustration of symmetrical difference

2.3.1.2   Buffer Operation

Buffers are polygons representing the area within a given distance of a geometric feature (e.g. point, line and polygon) that is used for the proximity analysis. It can be used to specify a buffer radius around each geometric feature. Buffering usually creates two areas:

(i) One area with the selected real-world features in a specified distance.
(ii) The other area that is beyond. The region beyond the specified distance is referred to as the buffer zone.

Buffer zones should often be represented as polygons with certain polygonal features, lines or points (Fig. 2.20).

The buffer distance or buffer size can depend on the numeric values provided for each feature in the vector layer attribute table. In map units, according to the coordinate reference system (CRS) which is used for details, numerical values are to be specified. To illustrate the noise level, a variable buffer can be used with a greater distance on roads and a shorter distance for more quiet roads (Fig. 2.21).

2.4   Voronoi Diagram

The diagram Voronoi is named after the Russian mathematician Georgy Fedoseevich Voronoi, who defined and investigated in 1908 the n-dimensional general case, which was popularized in the interpretation and presentation of given phenomena since the nineteenth century. Voronoi diagrams are also known as polygons of Thiessen, in which Delaunay criterion is mostly used to measure (Mu 2004) (Fig. 2.22). The polygons are convex and generate points closer to their polygon points than others. Such polygons represent a division of the plane through a certain distance to points in regions. The proximal Thiessen polygons are defined as follows:

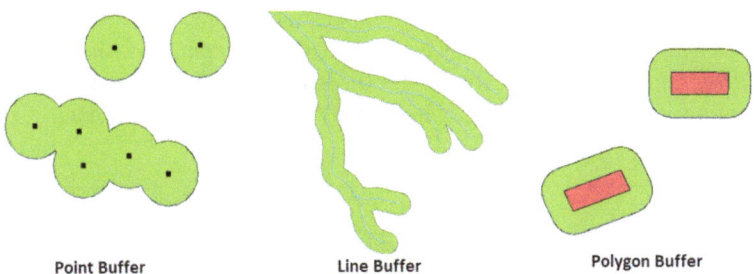

Point Buffer   **Line Buffer**   **Polygon Buffer**

Fig. 2.20 Illustration of buffer in GIS analysis

Fig. 2.21 Illustration of buffer analysis in GIS

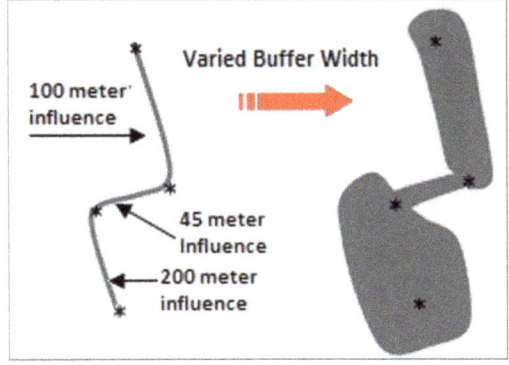

- Both objects are triangulated to the Delaunay criterion in a Triangulated Irregular Network (TIN).
- Any triangle edge is created by the perpendicular bisectors surrounding the edges of the Thiessen polygons. The location at which the cross-sectors determine the orientation of the polygonal vertices of Thiessen Polygon.

Voronoi diagrams were used to predict continuous 3D spatial data, while other techniques, such as spatial autocorrelation and kernel density models, have been typically used. Alternatively, Voronoi helps to discriminate human space. For example, the distribution of operation inside such stations uses rail stations as points to construct a space with different weights around an area and is then used to describe areas in which commuter traffic is present in relation to another neighbouring zone (Fig. 2.23).

2.5   Map Algebra

Map Algebra uses mathematical expressions that include raster data operators and functions. The relational, Boolean, logical, combined and bitwise Map Algebra operators develop new values by using one or more inputs. The concept was first presented by Dr. Dana Tomlin in his book entitled *Geographic Information Systems and Cartographic Modeling* (Tomlin 1990). It is a high-level computational language used for performing cartographic spatial analysis using raster data.

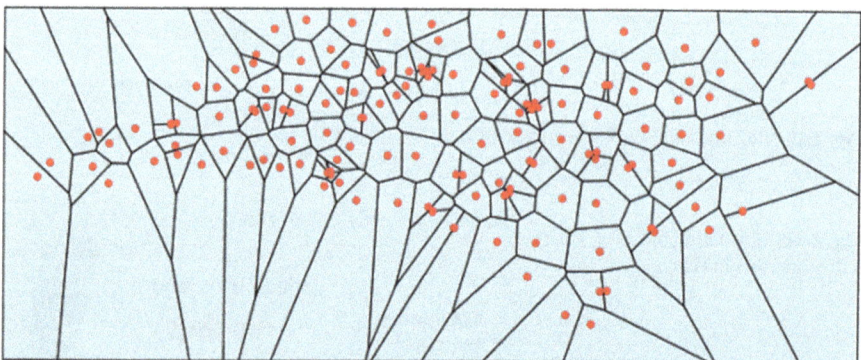

Fig. 2.22 Illustration of Voronoi diagram

Fig. 2.23 Addition
function of GIS analysis

1	4	5		5	1	3		6	5	8
5	3	2	**+**	1	2	1	**=**	6	5	3
2	5	2		1	4	2		3	9	4

This is a basic and efficient algebra that can carry out all tools, operators and spatial analysis functions of the spatial analyst. It does raster processing really faster, as raster data sets store coordinates implicitly (Karimipour et al. 2009). The matrix location, raster resolve and origin of a given cell must be determined in order to obtain the coordinate. Nevertheless, as long as we maintain the same cell location after processing (single-specific correspondence), a cell's geographic position is rarely important. Furthermore, if two or more raster data sets have the same size, scanning and resolution, they can be known as processing matrices (Liu and Mason 2009). Map algebra (or cartographic modelling) uses raster operations into four subclasses as follows.

2.5.1 Local Operations

Both cell-by-cell processes in one or more layers are included in local operations. Another traditional case of local operations is raster algebra. In this respect, two rasters are added, removed and square root. For examples, local operations between two layers can be used:

- Arithmetic operations: This allows addition (+), subtraction (−), multiplication (*), division (/), modulus (MOD) and unary minus (−).
- Statistical operations: For each cell, statistics can be calculated using a statistical feature like minimum, maximum, average and median.
- Relational operations/set algebra: This allows to build logical tests, returning values of true (as the value 1) and false (as the value 0). This type of operator can be used as equal (= =), not equal (^=, <>), less than (<), less than or equal (<=), greater than (>) and greater than or equal (>=) (Fig. 2.23).
- Trigonometric operations (sine, cosine, tangent, arcsine).
- Exponential and logarithmic operations (exponent, logarithm).

- *Boolean algebra* operators are allowed to chain logical test: AND (&), OR (|), XOR (!) and NOT (^).
- *Logical operator:* Every non-zero value for the map algebra is known as being logical (1), with zero being called a logical error (0). The logical operators DIFF (e.g. logical difference), IN (e.g. contained in list) and OVER (e.g. replace) allow to build logical tests on a cell-by-cell basis.
- *Combinational operator:* This combines several input raster attributes. All unique value mixes are found by these operators, and each has a unique ID, returned to the output grid.

2.5.2   *Focal Operation*

The main cell and its neighbours consider focal operations. Typical size of 3-by-3 cells is the kernel, filter or moving window, which are the central cell and its eight surrounding neighbours, but any other form (not necessarily rectangular) as defined by the user is possible. A focusing procedure uses the new value for the central cell; the aggregation function extends to all cells within the specified neighbourhood and then moves to the next central cell (Burrough et al. 2015).

The upper right-hand corner of the moving window will always be the minimum size. In the processing of images, focal roles play a key role. The medium feature for eliminating extremes is used for low-passing or fastening filters (Fig. 2.24). This can substitute the mean by the norm, the commonest value, for categorical results.

2.5.3   *Zonal Operation*

In several raster cells, zonal operations implement the aggregation functions. This will dissolve the cells of one raster using an aggregation feature, according to zones (categories) of another raster. In contrast to the neighbourhood window in case of focal operations, a second raster, usually a categorical raster, defines the zonal filters (or the 'zonal') for zonal operations. The product of a zonal operation is a descriptive list, collected according to zonal GIS world statistics.

Fig. 2.24 Focal operation in GIS

2.5.4 Global Operations

A global operation is a process that is performed on each output cell using all of the cells of the input raster. An example of a global function is the *Euclidean Distance* tool which computes the shortest distance between a pixel and a source location. This can also generate single-value outputs such as the overall pixel mean or standard deviation. Global operations and functions can also generate single-value outputs such as the overall pixel mean or standard deviation.

2.6 Data Formats

Spatial data forms provide the information a computer needs to digitally reconstruct the spatial data. We have grid cells that represent real-world characteristics in the raster world. We have points, lines and multiple paths in the vector world (Table 2.2). The main formats for these files are government mapping agencies such as the USGS or the National Geospatial Intelligence Agency, or GIS development agencies. In GeoComputation, the common file formats used are as follows:

2.7 Web Services and GeoComputation

Recent development of information technology has shaped the future evolution of GeoComputation technology. Diffusion of Web technology increases the dissemination of geographical information on the web. The Open Geospatial Consortium (OGC) has created a number of specifications for web services, namely, Web Feature Service (WFS), Web Map Service (WMS), the Web Map Tile Service (WMTS), the Web Coverage Service (WCS) and even a Web Processing Service (WPS). Subsequently, large-scale web mapping (e.g. OpenStreetMap, Google Maps API) and voluntary web mapping efforts create many opportunities for geographic research and applications. There are several open-source software programs which in the GIS community have gained enormous prominence. They include GIS Server, the GIS Database Management System, the GIS Research Tools (GeoTools), from Web (MapServer). A significant improvement to the GeoComputation infrastructure is being made with the introduction of Google Map Series products (Google Map, API and Google Earth). A dynamic satellite picture at various resolutions plus smooth zooming and panning gives some regional to street-level spatial data (Fig. 2.25). All the above improvements have made web services popular and have turned thousands into Internet mapping fans.

Many powerful support strategies can be used to motivate GeoComputation, such as grid computing; semantic cloud services; efficient, intelligent, secure web portals; and available on a highly convenient basis. Therefore, conventional GeoComputations are used in desktop-to-network delivery environments, with

Table 2.2 Data formats used in GIS

Extension	File type	Description
Vector formats		

Esri Shapefile	.SHP,.DBF,.SHX	The shapefile is BY FAR the most common geospatial file type you'll encounter. All commercial and open sources accept shapefile as a GIS format. It's so ubiquitous that it's become the industry standard
		But you'll need a complete set of three files that are mandatory to make up a shapefile. The three required files are:
		  SHP is the feature geometry
		  SHX is the shape index position
		  DBF is the attribute data
		You can optionally include these files but are not completely necessary.
		  PRJ is the projection system metadata
		  XML is the associated metadata
		  SBN is the spatial index for optimizing queries
		SBX optimizes loading times

Geographic JavaScript Object Notation (GeoJSON)	.GEOJSON.JSON	The GeoJSON format is mostly for web-based mapping. GeoJSON stores coordinates as text in JavaScript Object Notation (JSON) form. This includes vector points, lines and polygons as well as tabular information GeoJSON stores objects within curly braces { } and in general has less markup overhead (compared to GML). GeoJSON has straightforward syntax that you can modify in any text editor Webmaps browsers understand JavaScript so by default GeoJSON is a common web format. But JavaScript only understands binary objects. Fortunately, JavaScript can convert JSON to binary
Geography Markup Language (GML)	.GML	GML allows for the use of geographic coordinates extension of XML. And eXtensible Markup Language (XML) is both human-readable and machine-readable GML stores geographic entities (features) in the form of text. Similar to GeoJSON, GML can be updated in any text editor. Each feature has a list of properties, geometry (points, lines, curves, surfaces and polygons) and spatial reference system There is generally more overhead when comparing GML with GeoJSON. This is because GML results in more data for the same amount of information

(continued)

Table 2.2 (continued)

Extension	File type	Description
Google Keyhole Markup Language (KML/KMZ)	.KML.KMZ	KML stands for Keyhole Markup Language. This GIS format is XML-based and is primarily used for Google Earth. KML was developed by Keyhole Inc. which was later acquired by Google KMZ (KML-Zipped) replaced KML as being the default Google Earth geospatial format because it is a compressed version of the file. KML/KMZ became an international standard of the Open Geospatial Consortium in 2008 The longitude and latitude components (decimal degrees) are as defined by the World Geodetic System of 1984 (WGS84). The vertical component (altitude) is measured in metres from the WGS84 EGM96 Geoid vertical datum
GPS eXchange Format (GPX)	.GPX	GPS Exchange format is an XML schema that describes waypoints, tracks and routes captured from a GPS receiver. Because GPX is an exchange format, you can openly transfer GPS data from one program to another based on its description properties The minimum requirement for GPX is latitude and longitude coordinates. In addition, GPX files optionally store location properties including time, elevation and geoid height as tags
IDRISI Vector	.VCT.VDC	IDRISI vector data files have a VCT extension along with an associated vector documentation file with a VDC extension VCT format is limited to points, lines, polygons, text and photos. Upon the creation of an IDRISI vector file, it automatically creates a documentation file for building metadata Attributes are stored directly in the vector files. But you can optionally use independent data tables and value files

MapInfo TAB	.TAB.DAT.ID.MAP.IND	MapInfo TAB files are a proprietary format for Pitney Bowes MapInfo software. Similar to shapefiles, they require a set of files to represent geographic information and attributes.
		  TAB files are ASCII format that link the associated ID, DAT, MAP and IND files
		  DAT files contain the tabular data associated as a dBase DBF file
		  ID files are index files that link graphical objects to database information
		  MAP files are the map objects that store geographic information
		IND files are index files for the tabular data

(continued)

Table 2.2 (continued)

Extension	File type	Description
OpenStreetMap OSM XML	.OSM	OSM files are the native file for OpenStreetMap which had become the largest crowdsourcing GIS data project in the world. These files are a collection of vector features from crowdsourced contributions from the open community The GIS format OSM is OpenStreetMap's XML-based file format. The more efficient, smaller PBF Format ('Protocolbuffer Binary Format') is an alternative to the XML-based format The data interoperability in QGIS can load native OSM files. The OpenStreetMap plug-in can convert PBF to OSM, which then can be used in QGIS
Digital Line Graph (DLG)	.DLG	Digital Line Graph (DLG) files are vector in nature that were generated on traditional paper topographic maps. For example, this includes township and ranges, contour lines, rivers, lakes, roads, railroads and towns Much of the US Bureau of Census Topologically Integrated Geographic Encoding and Referencing (TIGER) data were generated using the standard DLG format
Geographic Base File-Dual Independent Mask Encoding (GBF-DIME)		The GBF-DIME file format was developed by the US Census Bureau in the late 1960s as one of the first GIS data formats to exist. It was used to store the US road network for major urban areas, which is a key factor in census information GBF-DIME supports choropleth mapping and also assisted in removing error for digitizing features. DIME was a key component to the current TIGER (Topologically Integrated Geographic Encoding and Referencing) system, which was produced by the US Census Bureau
ArcInfo Coverage		ArcInfo Coverages are a set of folders containing points, arcs, polygons or annotation. Tics are geographic control points and help define the extent of the coverage Attributes are stored in the ADF or INFOb tables. Each feature is identified with a unique number. These feature numbers are a way to link attribute data with each spatial feature Coverages were the standard format during the floppy disk era. But over time, this GIS format has become obsolete and mostly unsupported in GIS software

(continued)

Table 2.2 (continued)

Extension	File type	Description
Raster formats		
Tagged Image File Formats	TIFF	TIFF can use run length and other image compression schemes. It is not limited to 256 colours like a GIF
	GEO-TIFF	As part of a header in a TIFF format, it puts Lat/Long at the edges of the pixels
Graphic Interchange Format	GIF	A file format for image files, commonly used on the Internet. It is well-suited for images with sharp edges and relatively few gradations of colour
Joint Photograph Experts Group	JPEG	It uses a variable-resolution compression system offering both partial and full resolution recovery
USGS DEM		The USGS DEM standard is a geospatial file format developed by the US Geological Survey for storing a raster-based digital elevation model. It is an open standard and is used throughout the world
GTOPO30		Developed by US Geological Survey (USGS). It has a 30-arc second resolution (approximately 1 km) and is split into 33 tiles stored in the USGS DEM file format
DTED		DTED is a standard National Geospatial-Intelligence Agency (NGA) product that provides medium-resolution, quantitative data in a digital format for military system applications that require terrain elevation
Other common file formats		
Dual Independent Map Encoding	DIME	It is an encoding scheme developed by the US Bureau of the Census for efficiently storing geographical data. It was first coined by *George Farnsworth* in August 1967. The file format developed for storing the DIME-encoded data was known as Geographic Base Files (GBF). The Census Bureau replaced the data format with Topologically Integrated Geographic Encoding and Referencing (TIGER) in 1990
Geographic Data Files	GDF	GDF provides detailed rules for data capture and representation and an extensive catalogue of standard features, attributes and relationships. GDF is commonly used for data interchange in many industries such as automotive navigation systems, fleet management, dispatch management, road traffic analysis, traffic management and automatic vehicle location
GeoPackage	GPKG	It is an open, non-proprietary, platform-independent and standards-based data format for geographic information system implemented as an SQLite database container. Defined by the Open Geospatial Consortium (OGC) with the backing of the US military and published in 2014

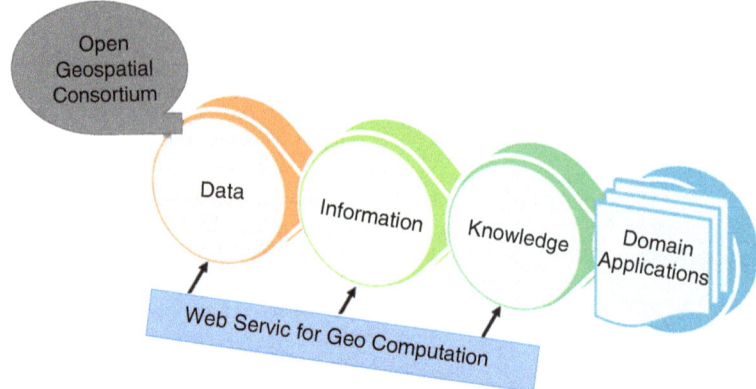

Fig. 2.25 Web service workflow for GeoComputation

major improvements in end-users, applications/research areas, data sources, layout complexity and methods for visualizing, as outlined below:

- End users are usually public and technical, although GeoComputations are traditionally mainly intended for technical uses.
- Technology and study areas shift from independent to interdisciplinary and integrated research systems.
- The visualization process evolves from basic image-based searching into three-dimensional, Internet-based, virtual globe representations that allow digital geospatial data to appear interactively and expressively.
- The data to be analysed today extend the static geographical data, which requires data obtained from globally dispersed sensors that are complex with continuous space and time.
- Complex GeoComputing and detailed simulation have become prevalent in routine geo-space research and decision-making.

References

Burrough PA, McDonnell R, McDonnell RA, Lloyd CD (2015) Principles of geographical information systems. Oxford university press, New York

Jitkajornwanich K, Elmasri R (2017, March) Proving completeness of OpenGIS SQL spatial relationships and operations. In: 2017 International Electrical Engineering Congress (iEECON). IEEE, pp 1–4

Karimipour F, Delava M, Frank A (2009) An algebraic approach to extend spatial operations to moving objects. World Appl Sci J 6(10):1377–1383

Liu JG, Mason PJ (2009) Essential image processing and GIS for remote sensing. Wiley-Blackwell, Chichester

Mu L (2004) Polygon characterization with the multiplicatively weighted Voronoi diagram. Prof Geogr 56(2):223–239

Tomlin CD (1990) Geographic information systems and cartographic modelling (No. 910.011 T659g). Prentice-Hall, Englewood Cliffs

Chapter 3
E-Research and GeoComputation in Public Health

Abstract Computer technology advancements are further shaping research, practice and related training in science and engineering. The chapter is seen as a key e-science and e-research study. With the rapid development of the Internet of Things (IoT) and wearable technology, huge numbers of health-related data are generated every moment. Recent technologies like grid and cloud computing have resulted in even more than an increase in treatment efficiency. The features of this new e-research approach are discussed using the related examples in this chapter. It is argued that in a geospatial sense, specific approaches can be beneficial. The last portion of the chapter focused on various ways of growing availability of powerful e-research resources for GeoComputing and expanding.

Keywords E-research · Internet of Things · Cloud computing · Fog computing · Spatial model · Cartographic confounding

3.1 Introduction

The authors have self-archiving policies that allow authors to post the final version of the publications in their papers on a personal website or on libraries and academic repositories, with many open-access research documents available in Elsevier, Springer, Wiley, Sage, Blackwell and Taylor and Francis. Open-access publishing has a far larger reading potential than the subscription-fee papers, both for academics and the public. E-science has done plenty with modern techniques for the developments of technology, such as the measurement of the grid and the distributed testing partnership. A significant amount of health-related data is generated using smart devices and environmental information from wearable devices and stationary sensors. The data are valuable resources for health, research and business applications. The correct allocation of these health data can contribute to progress in the public health system for all the relevant stakeholders. Up to the 1950s, scholars usually found crucial figures for non-infectious disease cross-sectional and time-series inquiries. In the second half of the twentieth century, funding allowed

© The Author(s), under exclusive license to Springer Nature
Switzerland AG 2021
G. S. Bhunia, P. K. Shit, *GeoComputation and Public Health*,
Springer Geography, https://doi.org/10.1007/978-3-030-71198-6_3

researchers to develop cohorts of individuals. In the twenty-first century, research supports and participation rates are deteriorating. Using electronic health records (EHRs) currently offers low-cost methods of access for epidemiological research to rich longitudinal data on large populations. In order to address questions concerning multifaceted causal connections, an EHR can be connected to contextual data via the geographical information system (GIS) and connected to self-reported data. EHR data sets also enabled environmental and social epidemiologists in a number of physical, constructed and social settings to impact data on patient spread.

A pioneering study by Deiner et al. (2016) in epidemiology, which showed that early binge of epidemics can be seen through operational inquiries on social media indicators such as Google Search and Twitter, has shown Big Data's potential. The findings of the study suggested that the monitoring of Internet questions could allow for early detection and the identification of biosecurity intimidation and epidemics. Late diagnosis and early response are clear benefits on which more data are reviewed, including paper records, pharmaceutically monitored transactions and patient references (Benke and Benke 2018).

The Internet of Things (IoT) is explosively grown during the present year and has significantly improved different aspects of the community, including healthcare (Gubbi et al. 2013). IoT has developed industry-agnostic jargon in which critical information is provided that supports healthcare professionals in their work and increases productivity and data analysis (Ahmadi et al. 2019). Advanced technologies have been widely used since recent decades such as telemedicine, digital hospital and online and mobile safety. The development of IoT has contributed to the exponential growth of smart systems such as physical and intelligent healthcare (Zheng and Rodríguez-Monroy 2015). E-science has done a great deal to establish methods for generating information such as grid computing and teamwork in remote science.

3.2   Usefulness in E-Research in Public Health

Statistic replications of large-scale experimental studies affect the volume and accuracy of the results and are subject to several epistemic inconsistencies. The present focus on big data over the Internet was linked to the rise of artificial intelligence and decision-making in diagnostic and technological development following recent advances. Passive data sets are not very important in data science in the preserved archives when information for decision taking, forecasting and data processing can be accessed. It was defined as the integration of IT, statistics and database management with increasingly rising use of IT and the introduction of advanced graphics. E-research have the following significant characteristics which can occur if different combinations:

- Variability (deficiency in form, coherence and context)
- Variety (imagery, digital data, text information)
- Speed (real-time and very fast communication processing)
- Truthfulness (precise details, sound and insecurity)
- Volume (very broad data sets)

Stan Openshaw was a leading author of the definition of the data-intensive spatial analysis system into a collection of publications from the late 1980s and 1990s (Openshaw et al. 1988) and the global explanation computer (Openshaw and Turtron 2001). The key concern of GC is to enhance geography with a toolbox of approaches to model and analyse a series of multifaceted, often deterministic problems (Gahegan 2000). The assemblage of GC tools has grown up in an unrestrained and miscellaneous means to address a varied range of practices and application fields from numerous viewpoints enthused by the statistics, pattern recognition and artificial intelligence (AI). GC has been primarily related to GIS and its capacity to deliver spatial data for many decades, contributing to developments in geostatistics and multilevel computing, network analysis methods, statistical modelling, geodemographics and simulation.

The charm of this research has been constantly developed and matured by the development of computational and e-research competences. For some fields this dream is already understood with the word cloud computing. For instance, Google Docs users, which is a service that allows people to create and save documents through a web browser, are unaware of the location of their materials. Users enter and use a platform that runs the templates for logging in and set-up of their simulation(s). Although the user can monitor the state of the simulations and obtain feedback from the database, without understanding where the simulations are being done, the models still run on the Leeds grid computer. Yet new cloud technologies are evolving which may make e-research obsolete for these institutional grids. Amazon's web services (http://aws.amazon.com/) and Google's database engine (http://cloud.google.com) include the ability to access simultaneous virtual computers or processors if needed. As a result, Amazon web services may be enabled. For the combination of its features, the platform the best be combined for web-based spatial decision support system by supplying the underlying framework with ready-to-use maps, templates, applications and data for the users. The usefulness of e-research in health-related application is described as below.

3.2.1 Internet of Things (IoT) and Public Health

With the growing advancement of Internet of Things (IoT) and wearable devices, vast volumes of health-related data are created at any moment. These enormous amounts of data comprise great value and can bring profit to all shareholders in the healthcare ecosystem Internet of Things at a Glance 2019). Presently, the bulk of the information in the public and private repositories is siloed and separated into separate structures. IoT connects all sorts of allied 'things' through a large network of interrelated knowledge without human interference. The introduction of IoT and the developments in technologies for wireless communication (WCT) enable patients to provide treatment in real time (Abidi et al. 2017). Subsequently, numerous sensors and portable devices help to determine heart rate (HR), respiration rate (RR) and blood pressure (BP) through a single touch. However, the integration of IoT in the healthcare system makes several challenges such as data management, storage,

security, privacy, exchange of data, ubiquitous access, etc. Presently, cloud computing allied with IoT provides access to common medical data and shared infrastructure universally (Fig. 3.1). Cloud computer distributes Internet-based computing
infrastructure that delivers fastened location and versatile tools and sizes, including
servers, storage, networking, applications and data processing. Furthermore, cloud
computing shifts from central paradigms to decentralized paradigm, i.e. fog computing. Fog computing performs edge data analytics that allows in-patient processing in real time, improves data privacy and reduces costs (Bhatia and Sood 2019).
IoT offers appropriate solutions for a number of health services applications.
Accordingly, 500 billion computers, equivalent to 58 intelligent devices per person
on Earth, will be connected by 2030 according to the CISCO report (2019).

For the last 4 to 5 years, a comprehensive IoT study on healthcare has been carried out and bestowed different facets of IoT in the medical field, such as technologies, systems and implementations, and as the security and standardization analysis,
etc. (Table 3.1).

Because it facilitates health technologies with the complete functionality of the
IoT and CC, the IoT in the healthcare framework (IoTHeF) is the primary IoT in
healthcare. There are three basic components of IoTHeF, including topology, architectures and networks, each of which has a particular role in IoT care. The IoTHeF
system is a common use of communications activities between intelligent devices
and demonstrates the vibrant role of gateways in the development of physical IoT

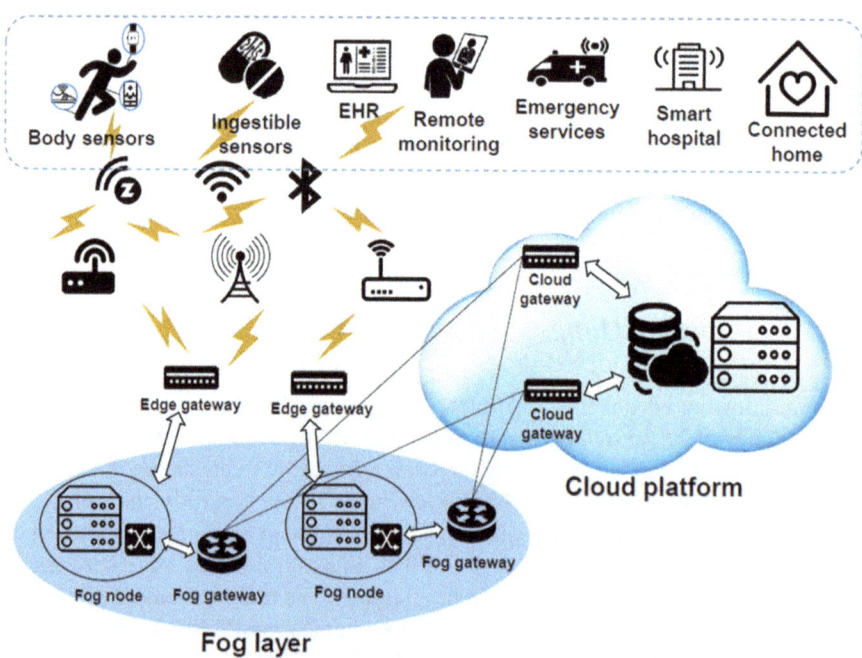

Fig. 3.1 An overview of a typical IoT and cloud computing-based healthcare system (Source:
Minh Dang et al. 2019)

Table 3.1 List of contribution from previous surveys on e-research in healthcare

Use of IoT/ cloud computing (CC)	Contribution	Year	References
Both IoT and CC	Three fundamental factors are indicated for effectively managing resources in cloud-based healthcare system Deriving information used for fog computing in healthcare system Analysing the limitations of recent methods, systems and frameworks	2019	Mutlag et al.
CC	Describing the challenges and opportunities for fog computing to healthcare For real-time applications, three-layer healthcare architecture has been introduced	2018	Kumari et al.
CC	Fog computing-based framework has been proposed to accelerate the response to mobile patients A prototype framework has been proposed that will reduce response by four times	2020	García-Valls et al.
IoT	Focus on recent IoT in e-health research Systematic IoT in e-health has been introduced Challenges and future direction for IoT in e-health List security issues in IoT devices and networks	2018	Farahani et al.
IoT and CC	Describe some aspects of IoT architecture in healthcare Cloud-based architecture role for IoT in healthcare has been investigated Critical issues and challenges of IoT in healthcare study have been discussed	2018	Ahmadi et al.
IoT	Basic elements used for IoT in healthcare system have been described Different types of sensors and communications methods have been analysed Introduce a framework applied in various IoT applications in healthcare Reviewing on cloud computing data storage	2017	Baker et al.
CC	Literature survey done on fog computing for healthcare Analysed network level used for fog computing tasks Fundamental components of fog computing have been discussed	2017	Kraemer et al.
IoT and CC	Review services and applications of IoT in healthcare system Security and privacy issues of IoT in healthcare frameworks Difficulties and future directions for IoT in healthcare	2015	Islam et al.

machines. Several sensor nodes have been used in this system to track patients, to gather data and then to pass all compiled data to a network of sink nodes. The suggested architecture includes a gateway to the back end node that transfers medical data from various devices, a tunnelling protocol that facilitates data transmission from IPv6 networks to low-power wireless personal-area networking (6LoWPAN) and a network using IPv4/Internet Protocol 6 (Ipv6) protocol and a socket that analyses and documented the protocol.

3.2.2   Cloud Computing for Healthcare

Cloud computing (CC) paradigm has become one of the hottest topics in information technology having the benefits of scalability, mobility and security. CC has recently emerged as a backbone of IoT healthcare systems (Sultan 2014). Subsequently, CC has the capability of sharing information among health professional, caregivers and patients in a more structured and organized way. This platform helped practitioners to observe and evaluate health conditions by transmitting raw sensor information from the end user to the cloud platform for processing. However, the majority of the cloud data centre are geographically centralized and located from end users (Corcoran and Datta 2016).

3.2.3   Fog Computing for Healthcare

Presently, healthcare applications are shifting from cloud computing to fog computing. Fog computing technology was deployed to connect the user's layer to the cloud layer (Fig. 3.2). Fog computing is heavily virtualised and offers software between end users and the cloud for processing, storage and networking (Bonomi et al. 2012; Kumari et al. 2018). The core principle of fog computing is to transfer data centre functions to the fog nodes on the network's bottom. We call the fog nodes the layer of cloud. Because the activities of these instruments are at the edge of the network, the communication rate is higher and the consumer reaction time is decreased. This layer analyses the data and information obtained by the edge tools. The server functions like this plate. This stage collects massive amounts of real-time sensor data. The fog layer then moves the work to various edge machines attached to the fog layer, and large quantities of information are then processed. The computational function needs to be performed using an efficient task programming algorithm.

The major uses for fog computation are time-sensitive systems in which vast quantities of data are processed. Chen et al. (2010) have implemented a prototype of an intelligent WSN gateway in the home healthcare system. The machine will produce information in a low-power distributed device in real time. Hong et al. (2013) introduced mobile fog as a programming model for geo- and latency-sensitive Internet applications (Table 3.2). In general, the healthcare system consists of four tiers, apart from the different levels mentioned above: the area; the organization; the clinic or the ambulatory hospital; and the doctor, the infirmary or the patient. The knowledge exchange between all four layers must be handled effectively. Consequently, there are various significant threats to privacy and security.

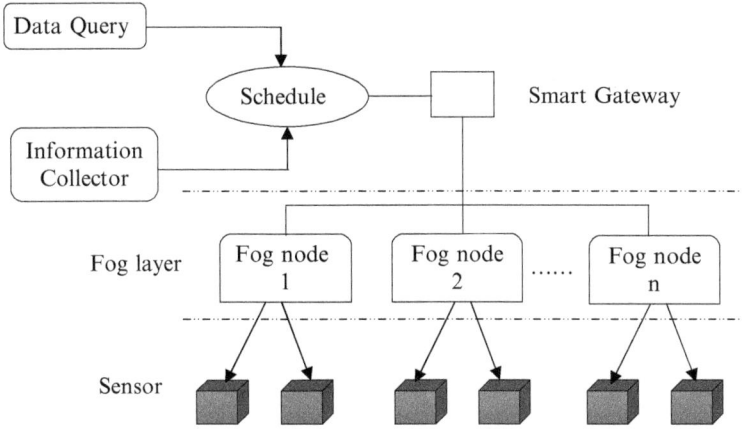

Fig. 3.2 Fog layer architecture in health management (Modified after Paul et al. 2018)

Table 3.2 Application of fog computing in public health (Source: after Minh Dang et al. 2019)

Field of study	Scope of study	Outcome	References
Smart health	Fog-enabled smart health has been developed to improve data sharing service Provide a privacy-preserving fog-assisted health data	Efficient data sharing service with privacy preservation	Tang et al. (2019)
	Establishment of fog-cloud framework for healthcare service in smart office Recommend a Severity Index to determine the adverse effects of different actions on personal health Implement an application set-up for prediction and alert generation of healthcare	BBN classifier/93.6%	Bhatia and Sood (2018)
	To expand network consistency and speed data processing system has been projected Set up a self-adaptive filter to reminisce absent or imprecise data routinely Recommend an RVNS queue to process pre-processed data	Reliable and faster processing speed	Wang et al. (2018)
	Project a fog-assisted cloud-based healthcare structure for early recognition of the virus outbreak Create alarms directly on the user's mobile phone in the fog layer On the cloud layer, the virus outburst is shown using sequential network analysis	J48 decision tree classifier/93.5%	Sood and Mahajan (2018)

(continued)

Table 3.2 (continued)

Field of study	Scope of study	Outcome	References
Acute illnesses	To provide high-level health services, e-Health gateway is designed Based on fog computing to generate a Geo-distributed intermediary layer between sensor nodes and cloud Generate a smart e-Health gateway prototype (UT-GATE) with high-level features	Enhance overall system efficiency	Rahmani et al. (2018)
Diabetes	Propose a fog-based health framework to diagnose and remotely monitor diabetic patients in real time Perform risk assessment of diabetes patients at regular intervals	J48Graft decision tree classifier/98.56%	Devarajan et al. (2019)
Hypertension attack	Propose an IoT-fog-based healthcare system to continuously monitor patients Predict the risk level of hypertension attack of users remotely Data collected from patients were saved on the cloud and shared with domain experts	ANN classifier/95.21%	Sood and Mahajan (2018)
Healthcare system privacy	Show research challenges in developing practical privacy-preserving analytics in healthcare systems Propose solutions to solve these challenges	Solutions for privacy issues	Sharma et al. (2018)
Wearable healthcare monitoring system	Suggest a cloud-based user verification system for secure authentication of medical data Generate a secret sitting key for upcoming secure communications Conduct a comprehensive virtual analysis on the system's communication and computation costs	System is resilient against known attacks	Corno et al. (2016)
Mosquito-borne diseases	Fog-based system has been announced to perceive and categorize different mosquito-borne diseases Implement social network analysis to demonstrate the outburst of the mosquito-borne disease on the cloud layer	FKNN classifier/95.9%	Vijayakumar et al. (2018)
Smart homes and hospitals	Introduce a cloud to fog U-healthcare monitoring system in smart homes and hospitals Investigate significant features of fog computing and extend to cloud computing		Nandyala and Kim (2016)

3.2.4   *Internet of m-Health Things (mIoT)*

mIoT is a real-time patient care surveillance system that uses smartphone and medical devices and cloud-based devices (Erdeniz et al. 2018) and the access to the cloud computing world of communications systems to forward applications/applications. This will become innovative IoT and CC substances in healthcare applications in the future as it delivers truly automated work. Recently, a platform (Cloud-MHMS) for an m-Health was introduced to observe healthcare investigations based on cloud computing (Xu et al. 2017). The mIoT law emphasizes the digitalization of medical products and associated care procedures by the use of intelligent medical tools and IT services (Web, Internet, applications, etc.). The findings of mIoT will be used by combinations of innovative equipment and resources to build entirely different forms of disease therapy (Fig. 3.3). A variety of smartphone applications, including myDario and SleepBot, have been developed that support system handling. Sensors may provide multiple details in order to promote therapeutic growth, but finding the right patients for the right clinical trials is especially critical. Body sensors are now fast approaching the general market from devices that were used only by athletes and racers, and customers and pharmacies will soon be able to access a range of statistics, including all pulses, blood pressure, ECGs and respiratory levels, and specialized data including inflammation, sleep cycles, etc.

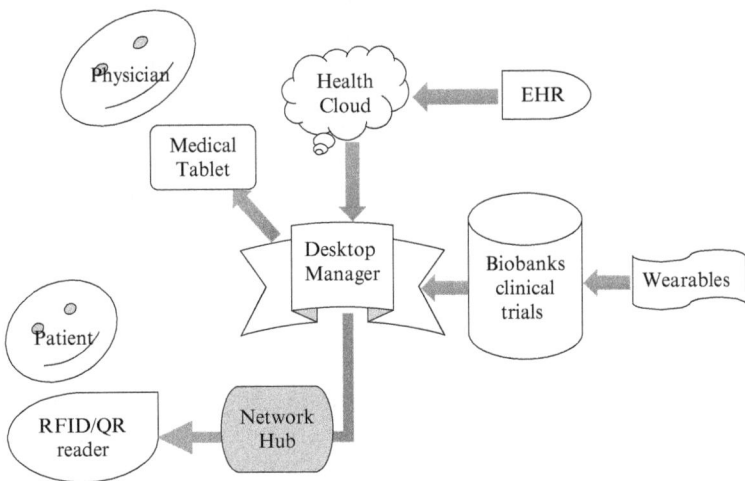

Fig. 3.3 Illustration of mIoT will took hospital in practice

3.2.5 Cognitive IoT (CioT)

The recent evolution of sensor technology not only decreases sensor price but also makes sensor smart. Cognitive computing refers to an intelligent computer, capable of imitating human brain problems (Zgheib et al. 2017). The CioT framework ropes data mining and works on storing and reviewing an immense volume of data. In a comprehensive CioT framework, all the sensors will operate together with other autonomous tools in order to accurately identify the health status of the user. The IoT and cloud computing in the context of healthcare rely on a perceptive approach that is smart enough to make appropriate decisions based on placid data.

3.2.6 Smartphone Solutions in Healthcare

New computing devices such as laptops, iPads and personal digital helpers (PDA) have had a new advancement and have had a significant impact on healthcare. The platform provided state-of-the-art innovations, including multimedia messaging devices, web browsing, video telephony, video recording, camera and an enormous number of applications. Innovation Single-chip Systems (SoC), extendable RAM capacity, large-scale storage, a highly customized smartphone operating system, wider screen size, and monitor resolution make smart telephones a handshake device. A huge number of mobile apps are now available in IOS and Android health-care applications. Few examples of healthcare apps used by physicians for the patient's treatment are described in Table 3.3. Diagnostic apps are, for instance, used to match the signs of patients with a broad medical symptoms database to enable them to support the medication most appropriate. A full list of medicament names, descriptions, side effects, interactions, dosage and characteristics usually includes medication reference apps. Education apps include comprehensive drug commands, video lessons on numerous events and medical student education activities. Apps for medical news suggest technologies that deliver the latest global medical coverage. Telemedicine devices enable patients who do not come to the hospital or clinic to get medical care remotely by phone or video. In addition, clinical collaboration systems offer a clear interface between doctors and operators.

(i) Due to the information technology revolution that began in the last century, the healthcare industry has changed dramatically. Advanced technology such as telemedicine, automated clinics, robotics, education and mobile health is incredibly useful in recent years, and now the rapidly increasing IoT industry promotes automated and smart education (Zheng et al., 2015).

(ii) Cloud technology should be applied by geospatial researchers and experts to have cost-efficient but technically sophisticated motives for geospatial decision-making.

Table 3.3 Healthcare apps used for the description purpose

Category			Description
Diagnosis	PEPID, UpToDate, Prognosis: Your Diagnosis, Diagnosis Medical App, Quick Medical Diagnosis & Treatment, InSimu Patient-Diagnose Virtual Clinical Cases, Ada-Your Health Guide, Doctor Diagnose Symptoms Check, Cardiac diagnosis, Eye Diagnosis, Self-Diagnosis, VisualDx	PPEPID, UpToDate, Prognosis: Your Diagnosis, Heart Rate and Pulse Monitor, Differential Diagnosis Guide, 5-Minute Clinical Consult, Ear Age Diagnosis, Emergency Central, Dem DX Diagnostic Reasoning, Diagnosaurus DDx, mRay	Evidence-based clinical and drug information resources for point-of-care decision-making
Drug references	Epocrates, Drugs.com Medication Guide, KnowDrugs Drug Checking, Drug Dictionary Offline, Drug INFO, Drugs Classifications	Epocrates, Drugs.com Medication Guide, Medicine Dictionary, Davis's Drug Guide	Mobile medical reference applications, look up drug information, identify pills
Education	Medscape, Muscle Trigger Point, Visual DX, PEPID	Medscape, 3D4Medical, Muscle Trigger Point, Visual DX, PEPID	Show practical clinical sources, including detailed guidelines on drugs, videos tutorials and educational exercises for students
Medical news	MedPage, Medscape, Newsfusion, Internal Medicine News	MedPage, Medscape, Newsfusion, Internal Medicine News, NEJM ThisWeek	Health apps support the flow of information in the health industry, becoming a new digital tool to help complement traditional searches for health news
Telemedicine	MDLIVE, LiveHealth Online Mobile, Express Care Virtual, Amwell, Lemonaid, I Online Doctor, Ask a Doctor, Doctor's Circle, MyLive Doctors, JustDoc Online Doctor, Carry doctor, Personal Doctor, Halodoc, My Swaath-Doctor consultation, doctor appointment	MDLIVE, LiveHealth Online Mobile, Express Care Virtual, Amwell, First Opinion, Doctor Pocket, Ask a Doctor, iCliniq, Ask Apollo, DocOnline, Continuous Care Health, Just Answer, OkaDoc	These applications make connecting with a doctor fast, easy and convenient from anywhere at any time

(continued)

Table 3.3 (continued)

Category	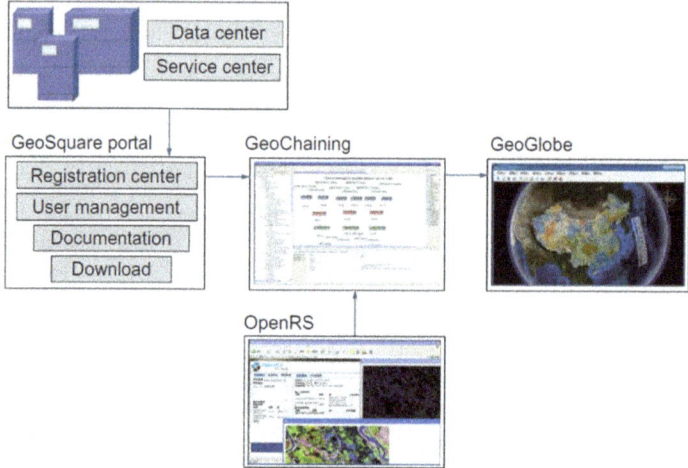 (Android)	(Apple)	Description
Medical calculator	MDCalc Medical Calculator, Medical Calculators, Calculate by QxMD, Caddy, MedicALC, Medical Formulas, Mediquations Medical Calculator, IV Infusion Calculator, eGFR Calculators	MDCalc Medical Calculator, Calculate by QxMD, MedicALC, Mediquations Medical Calculator, IV Infusion Calculator	Receive reliable clinical solutions fast and smoothly with fundamental input values

Fig. 3.4 GeoSquare architecture (Source: Gong et al. 2012)

(iii) Free GeoSquare has many uses, from work in the field of GIScience, to geo-science and technology training and to real-world applications (Fig. 3.4). Decision-makers can easily build comprehensive solutions for geospatial problems and work with different stakeholders using intelligent GIService chaining tools. In order to understand, create and exchange geospatial information, users are introduced to highly intuitive GeoCollaboration and Geovisualization environments.

(iv) In healthcare applications, wearable technologies have been used to perform different applications related to health such as remote diagnostics (Son et al. 2014) and disease monitoring (Zheng et al. 2019).

(v) Data mining machineries are playing an imperative role in serving geoscientists to resolve difficulties concerning some form of geographical and environmental data.

(vi) The Java GeoComputation (JGC) Application Programming Interface (API) allows java-based GC tools to be engineered to a single uniform interface to understand a variety of client application developers and end users. The JGC specification is created through a user-driven collaborative process in government, private industry and academia to work together to develop a draft specification that is subject to suggestion and input from all users and the general public.

3.3 Spatio-temporal Data Mining and Intelligent Service of Public Health

In the data mining world, over the last decade, spatio-temporal (ST) data mining approaches have been discussed.ST data mining studies uncover curious yet potentially valuable correlations in broad spatial and ST data sets that are previously unknown and differentiate between the necessities to mechanize ST information detection. Characteristically, this stage is to accurately make noise faults and missing data. Exploratory space-time analysis is steered in this phase to recognize the underlying ST distribution. After that, a suitable algorithm is nominated to run on the pre-processed data and generate output patterns. The process of spatial and spatio-temporal data mining is illustrated in Fig. 3.5. Huge amounts of ST records, such as climate change, social sciences, neurosciences, epidemiologist, travel, mobile health and earth sciences, are rapidly being gathered and analysed in different domains. Let us assume that behaviour taken by the human mind is stored in neuroimaging data along with the space from which the operation was measured and the time that the calculation was done. Dependent on how much time and location they have earned about Google's servers, online search criteria have to be fulfilled.

Spatio-temporal data mining (STDM) aids to consider novel formulation of the spatial and temporal information. One example involves the charge for space positions as objects and the use of spatial dimensions over time for the description of characteristics. For example, one of the goals of epidemiology is to group areas of elevated disease occurrence over time. Additionally, time points are viewed as artefacts, and characteristics are represented using measurements obtained from all geographical locations.

For instance, when applying disease distribution patterns to historical data, the goal is to identify the months/season of peak transmission of the disease in an area. There are also scenarios where events are treated as objects and features are defined based on the spatial and temporal information of events. As an example, malaria occurrence is viewed as an entity in combination with recognition and place and

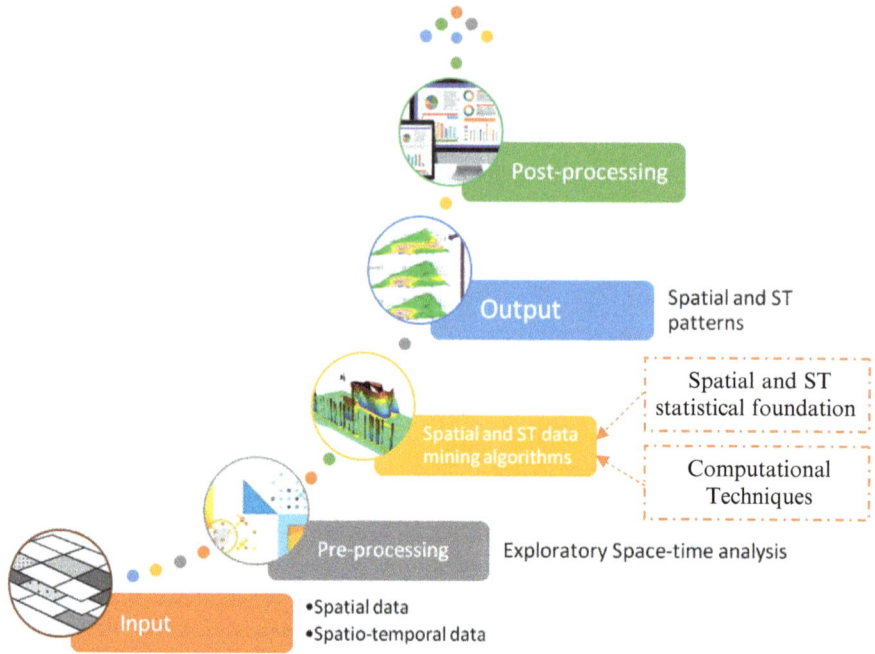

Post-processing

Output   Spatial and ST
                                    patterns

                                    ┌ ─ ─ ─ ─ ─ ─ ─ ─ ─ ┐
                                    │   Spatial and ST   │
Spatial and ST data   │ statistical foundation │
mining algorithms   └ ─ ─ ─ ─ ─ ─ ─ ─ ─ ┘
                                    ┌ ─ ─ ─ ─ ─ ─ ─ ─ ─ ┐
                                    │   Computational   │
                                    │   Techniques   │
                                    └ ─ ─ ─ ─ ─ ─ ─ ─ ─ ┘

Pre-processing   Exploratory Space-time analysis

Input   •Spatial data
                     •Spatio-temporal data

Fig. 3.5 The process of spatial and spatio-temporal data mining

time for the peak transmission within near proximity in space and time, as well as other features such as the complexity of the atmosphere and the number of environmental factors involved. Several literatures have been published and reported that the variation of sandfly density is associated with kala-azar transmission in India. An example of sandfly density for the lean season (January–March) and peak season (September–November) of 2014 has been collected from 140 locations of Muzaffarpur district (Bihar, India) illustrated in Fig. 3.5. The combination of spatial and time information in ST data therefore creates new challenges for STDM research, with a broad application in several fields of scientific and commercial significance. This approach includes a broad spectrum of research applications.

There are several literature methods that stretch through many decades of work in spatial statistics for purely spatial data mining (Cressie and Wikle 2015), spatial data mining (Aggarwal 2015) and spatial database management (Shekhar and Chawla 2003). In order to clarify the expertise and use of data mining, grouping, forecasting, detection of ab-norms and pattern mining, Shekhar et al. (2011) have been studying various types of spatial data and analysts have been exploring in the spatial data mining sector. Existing STDM work involves fundamental studies on

ST-point systems in the statistical society. ST clustering (Kisilevich et al. 2010) and trajectory pattern mining (Giannotti et al. 2007) have taken advantage of approaches to handling ST information.

3.3.1 Spatial and Spatio-temporal Data

The input data of public health is one of the important aspects of spatial and spatio-temporal data mining. There are three spatial data types, for example, the object model (e.g. point, line and polygon), the field model (e.g. grid consisting of areas covered by pixel) and the spatial network model (e.g. graph that indicates the interaction between vertical and edge spatial components) (Shekhar and Chawla 2003). In object model, the spatial objects are identified according to the application's context, whereas the filed model is useful to represent continuous data which is a function from spatial framework and partitioning of space to non-spatial attributes. The field model can be categorized into four groups: local (e.g. output of a given location based on the input at that location), focal (e.g. output at a given location based on the small neighbourhood of the location), zonal (e.g. operations work on inputs from a predefined zone) and the global operations (e.g. operations work based on the predefined global inputs). The spatial network model comprises a picture to illustrate the connection between spatial elements with vertices and edges.

In real-world applications, there are a wide variety of spatio-temporal data types that differ in time and place when data is assembled and have shown result in different groups of STDM concerns. Based on the temporal data, ST data can be divided into three groups – (i) temporal snapshot model (e.g. multi-temporal remote sensing data), (ii) temporal change model (e.g. sequential information of disease incidence with incremental changes occurring afterwards) and (iii) event or process model (e.g. peak and lean season of disease incidence with progressive changes) (Li et al. 2008). Based on the STDM formulation strategies, ST data are categorized into (i) data of events that reflect incidents that occur at points and times (e.g. location and time of disease events where the patients were first infected in an administrative unit), illustrated in Fig. 3.6; (ii) trajectory data, i.e. trajectory of moving bodies is being measured (e.g. spread of disease within an administrative unit); (iii) point comparison statistics for the measurement of a ST sector at travelling ST reference sites (e.g. monthly weather variable assessment in tropical and exotic areas), represented in Fig. 3.6; and (iv) raster data collected on fixed cells in the ST grid for observations of the ST field.

Standard deviation of ellipse (SDE) was used to create the spatial pattern of disease in an administrative unit which is based on the weighted field (e.g. disease incidence rate). The value in the output rotation field represents the rotation of the

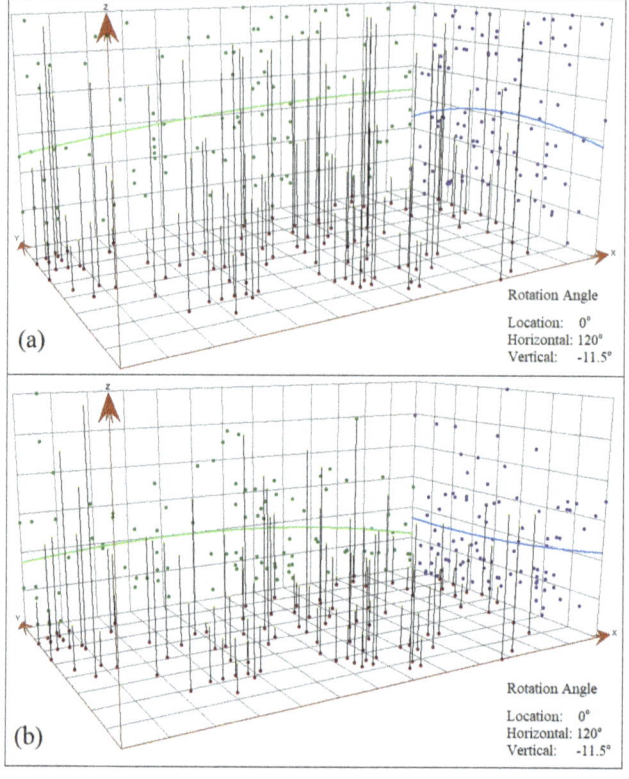

Trend analysis tools provides 3-D perspective of the data. Here, the location of sandfly density are plotted on x, y plane. The height of the stick in z-dimension represents the sandfly density. Subsequently, the sandfly density are projected onto the x,z plane and the y,z plane as scatterplots (*Figure 3.6*). This can be a thought of spatio-temporal views through 3-D data.

Fig. 3.6 Trend analysis of sandfly density in (**a**) peak (post-monsoon) and (**b**) lean season (winter season) in Muzaffarpur district (Source: Bhunia and Shit 2019)

long axis measured clockwise from noon, and distances are measured accurately using Euclidean distance (Fig. 3.7).

Indoor temperature (°C) and sandfly density (per trap/night) have been collected from the field from two different seasons (e.g. peak season and lean season). Based on the radial basis function interpolation technique, predicted surfaces of indoor temperature (°C) have been generated in two different seasons to investigate the autocorrelation of the data, making it less flexible and more automatic than kriging (Fig. 3.8). The sandfly density has been overlaid to make a spatial correlation between temperature and density (Fig. 3.9). Results showed suitable temperature for sandfly abundance varying from 25 °C to 28 °C in peak season, whereas, in lean season, suitable temperature for sandfly abundance ranges between 20 °C and 22 °C.

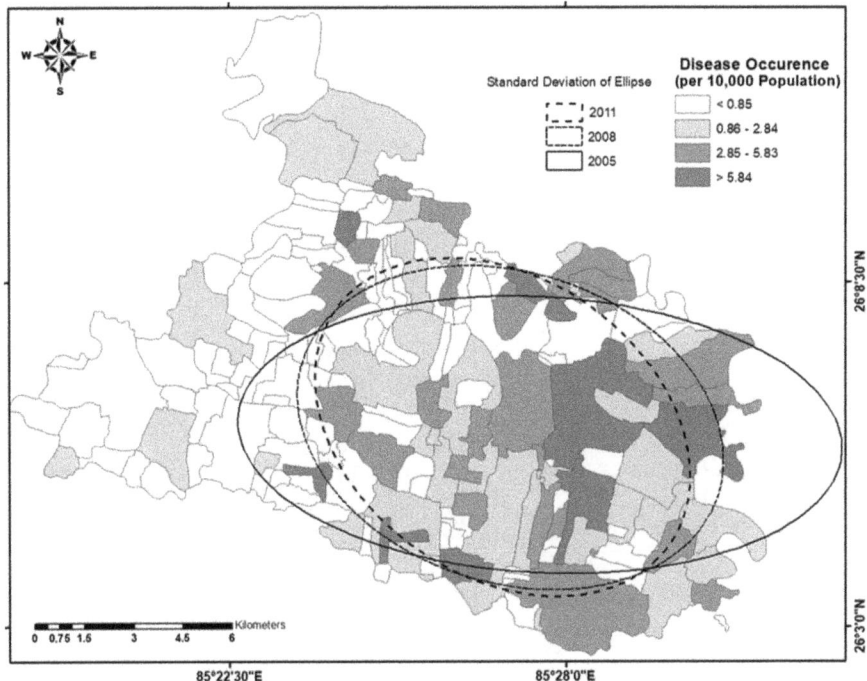

Fig. 3.7 Kala-azar transmission based on spatio-temporal data of Masaurhi block in Muzaffarpur district (Bihar, India)

3.3.2   Data Attributes and Relationships

Data attributes for spatial and spatio-temporal data can be categorized into three distinct types, namely, (i) non-spatio-temporal attributes, (ii) spatial attributes and (iii) temporal attributes. The spatial features describe the location of spatial diseases and/or vectors in spatial referral frame conditions (e.g. latitude, longitude and elevation), the area, the perimeter and spatial polyline shape. Non-spatial attributes are characterized by non-contextual objects, such as patient's name, age and occupational status. In addition to the scope of the process, temporal attributes consist of the time mark for a spatial layer (e.g. a raster layer). Non-spatial relationships are specifically formed by arithmetic, organization and identifying relationships.

In order to represent spatial-temporality in the informational system, each model concentrates on various aspects of frequency, time and space-time data. Moreover, models are evaluated that improvements can be correlated with a synchronous example of a changing in location of an object in the description of a spatio-temporal entity. However, a spatial-time object's morphology, topology and attributes may or may not change over time (Fig. 3.10). This factor allows numerous functions like *evolution, creation, fusion*, etc. to detect and define the movement or change of

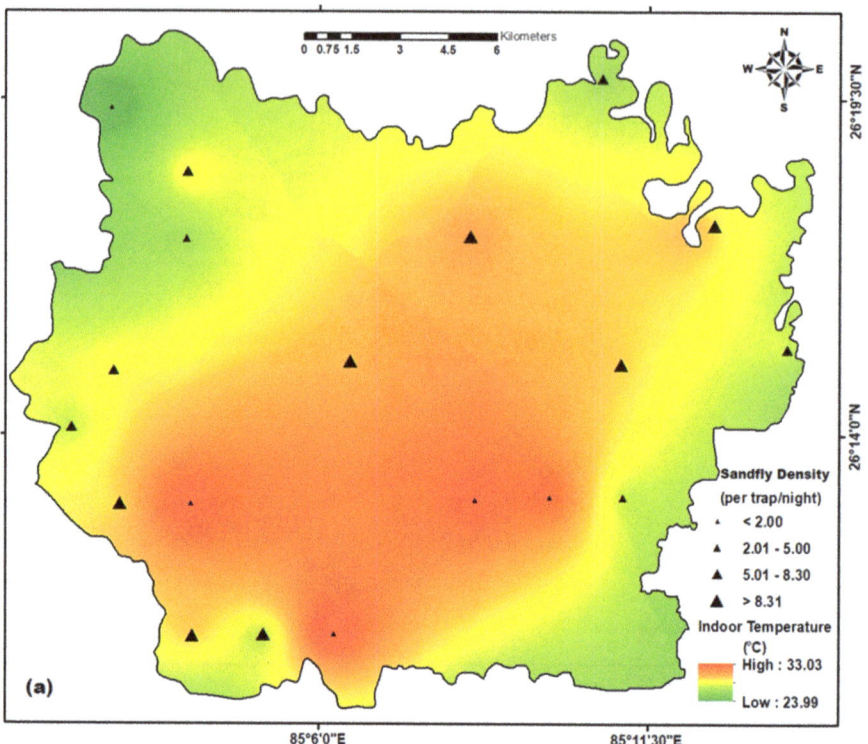

Fig. 3.8 Temperature distribution

objects in space, independently from their object identification. This norm can also be used to connect models for operations that can compute the speed and/or accel-eration of the spatio-temporal movement of objects. Space-time topologies, for instance, define basic metrics such as distance values, the position and the size of other artefacts. This also evaluates the capacity of the models to embody for a cer-tain time topological association among that spatial objects. As such, the identifica-tion of an object is able to determine the simulation potential of current space-time data structures. In some situations, the original instance of an object may best be destroyed and new versions recreated due to a substantial shift.

In terms of query capabilities, ST data models presented queries about locations and time, spatial and temporal properties and spatial and temporal relationships. Definitions of the trait are the characteristics of individuals irrespective of time and space (e.g. where is the place), distance (e.g. location-specific analysis), the nearest neighbour (e.g. the neighbouring disease regions affected) and topological queries (e.g. find the surface waterbodies for malaria breeding site of a particular area). The queries of ST database can be classified into three subcategories, such as (a) ST queries on discretely changing or moving reference objects (e.g. what is the disease location of a specific time t ?); (b) ST range queries including distance-join (e.g.

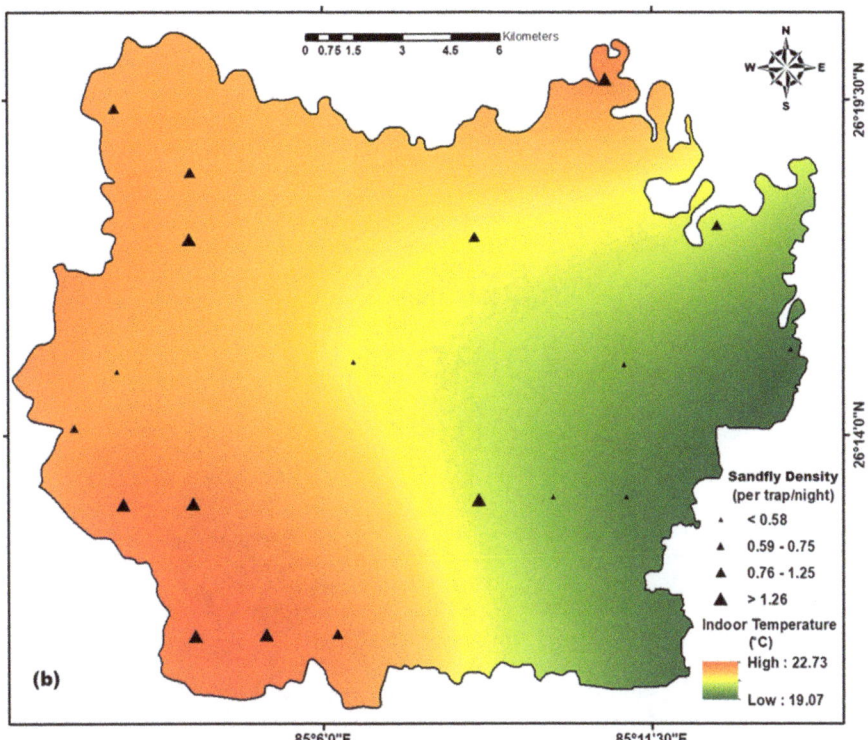

Fig. 3.9 Spatial association between indoor room temperature and sandfly density in (**a**) peak season and (**b**) lean season (December–February) of Baruraj block in Muzaffarpur district of Bihar State (India)

what happens to a region over a given period?); and (c) ST behaviour queries (e.g. when/where is the maximum occurrence of disease observed?).

3.3.3   Spatial and Spatio-temporal Statistics for Public Health Data

Spatial statistics can be categorized according to their underlying spatial data type (i.e. points with fixed locations and attribute values for the predictions in uncovered places) such as referenced data geostatistics, lattice statistics for areas data (e.g. to a countable collection of regular or irregular cells in a spatial framework), spatial point process for spatial point patterns (e.g. method is not focused on attribute value, but is based on specific statistics) and spatial network statistics (e.g. focuses on Euclidean space in public safety research applications).

Spatio-temporal statistics are a combination of spatial statistics and temporal statistics (Cressie and Wikle 2015). Analogous to spatial data, temporal data

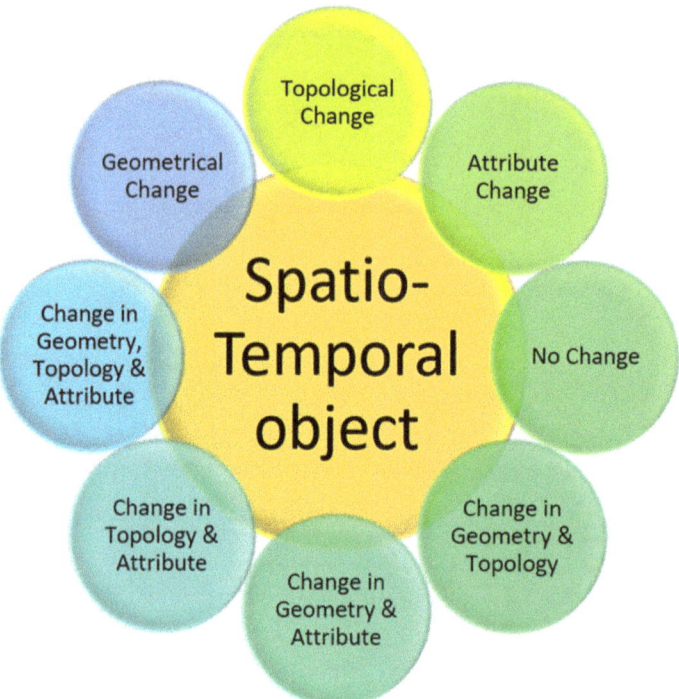

Fig. 3.10 Possible types of change of a spatio-temporal object

including spatial time series, spatio-temporal point process and time series of lattice data conserves inherent properties, for example, autocorrelation and heterogeneity. Spatio-temporal point process typically integrates time factor in the delivery process. The exploration of various patterns of space-time connection and teleconnection is important applications for public health. For instance, detecting space-time waterfall trends from health event datasets can help health authorities understand the spread of disease in one area and thus take appropriate steps to minimize disease events.

3.3.4   Data and GeoComputational Analysis in Public Health

3.3.4.1   Clustering

This refers to the classification of incidences of diseases that hold common beliefs, for example, clustering locations that are spatially contiguous, based on their time-series observations. Clustering ST points which have an unusually high density can be used to find out disease outbreaks, also called hotspots. For example, spatial scan statistics (Kulldorff 1997) and *Getis-ord G* statistics explore all possible regions as 'event detection for studying disease outbreaks' at various scales where the incidence of points is significantly higher than expected. Bhunia and Shit (2019) used spatial statistical approach (*Moran's I* and *Getis-Ord G$_i$*) to determine the spatial autocorrelation and cluster-outlier detection analysis of kala-azar transmission areas in India. CLARANS (Ng and Han 2002) is another clustering method that can be used in ST area for finding cluster of features in both space and time. Calinski-Harabasz pseudo F-statistic (Duque et al. 2007) is a ratio that reflects the similarity of group difference within the group, calculated as:

$$\frac{\left(R^2\Big/n_c - 1\right)}{\left(\left(1-R^2\right)\Big/\left(n-n_c\right)\right)}$$

where

$$R^2 = \frac{SST - SSE}{SST}$$

Yet SST reflects the disparity between the group, and SSE reflects similarities within the group:

$$SST = \sum_{i=1}^{n_c}\sum_{j=1}^{n_i}\sum_{k=1}^{n_v}\left(V_{ij}^k - V^{\bar{k}}\right)^2$$

$$SSE = \sum_{i=1}^{n_c}\sum_{j=1}^{n_i}\sum_{k=1}^{n_v}\left(V_{ij}^k - V_t^{\bar{k}}\right)^2$$

n = the number of features; n_i = the number of features in group i; n_c = the number of classes (groups); n_v = the number of variables used to group features; V_{ij}^k = the value of the kth variable of the jth feature in the ith group; $V^{\bar{k}}$ = the mean value of the kth variable; $V_t^{\bar{k}}$ = the mean value of the kth variable in group i (Fig. 3.11).

Subsequently, the ST points clusters often help to evaluate the related non-ST attributes. For example, a series of disease events helps to identify regions with related diseases in space and time (Glatman-Freedman et al. 2016). DBSCAN algorithm used a number of clustering techniques for ST data analysis based on the density of spatial points data (Ester et al. 1996). To find out group of trajectories, similar trajectories data have been used that follow a similar route or stream. Distance calculation selection and the choice of clustering technique are the most appropriate trajectory clustering methods (Morris and Trivedi 2009). TRACLUS is one of the most important solutions to trajectories by dividing into smaller segments based on the MDL theory and the DBSCAN-based method for community line segments (Lee et al. 2007). ST data can be analysed in series-time by clustering k-means (Mezer et al. 2009), by hierarchical clustering (Goutte et al. 1999), by mutual

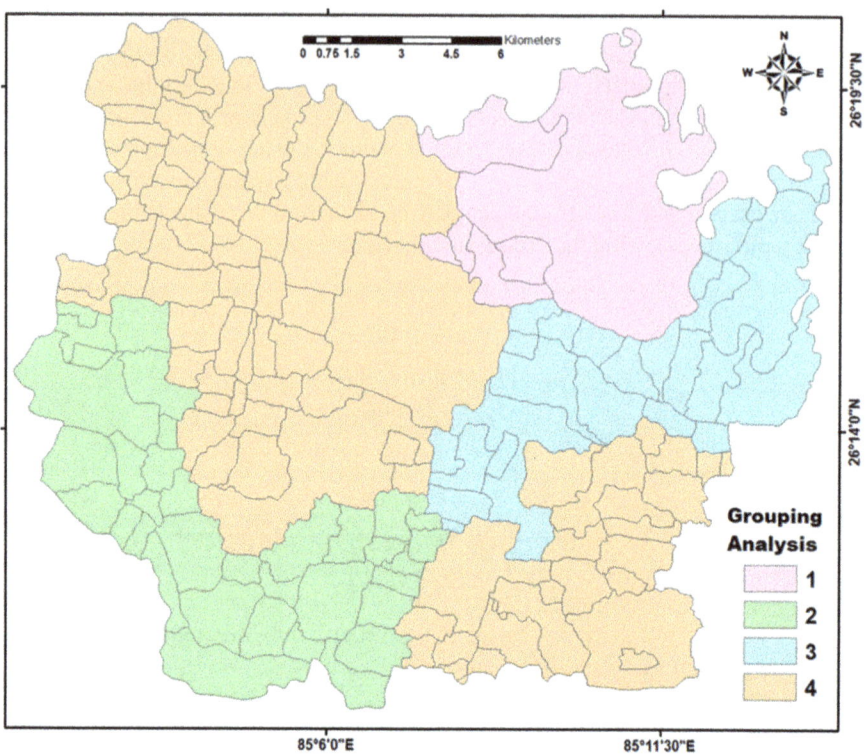

Fig. 3.11 Grouping analysis of kala-azar transmission villages of Baruraj block of Muzaffarpur district in Bihar (India). Ratio data is used to create the group based on the disease incidence for each village. Analysis tool assesses the effectiveness of dividing the features into a number of groups based on user specification. To create a group, tool will create a minimum spanning tree (e.g. connectivity graph for measuring adjoining relationships among features) reflecting both spatial structure and the associated attribute values of the feature. After that it will create groups by cutting the tree in best place by considering both the aspects, like within-group similarity and between-group difference

closest neighbouring (Steinbach et al. 2003) and by normalized-cuts spectral clustering (van Den Heuvel et al. 2008). Finally, the main goal is to locate groups of time stamps with identical charts while clustering spatial maps.

3.3.4.2   Predictive Learning

The main goal of predictive methods of learning is to map the dependent variables from the independent variables. In ST applications, the dependent and independent variables can belong to various kinds of ST data instances. The ST raster should be predicted by means of the spatial map to a scalar output variable. The spacious map used as input function involves the classification of images and objects, and the details delimited on spatial maps must be given an unconditional value for each image or subregion in an image. Recently, the approach of the convolutional neural networks (CNN) has been widely recognized in the computer community (Krizhevsky et al. 2012). An ecological indicator can be forecast in all places and periods using remote sensing measurements at local places and time stamps. For example, influenza outbreaks at an assumed location and time can be predicted in neighbouring places and times via web searches (Ginsberg et al. 2009) and Twitter messages (Culotta 2010). There are two main approaches for predicting ST points of reference, e.g. (i) temporal information usage knowledge to predict adjacent spatial values and (ii) spatial information estimates for neighbouring spatial values of the points. Most extensively used approaches for time-series predicting issues contain exponential smoothing techniques (Gardner 2006), ARIMA models (Box and Jenkins 1976), state space models (Aoki 2013) and Bayesian networks such as hidden Markov models and Kalman filters (Harvey 1990). Spatial autoregressive (SAR) models (Kelejian and Prucha 1999), geographically weighted models (GWR) (Brunsdon et al. 1998), Kriging (Oliver and Webster 1990) and Markov random field-based approaches (Zhao et al. 2007) are additionally used as prediction methods for spatial autocorrelation architectures.

3.3.4.3 Change Detection

In the situation of time-series data, the detection of changes was widely studied. Some similarity forms have been used in segments such as mean (Horváth et al. 2001), variance (Andreou and Ghysels 2002) and distribution statistics (Grosse et al. 2002) to define and alter time-series segments. In ST application, the spatial context of time series at every location is to identify changes in both space and time. The time-series framework can be defined by considering the time-series group that is identical to the given time series for a period of time or the time series observed at the location is near spatial proximity. Several change detection methods have been proposed by researchers and scientist such as iterative bottom-up approach (Boriah et al. 2010), Gaussian process-based approach (Karpatne et al. 2016), anomaly detection approach (Mithal et al. 2011), forecasting-based approach (Liang et al. 2014) and sub-sequence pattern matching approaches (Zhu et al. 2012) to define different types of changes in 'periodic' time-series analysis. The suitability of such approaches as point, line segment, polygon or network to space-time data and the time scale of the transition to be described as a point or time interval is specified.

3.3.4.4 Relationship Mining

A growing number of observational health and place studies have been using detailed, quantitative methods and techniques for statistical analysis of the general 'contextual' effect on public health. Recent developments in health science have nevertheless focused on formulating and establishing realistic empirical trends of indirect routes by which health may be affected. ST data mining will reveal correlations between time-series pairs with any similarity measures which are differentiated over time-series illustrations. The association between two different groups of contiguous places, not just a couple of places involved in a relationship, can be defined as similarity/discrepancy. The sets of regions along with their associations are therefore necessary to be determined simultaneously. Clustering algorithms independently discovered the relationships between regions. A noteworthy consideration, when mining relationships in ST data is to determine the strength of relationships among pairs of regions that may vary with time. As it is helps to determine the pair of interacting regions as well as the time window. Atluri et al. (2014) planned to use a mining pattern to explore the complex relationships between time series from many areas. Kawale (2011) used graph-based methods for climate science applications and the relations within a time interval are exposed from the equivalent graph. Another feature of the ST framework for data may be found in time lags in the discovery of relationships between time series (e.g. time taken to affect the system) (Lu et al. 2016). Dynamic relationships that are delayed are more complicated, with delayed relationships that last for a short period rather than the whole time. A major challenge in discovering delayed connections is the increase in the number and freedom to identify relationships that can lead to false detections without carefully conducting statistical correction for multiple hypothesis tests (Atluri et al. 2018). The Nobel prize winner Clive Granger has carried out causality research based on an autoregressive model to understand the cause (x) and effect (y) relationship. This view inspired the development of incidental maps for multivariate data from time series, which created probabilistic graphical models with unintentional node edges.

3.4   Theories and Models of GeoComputation for Public Health

The GeoComputing (GC) principle is a method for the study of spatial data. The computational dimension seems to be the joint denominator of the techniques. Faster and efficient computers and information technology advancements have had a profound influence on all statistical fields. Geospatial intelligence permits the approximation of parameters in richer and more truthful model-based illustrations of natural phenomenon, thus emancipating the fancy of the scientific community. The use of GC facilitates the study of social and medical patterns and the simpler accessible way to communicate spatial data within their space contexts. This makes it possible to display spatial data in a much more consumable way than data tables or texts. The epidemiology aims at advancing approaches and practices that allow the disease phenomenon in populations to be identified, clarified and envisaged, with an assessment of inhibition. Health status is a consequence of numerous elements. Distinct biology and behaviours, physical and social environments, strategies and interferences and access to eminent healthcare are influencing aspects that can subsidize the health of people and communities (Eberhardt et al. 2001). GC can assimilate statistical and geographic data that allow for the visualization of geographical relationships. Moreover, GC allows strategy makers to simply envisage difficulties concerning present health and social amenities and the natural environment and so more efficiently target resources. Hence, the investigation of geographical distribution of diseases, clustering analysis, range of service catchment, Euclidean distance between locations, etc. have paid to the construction of knowledge in public health. Subsequently, the ecological approach can emphasize on mapping the geographical circulation of diseases with the documentation of spatial clusters of cases and the investigation of relations between the disease incidence and environmental acquaintances interrelated to the united sphere.

The term GC applies to a variety of activities involving the use of modern computational techniques and methods to demonstrate geographical differences across the size of the phenomenon (Longley et al. 1998). It includes an extensive series of artificial intelligence and modern computational intelligence, comprising expert classifications, fuzzy sets, genetic algorithms, fractal modelling, neural networks, cellular automata, visualization and data mining. In recent research about user accessibility, the use of GC as a time-geographic tool is most evident (Weber 2003; Weber and Kwan 2003). They rely on the time-geographical structure of the potential path area, which is the geographical region within time limits resulting from some observable characteristics of this area. The GC algorithms offer an effective environment to exemplify and manage the enormous amount of geographic data in the real world. Numerous illustrations are discussed below to demonstrate the application of GC in time-space studies.

Documented public health hypotheses, models and/or contexts were classified according to their theory, model or structure and whether their relation was inferred or explicitly defined. For categorizing each theory, model and/or framework, Nilsen (2015) used:

- Theories include constructs or variables and predict the relationship between variables;
- Models are descriptive, simplify a phenomenon and may include phases or steps.
- The frameworks comprise concepts, constructs or categories and identify the relation between variables.

3.4.1 Spatial Microsimulation Model (SMM)

Space microsimulation is a large-scale simulated population-specific microsimulation (Ballas et al. 2006). The approach 'bottom-up' focuses on individuals and households, rather than 'top-down' methods that focus on aggregate statistics and flows. The inclusion of geographical data adds the "spatial" aspect to the system and enables rich data sets to be generated on a range of geographical dimensions. Previously, a number of health-related studies have been used in the

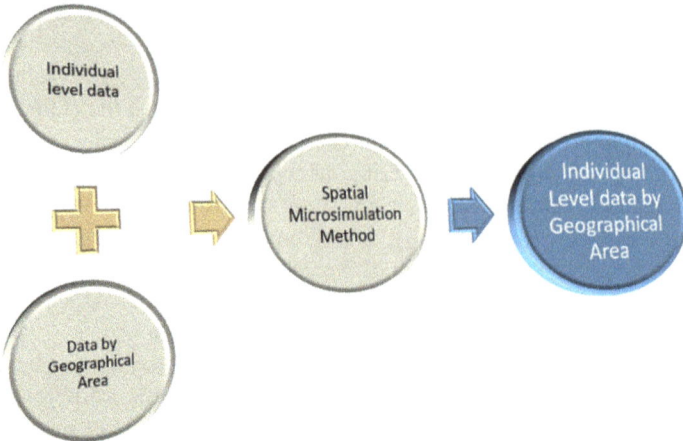

Fig. 3.12 Visual representation of the spatial microsimulation process

microsimulation process (Edwards and Clarke 2009; Morrissey et al. 2010; Burden and Steel 2016). Figure 3.12 provides a visual demonstration of the method of spatial microsimulation. In order to 'contain' data through sampling or weighting, the combination of shared variables of census and survey data is used to create a single dataset containing all interest variables. The final data set results in the formation of a population of individuals, distributed across the geographical areas of the census, and their individual characteristics are preserved by the survey data. Such variables are referred to as the 'target variable(s)' and are the findings of interest for the analysis that were not available at geographically disaggregated scales previously.

The key benefit of this approach is that it is able to create new personalized datasets from readily accessible secondary data sources using robust population synthesis techniques. Spatial microsimulation, by synthesizing new variables, may therefore address those areas where there are limited or no data available. In addition, spatial microsimulation is more readily accessible than ever, as the technology is being attempted to make new and current users more effective. The statistical nature of spatial microsimulation models can also be seen as a positive because, while not possessing the interactivity of more dynamic simulation techniques. Spatial microsimulation may therefore be particularly useful in the application of theoretical frameworks. This can affect the amount and form of variables available to limit data as well as the simulating target variables.

3.4.2 Agent-Based Model (ABM)

Agent-based modelling is important to understand and generate decisions in these highly compound, interlaced and disseminated environments (Shook et al. 2013). This is a 'bottom-up' approach and is a computer simulation of real-world environments. ABM is a patch-based simulation, derived from the 1970s cellular automata model. ABM was identified as the most appropriate tool to investigate complex health inequalities (Speybroeck et al. 2013). The simplification of behaviours before operational models to identify the embryonic phenomenon arising from these is a central concept of ABM. These behaviours are implemented by rules that monitor the model's operation and are based on theory or dataset patterns. ABM was most widely used in the general field of public health with examples of applications including an analysis of the influenza epidemic aggregation effects and cholera (Potter et al. 2012) and spates of other infectious diseases (Grefenstette et al. 2013).

The primary advantage of ABM is the ability to integrate research interactions and response mechanisms. From a spatial perspective, the ability to precisely mimic specific geographical areas in a complex modelling setting by means of a combination of GIS offers an opportunity to better understand the surrounding local area results in health. ABM can also be used to test future scenarios in order to obtain patterns of data for future years that has been fully verified, calibrated and validated. Finally, ABM can also be highly sensitive to its initial configuration and to slight adjustments in the interaction laws, which increase the need for sensitivity analyses.

The computational and data sciences, particularly GC, find significant demands in *health-GIS* (Shi and Wang 2015). Over the last few decades, health-GIS has been a major driver of spatial analyses, mainly spatial statistics. Model-based approaches presume that a complete model with an integrated structure will reflect the geographic characteristics of the disease (Lawson and Bertollini 1999). Typical illustrations comprise the Geographical Analysis Machine (Openshaw et al. 1987), the spatial filtering method (Rushton and Lolonis 1996) and the k-nearest neighbour methods (Besag and Newell 1991). Such models may include probability models, Bayesian models, event control points and sample counts (Lawson 2001). Both these are, in several ways, different cases for estimating the kernel density (Shi et al. 2013). The Monte Carlo process is a key tool for the assessment of statistical importance in these systems.

3.4.3   Local Intensity Estimation

Kernel-based methods have been widely employed in health research. Estimating kernel ratios (KRE) in different ways can be carried out, such as (i) the bandwidth can be set in the entire region (e.g. populations at risk or disease events) or adaptable with local situation and (ii) the kernel can be focused at the place (e.g. where the risk of diseases should be estimated) or the place of an event point (e.g. where the disease occurs). The broad bandwidth of the kernel produces the approximate value with a great spatial uncertainty. Not only does KRE alleviate problems related to areal units; it also manages less aggregated patient and population information in health study (Shi 2010). However, KRE is a highly computationally concentrated process since it wants to accomplish a kernel calculation. KRE plays an important role in detecting spatial clusters (Carlos et al. 2010; Berke et al. 2010).

3.4.4 Restricted and Unrestricted Monte Carlo Process

For disaggregating data, *restricted and controlled Monte Carlo* (RCMC) process has been used (Shi et al. 2009). Jacquez and Jacquez (1999) have established a method for breaking down areal data by transporting arbitrary polygon positions. Fortunately, the health-GIS has to deal regularly with data in a polygonal form. RCMC assumes that the distribution of subjects within a polygon is mainly followed by the circulation of the background. The more accurate and precise the context layer, the better the distribution of subjects is shown based on this principle. RCMC disaggregates the areal level evidence, so that the difficulties related to extremely combined data can be eased. Additionally, governing the randomization with the geographically comprehensive contextual evidence associated to sites of patients and/or populations and accordingly lessens the spatial ambiguity. Moreover, several rehearsals of the randomizations permit numerical assessment of the spatial vagueness obtained from the geographical fuzziness.

3.4.5 Unrestricted and Controlled Monte Carlo

Unrestricted and controlled Monte Carlo (UCMC) method produces a situation of nullity and is not restricted by the polygons that accompanied the cases of disease. By comparing the real local intensity value to several random intensity values at the same location produced by UCMC, you can determine the statistical meaning of the local intensity value of a disease. For example, when 99 random values have been generated and the current intensity value is higher than 98, then the probability of having a high intensity value simply by probability is not greater than $P = 0.02$. UCMC may generate artificial controls to compare environmental exposure to disease cases. Dispersion of the UCMC points matches the distribution of the at-risk population or predicted number. Eventually, this cycle imitates the epidemiology of the case control system. The variation in the UCMC iterations then represents the uncertainty with regard to the sampling of the population at risk.

3.4.6 Geographic Machine Analysis

Geographic Machine Analysis (GMA) is accomplished by revealing clusters of events/diseases and creating maps for detecting precedence areas for public health interferences and would not intend at serving to elucidate the existence of phenomenon. In 1987, Stan Openshaw and others invented the name and its acronym. The approach was developed for assessing whether spatial clustering was evident in different forms of cancer in children in Northern England at the Department of Geography at the University of Newcastle in the mid-1980s (Openshaw et al. 1987). A GMA provides an imaginary new approach to point pattern data analysis based on a fully automated method, which explores a point data set for pattern evidence without unduly affecting specified areal units or data error. There is no need for detailed knowledge or definition of particular location-specific assumptions. If there is clear proof of the geographical space pattern in geographic data, the GMA finds it. The GMA method was mainly computational, based on GIS technology developed, and data were used in the distribution of leukaemia among children across Northern England. GMA was a cluster automation detector for dot patterns which included geovisualization elements, native kernel density mappings with different bandwidth and significance tests for Monte Carlo. The fundamental GMA performs an exhaustive search using an estimate of the entire research field to all possible clusters.

3.4.6.1   Moran's Index

To detect dependence between geographically proximate events, Moran coefficient or semi-variogram model has been used to measure the spatial autocorrelation (Anselin 1995). This tool calculates the mean and variance for the attribute being estimated. This deducts the mean for each function value, creating an aberration from the mean. Deviation values are replicated to generate a cross-product for all neighbouring functions. *Moran's I* statistic for spatial autocorrelation is given as:

$$I = \frac{n}{S_0} \frac{\sum_{i=1}^{n} \sum_{j=1}^{n} w_{i,j} z_i z_j}{\sum_{i=1}^{n} z_i^2}$$

where Z_i is the deviation of an attribute feature I from its mean $(xi - \bar{x})$, $w_{i,j}$ is the spatial weight between feature I and j, n is equal to the total number of features and S_o is the aggregate of all the spatial weights:

$$S_o = \sum_{i=1}^{n} \sum_{j=1}^{n} W_{i,j}$$

The cross-product is positive when the values of the adjacent characteristics are both higher than the mean and both smaller than the mean. The cross-product will be negative if one value is smaller than the mean and the other is larger than the mean. A positive value for 'Moran's Index' shows that a feature has neighbour features which are equally high or low, which is part of a cluster. A negative value for 'Moran's Index' means that a function has adjacent appearances with disparate values; this is an outlier.

3.4.6.2 CUSUM Chart

Cumulative sum (CUSUM) graphs are a type of control diagram used for tracking small process shifts. CUSUM chart extensively used by public health departments was established to perceive the variations in patterns over time (Page 1954). These are created by cumulative recording of events over time. The CUSUM techniques notice shifts in single or multiple parameters while typically assuming the target parameters are constant. The MINITAB 15 program has been run on the CUSUM chart. In this graph, the variations between X and the target value μ (or an average regulated value) are obtained as the samples are taken, calculated as

$$S = \sum_{j=1}^{i}\left(X_j - \mu\right)$$

where X = mean of the jth sample size ≥ 1.

3.4.6.3 Space-Time Accessibility

The first effort made by Kwan (1998) for the measurement of space-time accessibility. This study scrutinized specific access of urban opportunity based on three major sources, i.e. (i) activity travel of diary dataset collected through mail, (ii) digital geographic database of land parcel and (iii) detailed digital street network. These space-time statistics were measured by an ARC Macro-Language (AML) algorithm for each individual and carried out on the ARC/INFO GIS. Such assessments evaluate the room, a number of opportunities and the attraction of the opportunities to be achieved. The findings of the analysis indicate that space-time inventions are conducted to illustrate human differences that cannot be differentiated by standard comfort scales. By using the geographical database with the improved geocomputing algorithm to evaluate accessibility space-time steps, Weber and Kwan (2003) developed a second-generation algorithm. The research approaches provide a practical picture of the transportation network's time characteristics and local opportunities.

3.4.6.4 Spatio-temporal Conditional Autoregressive (STCAR)

An Artificial Neural Network (ANN) can be used as a discovery platform inside data-rich ecosystems and combined in one spatial database with GIS technologies, with numerous kinds of nature. ANN is enthused by the technique the brain progresses information. A computing device with a large number of closely connected components in equilibrium to solve specific problems is the main component of this architecture. An ANN is optimized into a learning method for a particular task, such as pattern recognition or data classification (Gopal et al. 1999). In practice, ANNs are particularly helpful for grouping, calculating and mapping problems that are indulgent and have a lot of training data at their fingertips.

3.5   Accuracy and Uncertainty of GeoComputational Models for Public Health

Uncertainty refers to the dearth of detailed knowledge as to what the reality is whether qualitative or quantitative. If the complexities are high, the two parts of the set of possible outcomes can be adjusted immensely. Uncertainty analysis should be an iterative method and must challenge the discrepancy between variability and true uncertainty portraying probable outcome. Uncertainty assessment was commonly used in the safety risk evaluation in the 1980s (Bogen and Spear 1987). By measuring the scale of individual amounts and finding the breaches in science theory essential to prophecy based on the underlying corollaries, uncertainty can be established. Since the last decades, the geographic approachability of service is an imperative topic in health geography, spatial epidemiology and public health. The emphasis in most epidemiological research is on the health effects of humans. The inconsistencies in this study may be representative of age, aetiology, mystifying influences and knowing the interconnections between mixed individuals. The most important aspects influencing the uncertainty of public health analysis through GeoComputational analysis are illustrated in Fig. 3.13. This section describes our knowledge, accuracy and uncertainty of GeoComputation analysis in public health research.

3.5.1   Data Error

In geographical context, epidemiological analysis to be carried out at an ecological level, for detecting association between exposure distribution and disease occurrences. Accuracy defines the agreement of a measurement with a putative location or exact value. Therefore, if the population and health statistics are unreliable and inaccurate, the analysis results may be exaggerated via the selection bias. Census statistics are not accurate in many countries and should be used on a comprehensive basis. These data will usually be gathered every decade for a single snapshot. The variations in demographics between the censuses also raise the vagueness and unreliability of these results. Health Registers is an another source of deformations, shows spatial and temporal variation (Forand et al. 2002).

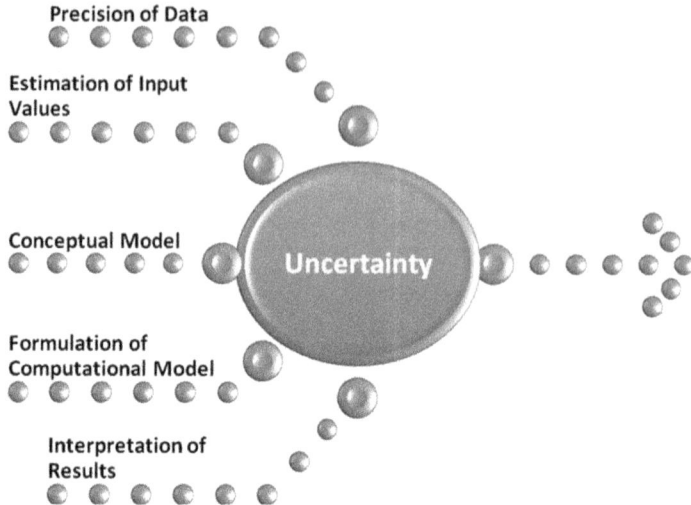

Fig. 3.13 Schematic diagrams of major components enhancing uncertainty in GeoComputational analysis on public health

3.5.2   Positional Error

Geocoding is the procedure of enchanting address/evidence and altering it into geographic coordinates beneficial for health analysis. This is to be done based on a reference dataset from which the geographic coordinates can be calculated. This measurement is carried out by knowing the from and end points of geographic coordinates through interpolation along a street segment and/or a real interpolation within a small administrative unit. The accuracy of the geocoding process is based on two aspects: (i) completeness (e.g. proportion of successfully geocoded reference information) and (ii) positional accuracy (e.g. adjacence between real coordinates and geocoded coordinates). Zandbergen et al. (2011) have abridged the characteristics of positional error in four generalized ways:

(i) Precision depends on rural areas' extent compared to the urban area which reduces inconsistency on street segments on the façade.

(ii) Sampling distribution, which is not always Gaussian, e.g. a mixture of Gaussian and t-distribution, given a good match for positional error distributions observed (Zimmerman et al. 2007).

(iii) Displacement direction of positional errors is not symmetrical and is determined by rectilinear conformation (e.g. cardinal direction) running north-south and east-west; hence, geocoding positional errors along these axes would be greater.

(iv) Geocoding positional errors are spatially autocorrelated.

3.5.3   Cartographic Confounding

Cartographic confounding takes place when spatial accuracy is a location attribute and is generally accepted as basic aspects of geocoded data (Oliver et al. 2005). In classical epidemiological terms, one factor can be calculated by its interaction with others (confounder) and disease risk because of its impact. Also, if the factor of interest is regional, the location of maps of the disease can be confused by a disease factor not uniformly distributed around the study region. Therefore, an accurate analysis of maps needs an understanding of this phenomenon that we refer to as 'cartographic confounding'. Therefore, spatial trends of the incidence of disease may confound the location and the socio-demographic risk factors for the disease. Geographically based, i.e. location related, is the cartographic confounding.

Many examples of cartographic confounding in geocoded cancer studies have been shown with spatial bias regression (Griffith et al. 2007), leading to inaccurate estimates in exposure (Zandbergen 2007), health-environment relationship in cancer epidemiology (Mazumdar et al. 2008) and rural settings (Cayo and Talbot 2003). The idea that geographical location can serve to take account of unmeasured or unmeasurable risk from the disease is a central contribution from spatial disease analyses. But when the risk factors for missed sites are some of the same risks as for the disease being examined, this confounding between the risk factor and the site makes it harder to obtain an unbiased spatial analysis (Gregorio et al. 1999). Methods designed to make possible unrepresentative data accounts should therefore also take this geographical component into account. One way to deal with this problem is through imputation of geocodes to minimize the lack of data.

3.5.4   Misclassification

Misclassification can possibly familiarize spurious spatio-temporal patterns in risk which would directly mark the cogency of any epidemiological study (Oliver et al. 2005). Misclassification may change the odds (OR) ratio of exposure to disease in the population. This inconsistency can sometimes pose major problems when determining the existence and intensity of the interaction exposure-disease, as the reasons for the variance are uncertain and the magnitude of the variation is high. In an exposure-disease association analysis, two forms of misclassification may occur: non-differential and differential. But differential misclassification can have indeterminate, very far null, impacts on null or even on the opposite side of the null. There is no differential error because neither disease cluster susceptibility nor disease specificity differs according to exposure level. Difference misclassification, on the other hand, occurs when the disease misclassification differs by exposure type. The effect of geocoding errors is based on the spatial variations in population or risks

(Ward et al. 2005). Subsequently, the migration of population will also introduce errors in spatial and temporal risk pattern (Arnold 1999). This is particularly challenging for consequences with long dormancy periods between exposure provocation and disease onset, but the dormancy phases, migration and germane contact matrices are not well categorized.

3.5.5 Spatial Extent

The administrative limitations of public health have also changed over time, making it more difficult to develop various coding system administrations for the geographic sector by different departments. Inconsistent geography is troublesome for any space-time study that involves time periods when limit changes occur and is a major challenge in the event that evocative figures become distracting and correlated over time. Health threats are often associated with fairly arbitrary functional areas, but risk scans are prone to changes in the issue of affected areas (Openshaw 1984). The accumulation of data at various spatial resolution levels is unmitigated to skew results that can perturb the interpretation of the results.

The findings obtained solely by clustered data should not be used to create assumptions about the existence of a particular connotation. Some of the ecologic prejudices are generated in the region by using the data at the micro level. Small area studies allow local effects to be investigated (Elliott and Wartenberg 2004). Advances in statistical approaches in the last decade contain the allowance of spatial disease mapping models to integrate the time dimension. Ecological investigations promoted from the amalgamation of individual and areal level evidence recently projected by Jackson et al. (2006) may be subject to ecological bias.

3.5.6 Spatial Weights

To quantify the impacts of spatial weights in potential cluster membership, interpolation and predictive models, a formal understanding of positional uncertainty has been described. In public health application, spatial weights have been used: (i) in exposure job to regulate connection in buffer zone, (ii) in spatial regression to enter immediacy between duos of sites and (iii) in geostatistics to compute variograms that support the spatial interpolation. For instance, the effects of positional error on nearest neighbour relationships have been evaluated by Jacquez and Rommel (2009) and measured the amount of bias that would be presented by positional errors for places of residence.

3.5.7   Spatial Modelling

Interpretation and decision building with geographical data should be completed with knowledge of their nature and reliability. 'Uncertainty' is an explanation of the deficient information of the true value of a specific variable, or its real unpredictability in an individual or a group. It can be introduced into a health risk assessment at every step of the process. Finkel (1990) and US EPA (1999) have classified uncertainty into (i) variable uncertainty, (ii) model uncertainty, (iii) decision-rule uncertainty and (iv) variability. When the variables seeming in equations cannot be measured precisely, uncertainty occurs due to equipment limitations and ST variances, whereas the model uncertainty occurs due to increased complexity and lack of data to predict the model. The uncertainty of decision rule arises when determining an acceptable level of risk to balance between different social concerns, single point estimation instead of joint probability.

The quality assessment of a research work to be estimated based on the methodological quality and data validation of articles. Caldwell et al. (2011) and Creswell (2008) considered several questions for quality assessment of the research work, as follows:

- Is the methodology identified and acceptable?
- Is a theoretical lens or perspective used for research guidance, with a reference?
- Is the theoretical context defined?
- Is the theoretical structure simply tied to the problem?
- By using a conceptual structure, are the definitions properly demarcated?
- Are the associations among concepts clearly acknowledged?

A positive and strong scoring meant that information was provided enough to score a yes. The workout with less score meant some information was given but not enough to score yes. The relative risk is interpreted or mapped in epidemiology without any attempt to disclose the uncertainty in the risk estimates. The Bayesian methods have a powerful output in terms of relative risk-related uncertainty metrics, such as 95% credibility intervals and risk ranks (Jarup et al. 2002). When there is knowledge to measure ambiguity, many methods may be used to combine spatial and aspatial data to promote clarification. A single choropleth map can, for example, show regional variability on risk assessments combined with their added vulnerability (Monmonier 2006). De Cola (2002) proposed that unsaturated colour or decreased crispness in borders increases uncertainty. Goovaerts (2006) has reported that multiple maps can be used to sneakily disclose uncertainty either statically or in a simulation geographically of risk values. Subsequently, a geographical obscurity may be covered by a locational error or the collection of data.

3.6   A Tiered Approach to Accuracy and Uncertainty

The accuracy and uncertainty assessment are an important step in the risk classification process. Three tiers are involved and may be useful discretely to estimate the accuracy and uncertainty, i.e.,

(i) For acceptable measurement scales, the adjustment and co-variation of all input values must be clearly demonstrated.
(ii) Sensitivity analysis can be used to determine how differences in inputs are influenced by models' predictions, for example, to distribute input values based on their position in the model performance.
(iii) The method of differentiation propagation should be considered to map the relation between general accuracy and uncertainty in risk estimates.

3.7   Conclusion

Within Health-GIS, the computational and data sciences, particularly cyberGIS, find significant demands. Health-GIS has emerged over the past several decades as a major driving force underpinning the evolution of spatial analytics, particularly spatial statistics. Highly sophisticated models and procedures have been developed to characterize spatial disease distribution and to predict ill-health risks based on environmental and socio-economic factors. The availability of data, methods and technology is challenging for the future of spatial epidemiology. Multidisciplinary teams are involved to find out approaches which are multiplying errors and complement each other. Computer resources are made available in order to promote the analysis process to manage environmental public health data. The spatial statistical approach was used in terms of information and time without arbitrary criteria. Consequently, understanding the issues on local data and the explanation of the systematic output can persist a vital constituent of the task, rather than being diluted by the overwhelming anxieties on time and cost of undertaking the investigation. However, geocoded data's transparency provides excellent opportunities for epidemiological study and allows large-scale work to be assessed over a long period.

However, complex spatial variations and trends, complicated pertinent elements and heterogeneous populations at risk have presented challenges for the formulation of stable, general and yet precise modelling solutions, small number difficulties and aggregate data, uncertain locational details and the random effects. The data mining refers to define and assess links between disease/health and environment based on complex and vast spatial data on human and natural systems. With mobile devices, sensor networks, remote sensing, social media and any source of spatial data, this Big Data issue is especially important in the context of health-GIS. In the context of contemporary globalization and global climate change, health-GIS studies include more and more different international and interdisciplinary stakeholders. In order to understand and make decisions in this highly complex, interwoven and distributed world, cybric network simulation (e.g. agent-based modelling) is essential. Stochastic research adapts to accommodate health conditions with subjectivity and uncertainty. Problems like site versus place have recently been created, as well as an unclear background in health research studies. The rapid development in digital technology turns many fields of study – from science and engineering to humanities – into the GeoComputation and data sciences.

References

Abidi B, Jilbab A, Haziti ME (2017) Wireless sensor networks in biomedical: wireless body area networks. In: Europe and MENA cooperation advances in information and communication technologies. Springer, Cham, pp 321–329

Aggarwal CC (2015) Mining spatial data. In: Data mining. Springer, Cham, pp 531–555

Ahmadi H, Arji G, Shahmoradi L, Safdari R, Nilashi M, Alizadeh M (2019) The application of internet of things in healthcare: a systematic literature review and classification. Univ Access Inf Soc 18:837–869

Andreou E, Ghysels E (2002) Detecting multiple breaks in financial market volatility dynamics. J Appl Econ 17:579–600

Anselin L (1995) Local indicators of spatial association—LISA. Geogr Anal 27(2):93–115

Aoki M (2013) State space modelling of time series. Springer Science & Business Media

Arnold R (1999) Small area health statistics unit procedures for estimating populations in small areas. Studies on medical and population subjects-Off Popul Census & Surv 62: 10–24

Atluri G, Steinbach M, Lim KO III, MacDonald A, Kumar V (2014, April) Discovering groups of time series with similar behaviour in multiple small intervals of time. In: Proceedings of the 2014 SIAM International Conference on Data Mining. Society for Industrial and Applied Mathematics, pp 1001–1009

Atluri G, Karpatne A, Kumar V (2018) Spatio-temporal data mining: a survey of problems and methods. ACM Computing Survey (CSUR) 51(4):1–41

Baker SB, Xiang W, Atkinson I (2017) Internet of things for smart healthcare: technologies, challenges, and opportunities. IEEE Access 5:26521–26544

Ballas D, Clarke G, Dorling D, Rigby J, Wheeler B (2006) Using geographical information systems and spatial microsimulation for the analysis of health inequalities. Health Informatics J 12(1):65–79

Benke K, Benke G (2018) Artificial intelligence and big data in public health. Int J Environ Res Public Health 15(12):2796

Berke EM, Tanski SE, Demidenko E, Alford-Teaster J, Shi X, Sargent JD (2010) Alcohol retail density and demographic predictors of health disparities: a geographic analysis. Am J Public Health 100(10):1967–1971

Besag J, Newell J (1991) The detection of clusters in rare diseases. Journal of the Royal Statistic Society A 154(154):143–155

Bhatia M, Sood SK (2019) Exploring temporal analytics in fog-cloud architecture for smart office healthcare. Mobile Network & Application 24(4):1392–1410

Bhunia GS, Shit PK (2019) Geospatial analysis of public health. isbn:978-3-030-01680-7

Bogen KT, Spear RC (1987) Integrating uncertainty and Interindividual variability in environmental risk assessment. Risk Anal 7(4):427–436

Bonomi F, Milito R, Zhu J, Addepalli S (2012, August) Fog computing and its role in the internet of things. In: Proceedings of the first edition of the MCC workshop on Mobile cloud computing, pp 13–16

Boriah S, Mithal V, Garg A, Kumar V, Steinbach MS, Potter C, Klooster SA (2010, October) A comparative study of algorithms for land cover change. In: CIDU, pp 175–188

Box GE, Jenkins GM (1976) Time series analysis: forecasting and control. Holden-Day, San Francisco

Brunsdon C, Fotheringham S, Charlton M (1998) Geographically weighted regression. J R Stat Soc Ser D Stat (The Statistician) 47(3):431–443

Burden S, Steel D (2016) Constraint choice for spatial microsimulation. Popul Space Place 22(6):568–583

Caldwell K, Henshaw L, Taylor G (2011) Developing a framework for critiquing health research: an early evaluation. Nurse Educ Today 31(8):e1–e7

Carlos HA, Shi X, Sargent J, Tanski S, Berke EM (2010) Density estimation and adaptive band-widths: a primer for public health practitioners. Int J Health Geogr 9(1):1–8

Cayo MR, Talbot TO (2003) Positional error in automated geocoding of residential addresses. Int J Health Geogr 2(1):10

Chen Y, Shen W, Huo H, Xu Y (2010, July) A smart gateway for health care system using wire-less sensor network. In: 2010 Fourth International Conference on Sensor Technologies and Applications. IEEE, pp 545–550

Corcoran P, Datta SK (2016) Mobile-edge computing and the Internet of Things for consumers: extending cloud computing and services to the edge of the network. IEEE Consum Electronic Mag 5(4):73–74

Corno F, De Russis L, Roffarello AM (2016, June) A healthcare support system for assisted liv-ing facilities: an IoT solution. In: 2016 IEEE 40th annual computer software and applications conference (COMPSAC), vol 1. IEEE, pp 344–352

Cressie N, Wikle CK (2015) Statistics for spatio-temporal data. John Wiley & Sons, New York

Creswell JW (2008) Chapter 3 the use of theory, Thousands Oaks

Culotta A (2010, July) Towards detecting influenza epidemics by analyzing Twitter messages. In Proceedings of the first workshop on social media analytics (pp. 115–122)

Dang LM, Piran M, Han D, Min K, Moon H (2019) A survey on internet of things and cloud com-puting for healthcare. Electronics 8(7):768

de Cola L (2002) Spatial forecasting of disease risk and uncertainty. Cartogr Geogr Inform Sci 29:363–380

Deiner MS, Lietman TM, McLeod SD, Chodosh J, Porco TC (2016) Surveillance tools emerg-ing from search engines and social media data for determining eye disease patterns. JAMA Ophthalmol 134(9):1024–1030

Devarajan M, Subramaniyaswamy V, Vijayakumar V, Ravi L (2019) Fog-assisted personalized healthcare-support system for remote patients with diabetes. J Ambient Intell Humaniz Comput 10(10):3747–3760

Duque JC, Ramos R, Suriñach J (2007) Supervised regionalization methods: A survey. Int Reg Sci Rev 30(3):195–220

Eberhardt MS, Ingram DD, Makuc DM (2001) Urban and rural health chartbook: health, United States. National Center for Health Statistics, Hyattsville

Edwards KL, Clarke GP (2009) The design and validation of a spatial microsimulation model of obesogenic environments for children in Leeds, UK: SimObesity. Soc Sci Med 69(7):1127–1134

Elliott P, Wartenberg D (2004) Spatial epidemiology: current approaches and future challenges. Environ Health Perspect 112:998–1006

Erdeniz SP, Maglogiannis I, Menychtas A, Felfernig A, Tran TNT (2018, May) Recommender systems for IoT enabled m-health applications. In: IFIP International conference on artificial intelligence applications and innovations. Springer, Cham, pp 227–237

Ester M, Kriegel HP, Sander J, Xu X (1996, August) A density-based algorithm for discovering clusters in large spatial databases with noise. In Kdd (Vol. 96, No. 34, pp. 226–231)

Farahani B, Firouzi F, Chang V, Badaroglu M, Constant N, Mankodiya K (2018) Towards fog-driven IoT eHealth: promises and challenges of IoT in medicine and healthcare. Futur Gener Comput Syst 78:659–676

Finkel AM (1990) Confronting uncertainty in risk management. In: Center for Risk Management, resources for the future, Washington, DC

Forand SP, Talbot TO, Druschel C, Cross PK (2002) Data quality and the spatial analysis of disease rates: congenital malformations in New York state. Health Place 8(3):191–199

Gahegan M (2000) What is GeoComputation? A history and outline. Available at: http://www.geocomputation.org/what.html.

García-Valls M, Calva-Urrego C, García-Fornes A (2020) Accelerating smart eHealth services execution at the fog computing infrastructure. Futur Gener Comput Syst 108:882–893

Gardner ES Jr (2006) Exponential smoothing: the state of the art—part II. Int J Forecast 22(4):637–666

Giannotti F, Nanni M, Pinelli F, Pedreschi D (2007, August) Trajectory pattern mining. In Proceedings of the 13th ACM SIGKDD international conference on Knowledge discovery and data mining (pp. 330–339)

Ginsberg J, Mohebbi MH, Patel RS, Brammer L, Smolinski MS, Brilliant L (2009) Detecting influenza epidemics using search engine query data. Nature 457(7232):1012–1014

Glatman-Freedman A, Kaufman Z, Kopel E, Bassal R, Taran D, Valinsky L et al (2016) Near real-time space-time cluster analysis for detection of enteric disease outbreaks in a community setting. J Infect 73(2):99–106

Gong J, Wu H, Zhang T, GuiZ LZ, You L, Shen S, Zheng J, GengJ QK, Yang W, Li Z, Yu J (2012) Geospatial Service Web: towards integrated cyberinfrastructure for GIScience. Geo-spatial Inf Sci 15(2):73–84

Gopal S, Woodcock CE, Strahler AH (1999) Fuzzy neural network classification of global land cover from a 1° AVHRR data set. Remote Sens Environ 67(2):230–243

Goovaerts P (2006) Geostatistical analysis of disease data: visualization and propagation of spatial uncertainty in cancer mortality risk using Poisson kriging and p-field simulation. Int J Health Geogr 5(7). https://doi.org/10.1186/1476-072X-5-7

Goutte C, Toft P, Rostrup E, Nielsen FA, Hansen LK (1999) On clustering fMRI time series. Neuroimage 9(3):298–310

Grefenstette JJ, Brown ST, Rosenfeld R, DePasse J, Stone NT, Cooley PC, Guclu H (2013) FRED (A Framework for Reconstructing Epidemic Dynamics): an open-source software system for modeling infectious diseases and control strategies using census-based populations. BMC Public Health 13(1):1–14

Gregorio DI, Cromley E, Mrozinski R, Walsh SJ (1999) Subject loss in spatial analysis of breast cancer. Health Place 5(2):173–177

Griffith DA, Millones M, Vincent M, Johnson DL, Hunt A (2007) Impacts of positional error on spatial regression analysis: A case study of address locations in Syracuse, New York. Trans GIS 11(5):655–679

Grosse SD, Matte TD, Schwartz J, Jackson RJ (2002) Economic gains resulting from the reduction in children's exposure to lead in the United States. Environ Health Perspect 110:563–569

Gubbi J, Buyya R, Marusic S, Palaniswami M (2013) Internet of Things (IoT): A vision, architectural elements, and future directions. Futur Gener Comput Syst 29(7):1645–1660

Harvey AC (1990) Forecasting, structural time series models and the Kalman filter. Cambridge University Press, Cambridge

Hong K, Lillethun D, Ramachandran U, Ottenwälder B, Koldehofe B (2013, August) Mobile fog: A programming model for large-scale applications on the internet of things. In Proceedings of the second ACM SIGCOMM workshop on Mobile cloud computing (pp. 15–20)

Horváth L, Kokoszka PS, Teyssière G (2001) Empirical process of the squared residuals of an ARCH sequence. Ann Stat 29:445–469

Internet of Things at a Glance. Available online: https://www.cisco.com/c/dam/en/us-/products/collateral/se/internet-of-things/at-a-glance-c45-731471.pdf. Accessed on 23 Feb 2019

Islam SR, Kwak D, Kabir MH, Hossain M, Kwak KS (2015) The internet of things for health care: a comprehensive survey. IEEE Access 3:678–708

Jackson C, Best N, Richardson S (2006) Improving ecological inference using individual level data. Stat Med 25:2136–2159

Jacquez GM, Jacquez JA (1999) Disease clustering for uncertain location. In: Lawson A, Bertollini R (eds) Disease mapping and risk assessment for public health decision making. Wiley, London

Jacquez GM, Rommel R (2009) Local indicators of geocoding accuracy (LIGA): theory and application. Int J Health Geogr 8:60

Jarup L, Best N, Toledano MB, Wakefield J, Elliott P (2002) Geographical epidemiology of prostate cancer in Great Britain. Int J Cancer 97(5):695–699

Karpatne A, Khandelwal A, Chen X, Mithal V, Faghmous J, Kumar V (2016) Global monitoring of inland water dynamics: state-of-the-art, challenges, and opportunities. In: Computational sustainability. Springer, Cham, pp 121–147

Kawale P, (2011). Determinants of use of health information in Nathenje health area of Lilongwe District. A Dissertation Submitted in Partial Fulfilment of the Requirements of the Master of Public Health Degree, University of Malawi

Kelejian HH, Prucha IR (1999) A generalized moment's estimator for the autoregressive parameter in a spatial model. Int Econ Rev 40(2):509–533

Kisilevich S, Mansmann F, Nanni M, Rinzivillo S (2010) Spatio-temporal clustering. In: Data mining and knowledge discovery handbook. Springer, Boston, pp 855–874

Kraemer FA, Braten AE, Tamkittikhun N, Palma D (2017) Fog computing in healthcare–a review and discussion. IEEE Access 5:9206–9222

Krizhevsky A, Sutskever I, Hinton GE (2012) Imagenet classification with deep convolutional neural networks. Adv Neural Inf Proc Syst 25:1097–1105

Kulldorff M (1997) A spatial scan statistic. Commun Statist – Theory Meth 26(6):1481–1496

Kumar T, Ramani V, Ahmad I, Braeken A, Harjula E, Ylianttila M (2018) Blockchain utilization in healthcare: key requirements and challenges. In 2018 IEEE 20th International Conference on e-Health Networking, Applications and Services (Healthcom) (pp. 1–7). IEEE

Kumari A, Tanwar S, Tyagi S, Kumar N (2018) Fog computing for healthcare 4.0 environment: opportunities and challenges. Comp & Electric Eng 72:1–13

Kwan MP (1998) Space-time and integral measures of individual accessibility: a comparative analysis using a point-based framework. Geograph Analys 30(3):191–216

Lawson AB (2001) Statistical methods in spatial epidemiology. John Wiley & Sons, London

Lawson A, Bertollini R (eds) (1999) Disease mapping and risk assessment for public health. John Wiley & Sons, London

Lee JG, Han J, Whang KY (2007, June) Trajectory clustering: a partition-and-group framework. In Proceedings of the 2007 ACM SIGMOD international conference on Management of data (pp. 593–604)

Li Z, Chen J, Baltsavias E (eds) (2008) Advances in photogrammetry, remote sensing and spatial information sciences: 2008 ISPRS congress book, vol 7. CRC Press, Hoboken

Liang L, Chen Y, Hawbaker TJ, Zhu Z, Gong P (2014) Mapping mountain pine beetle mortality through growth trend analysis of time-series Landsat data. Remote Sens (Basel) 6(6):5696–5716

Longley PA, Brooks S, Macmillan W, McDonnell RA (1998) Geocomputation: a primer. Wiley, Chichester

Lu C, Mejia-Guevara I, Hill K, Farmer P, Subramanian SV, Binagwaho A (2016) Community-based health financing and child stunting in rural Rwanda. Am J Public Health 106:49–55. https://doi.org/10.2105/AJPH.2015.302913

Mazumdar S, Rushton G, Smith BJ, Zimmerman DL, Donham KJ (2008) Geocoding accuracy and the recovery of relationships between environmental exposures and health. Int J Health Geogr 7(1):13

Mezer A, Yovel Y, Pasternak O, Gorfine T, Assaf Y (2009) Cluster analysis of resting-state fMRI time series. Neuroimage 45(4):1117–1125

Mithal V, Garg A, Boriah S, Steinbach M, Kumar V, Potter C, Klooster S, Castilla-Rubio JC (2011) Monitoring global forest cover using data mining. ACM Trans Intelligent Syst & Technol (TIST) 2(4):1–24

Monmonier M (2006) Cartography: uncertainty, interventions, and dynamic display. Prog Hum Geogr 30(3):373–381. https://doi.org/10.1191/0309132506ph612pr

Morris B, Trivedi M (2009, June) Learning trajectory patterns by clustering: experimental studies and comparative evaluation. In 2009 IEEE Conference on Computer Vision and Pattern Recognition (pp. 312–319). IEEE

Morrissey K, Hynes S, Clarke G, O'Donoghue C (2010) Examining the factors associated with depression at the small area level in Ireland using spatial microsimulation techniques. Ir Geogr 43(1):1–22

Mutlag AA, Abd Ghani MK, Arunkumar NA, Mohammed MA, Mohd O (2019) Enabling technologies for fog computing in healthcare IoT systems. Futur Gener Comput Syst 90:62–78

Nandyala CS, Kim HK (2016) From cloud to fog and IoT-based real-time U-healthcare monitoring for smart homes and hospitals. Int J Smart Home 10(2):187–196

Ng RT, Han J (2002) CLARANS: A method for clustering objects for spatial data mining. IEEE Trans Knowl Data Eng 14(5):1003–1016

Nilsen P (2015) Making sense of implementation theories, models and frameworks. Implement Sci 10(1):53

Oliver MA, Webster R (1990) Kriging: a method of interpolation for geographical information systems. Int J Geogr Inf Syst 4(3):313–332

Oliver MN, Matthews KA, Siadaty M, Hauck FR, Pickle LW (2005) Geographic bias related to geocoding in epidemiologic studies. Int J Health Geogr 4(1):29

Openshaw S (1984) Ecological fallacies and the analysis of areal census data. Environ Plan A 16(1):17–31

Openshaw S, Turton I (2001) Using a geographical explanations machine to explore spatial factors relating to primary school performance. Geogr Environ Model 5(1):85–101

Openshaw S, Charlton ME, Wymer C, Craft A (1987) A mark 1 geographical analysis machine for the automated analysis of point data sets. Int J Geogr Inf Syst 1:335–358

Openshaw S, Charlton M, Craft AW, Birch JM (1988) Investigation of leukaemia clusters by use of a geographical analysis machine. Lancet 331(8580):272–273

Page ES (1954) Continuous inspection schemes. Biometrika 41(1/2):100–115

Paul A, Pinjari H, Hong WH, Seo HC, Rho S (2018) Fog computing-based IoT for health monitoring system. J Sensor 2018:1–7

Potter MA, Brown ST, Cooley PC, Sweeney PM, Hershey TB, Gleason SM, Lee BY, Keane CR, Grefenstette J, Burke DS (2012) School closure as an influenza mitigation strategy: how variations in legal authority and plan criteria can alter the impact. BMC Public Health 12(1):977

Rahmani AM, Gia TN, Negash B, Anzanpour A, Azimi I, Jiang M, Liljeberg P (2018) Exploiting smart e-Health gateways at the edge of healthcare Internet-of-Things: A fog computing approach. Futur Gener Comput Syst 78:641–658

Rushton G, Lolonis P (1996) Exploratory spatial analysis of birth defect rates in an urban population. Stat Med 15:717–726

Sharma S, Chen K, Sheth A (2018) Towards practical privacy-preserving analytics for IoT and cloud based healthcare systems. IEEE Internet Computing (99):1–1. https://doi.org/10.1109/MIC.2018.112102519

Shekhar S, Chawla S (2003) Introduction to spatial data mining. A Tour, Spatial Databases, pp 21–44

Shekhar S, Evans MR, Kang JM, Mohan P (2011) Identifying patterns in spatial information: a survey of methods. Wiley Interdiscip Rev Data Min Knowl Discov 1:193–214

Shi X (2010) Selection of bandwidth type and adjustment side in kernel density estimation over inhomogeneous backgrounds. Int J Geogr Inf Sci 24(5):643–660

Shi X, Wang S (2015) Computational and data sciences for health-GIS. Ann GIS 21(2):111–118

Shi L, Tsai J, Kao S (2009) Public health, social determinants of health, and public policy. J Med Sci 29(2):43–59

Shi X, Miller S, Mwenda K, Onda A, Rees J, Onega T et al (2013) Mapping disease at an approximated individual level using aggregate data: a case study of mapping New Hampshire birth defects. Int J Environ Res Public Health 10(9):4161–4174

Shook E, Wang S, Tang W (2013) A communication-aware framework for parallel spatially explicit agent-based models. Int J Geogr Inf Sci 27(11):2160–2181

Son D, Lee J, Qiao S, Ghaffari R, Kim J, Lee JE, Song C, Kim SJ, Lee DJ, Jun SW, Yang S (2014) Multifunctional wearable devices for diagnosis and therapy of movement disorders. Nat Nanotechnol 9(5):397

Sood SK, Mahajan I (2018) A fog-based healthcare framework for chikungunya. IEEE Internet Things J 5(2):794–801

Speybroeck N, Van Malderen C, Harper S, Müller B, Devleesschauwer B (2013) Simulation models for socioeconomic inequalities in health: a systematic review. Int J Environ Res Public Health 10(11):5750–5780

Steinbach M, Tan PN, Kumar V, Klooster S, Potter C (2003, August) Discovery of climate indices using clustering. In Proceedings of the ninth ACM SIGKDD international conference on Knowledge discovery and data mining (pp. 446–455)

Sultan N (2014) Making use of cloud computing for healthcare provision: opportunities and challenges. Int J Inf Manag 34(2):177–184

Tang W, Zhang K, Zhang D, Ren J, Zhang Y, Shen XS (2019) Fog-enabled smart health: toward cooperative and secure healthcare service provision. IEEE Commun Mag 57(5):42–48

Van Den Heuvel M, Mandl R, Pol HH (2008) Normalized cut group clustering of resting-state FMRI data. PLoS One 3(4):e2001

Vijayakumar V, Malathi D, Subramaniyaswamy V, Saravanan P, Logesh R (2018) Fog computing-based intelligent healthcare system for the detection and prevention of mosquito-borne diseases. Comput Hum Behav 100:275–285

Wang K, Shao Y, Xie L, Wu J, Guo S (2018) Adaptive and fault-tolerant data processing in healthcare IoT based on fog computing. IEEE Trans Netw Sci Eng 7:263–273

Ward MH, Nuckols JR, Giglierano J, Bonner MR, Wolter C, Airola M, Mix W, Colt JS, Hartge P (2005) Positional accuracy of two methods of geocoding. Epidemiology 16(4):542–547

Weber J (2003) Individual accessibility and distance from major employment centers: an examination using space-time measures. J Geogr Syst 5(1):51–70

Weber J, Kwan MP (2003) Evaluating the effects of geographic contexts on individual accessibility: a multilevel Approach1. Urban Geogr 24(8):647–671

Xu B, Xu L, Cai H, Jiang L, Luo Y, Gu Y (2017) The design of an m-Health monitoring system based on a cloud computing platform. Enterp Inf Syst 11(1):17–36

Zandbergen PA (2007) Influence of geocoding quality on environmental exposure assessment of children living near high traffic roads. BMC Public Health 7(1):37

Zandbergen PA, Hart TC, Lenzer KE, Camponovo ME (2011) Error propagation models to examine the effects of geocoding quality on spatial analysis of individual-level datasets. Spat Spatio-temporal Epidemiol 3(1):69–82

Zgheib R, Conchon E, Bastide R (2017) Engineering IoT healthcare applications: towards a semantic data driven sustainable architecture. In: eHealth 360°. Springer, Cham, pp 407–418

Zhao Y, Zhang L, Li P, Huang B (2007) Classification of high spatial resolution imagery using improved Gaussian Markov random-field-based texture features. IEEE Trans Geosci Remote Sens 45(5):1458–1468

Zheng X, Rodríguez-Monroy C (2015) The development of intelligent healthcare in China. Telemed & e-Health 21(5):443–448

Zheng X, Vieira A, Marcos SL, Aladro Y, Ordieres-Meré J (2019) Activity-aware essential tremor evaluation using deep learning method based on acceleration data. Parkinsonism Relat Disord 58:17–22

Zhu Z, Woodcock CE, Olofsson P (2012) Continuous monitoring of forest disturbance using all available Landsat imagery. Remote Sens Environ 122:75–91

Zimmerman DL, Fang X, Mazumdar S, Rushton G (2007) Modeling the probability distribution of positional errors incurred by residential address geocoding. Int J Health Geogr 6(1):1

Chapter 4
GeoComputation and Geo-visualization in Public Health

Abstract Geo-visualization is a subfield of Geoinformatics that draws attention both on visualization research, spatial algorithm research in data mining and machine learning. The chapter investigates how visualization can be successfully applied to all phases of problem-solving in geographical analysis, from initial hypothesis development through knowledge discovery, analysis, presentation and evaluation. This chapter also supports the application of methods of geovisualization to the major problems in GIScience. Several graphical representations (information spaces) are used to describe the general structure and clustering of the data and to gain insight into the relationships between the various variables.

Keywords Exploratory analysis · Disease cluster · Space-time cluster · Mashups · GeoComputation vs. GeoVisualization

4.1 Introduction

Visual analytic is the science of analytical reasoning facilitated by interactive visual interfaces.–Jim Thomas and K A. Cook (2005)

Geographical data are special regarding the multitude of vicissitudes that have influenced the assortment of public health data since the last decade. Progressively, manually collected and prudently plaid high-quality public health data are being amplified with mechanically collected information, where completeness and correctness can no longer be engaged for approval, for example, those gathered by Global Positioning System (GPS) devices or smartphones or extracted from satellite imagery. Visualization practices offer the potential to fetch the human being into the loop and to practice the human's ruling competence to distinct the eloquent from the irrelevant, subsequently freeing the human from dull handing out of huge volume of data by merging in computerized analysis algorithms. For space-time analysis, convergence of visual and computational methods is important and working and serviceable. Geo-visualization tools are highly mandatory. Such paraphernalia may support not only the skilled analyst but also the public service engaged in the analysis of knowledge in time and space.

© The Author(s), under exclusive license to Springer Nature
Switzerland AG 2021
G. S. Bhunia, P. K. Shit, *GeoComputation and Public Health*,
Springer Geography, https://doi.org/10.1007/978-3-030-71198-6_4

Geo-visualization (or *geographic visualization*) is a discipline in which human and machine intermediaries are important, by simplifying the process for data screening and optimizing the interface between human and device. MacEachren and Kraak (2001) described that geo-visualization is a procedure and study with visual geospatial presentations to support exploration, investigation, synthesis and demonstration of geospatial information. Visualization can play an important role in, e.g. 'process pattern developing' (visual illustration showing key features of a process as it unfolds) and, e.g. 'processing guidance' (e.g. interactive environments providing regulatory parameters of a knowledge-building process to form and alter its behaviour). The term Geo-visualization has been taken to mean quite various aspects by different researchers. It accentuates plans to prompt anthropological visual thinking through interface with maps to determine forms and associations in spatial data. Most of the researchers recommend that the geographic data are a vital emphasis to develop approaches that identify and influence the inimitable physiognomies innate in geographic data. In addition, the focus of logical methods is also on position and space and consequently on understanding the intricacy of particular communities and/or patterns of understanding and relationships in spatial space and probably their variability over time which is accountable for them. Researchers on geographical background have a tendency to sight geo-visualization 'as a tactic to deal with the intricacy of the ecosphere and often challenge to pulverize their geo-visualization exploration in geographic concepts and theory'. Geo-visualization is critical for dealing with today's datasets, which are increasing rapidly in scope and complexity in order to gain understanding, to decide designs and to maximize and navigate Byzantine developments.

Geo-visualization is a subfield of Geoinformatics that draws attention both on visualization research, spatial algorithm research in data mining and machine learning. This has the ability to effectively transfigure how geographic data is represented, acquire knowledge, draw inferences and create decisions. Subsequently, spatial data analysis techniques exposed their significance in evaluating patterns and models in terrestrial data using the GeoComputational method almost mechanically. Since the last 5 years, the growth of tightly unified data investigation and visualization approaches is still in the commencement, and supplementary study is required to make development. The instinctive data research group through machine learning, computational algorithm, statistics, etc. illustrated geo-visualization approaches within the public health research. For these instances, geo-visualization is adopted as part of the technology field and is accepted. However, the amalgamation between geo-visualization and geo-computation is still in the beginning, and more investigation is required to make evolution in this respect. Combining these two approaches can potentially improve the supremacy of each method exponentially.

Data is the heart of Geo-visualization system. Information discovery by data mining, spatial query, GeoComputing and related fields takes over the exploration engine and allows for geo-visualization (Fig. 4.1). However, the basic component of Geo-visualization is summarized below:

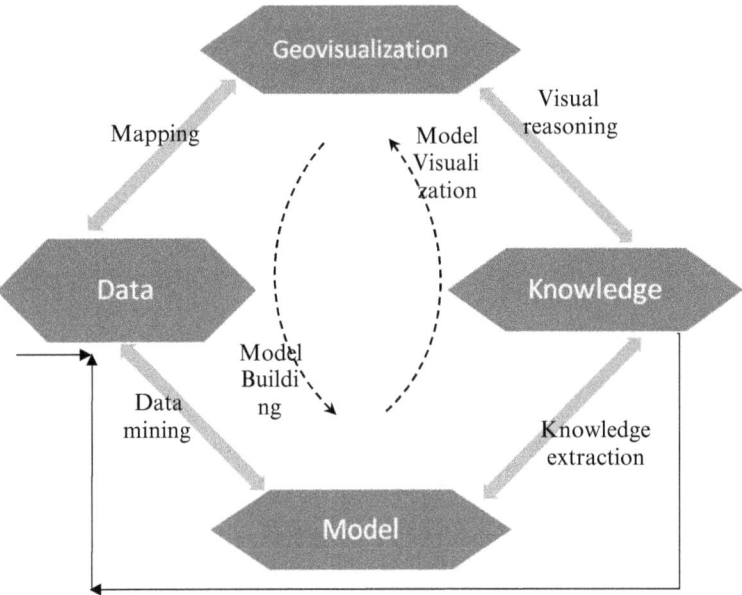

Fig. 4.1 Components of Geo-visualization

- *Data* – Geographic data background information influences the effectiveness of visualization methods and metaphors. The huge dimensions of geospatial data are now accessible, and the highly multivariate nature of some of the integral trends and patterns stands supplementary challenges when conniving valuable visualization techniques. Mining knowledge from progressive event sequences is one of the essential aspects in data mining, and frequent pattern mining (FPM) is a vital issue in chronological data mining (Gotz et al. 2014). In general, 'frequent' is demarcated with a pre-specified *minimal support* value that denotes to the mined pattern should seem in at least a certain percentage of the *event sequences*. The chronological properties of these events, for example, sequence and timing, are then analysed to determine relations with patients' eventual consequences. A number of systems have been used to attain acumens from disease event categorization data, vacillating from data mining system to interactive visualization (Wang et al. 2013a, b; Gotz et al. 2012).

- *Knowledge* – The illustration of geographic knowledge is a major concern today. At present, geographic data are not represented in common languages. Ad hoc models established by the GIS platform can represent and operate the data. In addition, the object-based data model can include hierarchies of generalization and component relationships to define a set of geographically oriented concepts. Exploration of information in libraries has created a platform for the exploration of spatial knowledge. One approach to incorporate information creation in spatial data exploration is to merge the techniques of numerical research with visual

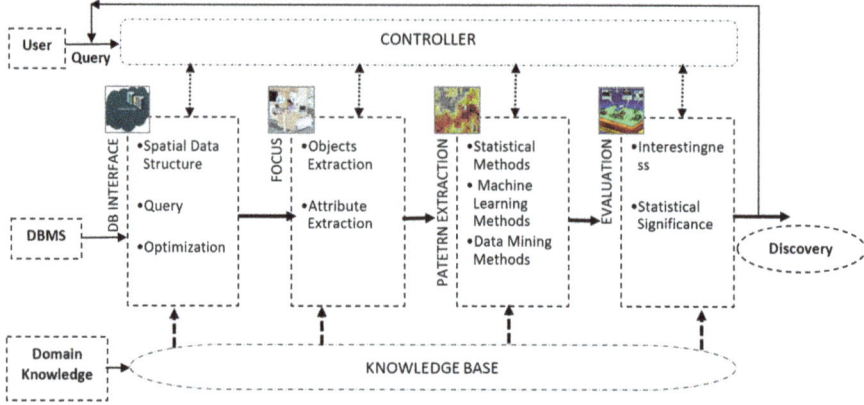

Fig. 4.2 Architecture of knowledge discovery in Geo-visualization and GeoComputation. (Modified after Koperski and Han 1999)

analysis in a framework which can facilitate exploratory analysis. Figure 4.2 illustrated architecture of knowledge discovery of Geo-visualization.

- *Visualization environment* – The simulation environment would be a requirement for interconnecting spatial significance and enabling the user to interrelate with the paraphernalia that create context. A thorough analysis of symbols, exhibits, stories, collective activities and so on would be necessary (MacEachren 1995). The use of graphical representation plays a key role in providing visual description of the data. This suggests a visualization of the sample assembly in its multidimensional space in order to allow for positions for interaction and association research.

These methods usually hierarchically create data to affirm the multi-layered collection of values correlated with a single case. More recently, simulation of patient cohorts has loosened interest. This provides a range of resources for the estimation, review and cataloguing of case data across patient groups (Wongsuphasawat and Gotz 2011). Probably the most important is outflow, a method for visualizing common modes of patient delivery in terms of procedures, symptoms or all other defined categories of historical events (Gotz et al. 2012). Those numerical cohort simulation methods are very effective and typically only manage a small range of categories of events. Wongsuphasawat (2011) used colour coding to discriminate various event types which can lead to an awfully multifaceted web of event pathway.

4.2   Exploratory Visual Analysis of Public Health Data

Geo-visualization is an emergent tool for probing through massive dimensions of data, for collaborating multifaceted patterns, for delivering a prescribed context for data appearance and for exploratory geographical data analysis (Gahegan 2000).

Coppock (1995) made Geo-visualization renowned not only in the growth of GeoComputation but also as a tool to recover dependability of hazard assessment and consequently decision support and also to progress the capability of non-experts to make benefit of the information presented.

4.2.1   Visual Query

The visual query can be done through an easy-to-use user interface component and a query engine. The key aspect of the image application is the episode classification, which covers all individual occurrences from the early onset of the illness and a potential diagnosis. There are three elements in the episode structure: (i) milestones (e.g. start and end of the episode), (ii) preconditions (e.g. a set of limitations which must be met before the starting point) and (iii) an outcome measure (e.g. way to evaluate the eventual result of an episode). An example of disease incidence rate for 5 consecutive years is illustrated in Fig. 4.3.

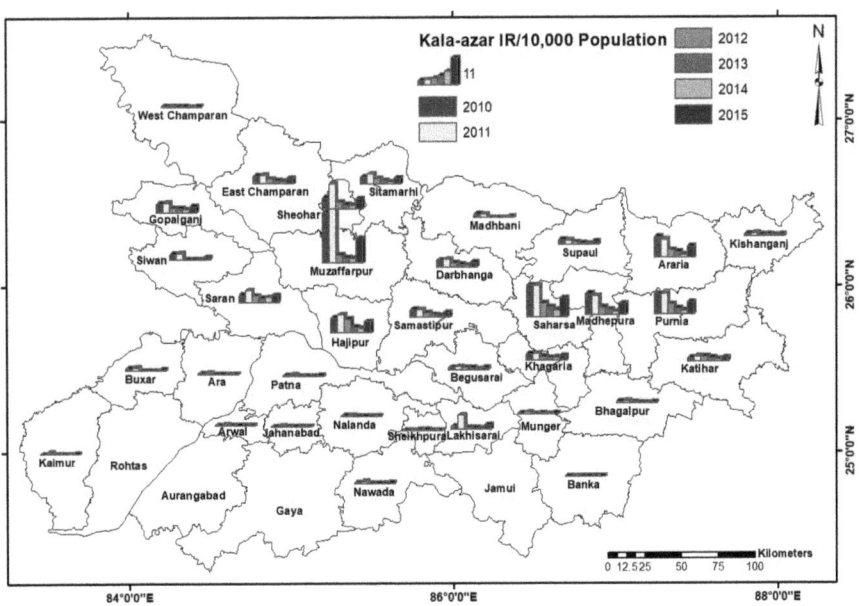

Fig. 4.3 Shows the kala-azar disease incidence rate per 10,000 population at district level. This output helps domain experts in their yearly surveillance tasks, allowing them to extract in timely manner accurate information on disease outbreaks

4.2.2   Temporal Pattern Mining

Temporal pattern mining (TPM) refers to a large number of organized events. TPM comprises two key elements, i.e. regular mine pattern and analyser of statistical models.

4.2.2.1   Frequent Pattern Miner

The key problem in time-based data mining is frequent pattern mining (FPM). FPM aims to find a number of subsequences in a collection of longer sequences of events. The FPM is responsible for the observation of frequent patterns which occur in a series of input systems. 'Frequent' in general is characterized with a minimum support value that is pre-specified, meaning that the extracted pattern is to occur in at least one percent of the event sequences. We use the term sequence of events here, which is the sequence of time-ordered events. Ayres et al., (2002) used bitmap based Sequential pattern Miner (SPAM) algorithm that assimilates a depth-first traversal of the search space with operative clipping instruments. SPAM adopts a pattern growing strategy that starts with an empty frequent pattern set F. Increasing F pattern with length 1 is developed by SPAM and tested if the original patterns surpass the sustenance threshold. If indeed, it is added to F. If yes, they will be added to F and SPAM will grow them with S- and 1-extensions. This procedure is repeated until no patterns can be identified which cross the threshold.

4.2.2.2   Statistical Pattern Analyser

Statistical pattern analyser (SPA) guises for links between the extracted patterns and the episode condition's consequence measure. The first hypothesis of the pattern analyser is that each episode is represented by a bag of pattern (BoP). The BoP illustration is an n-dimensional vector for each patient, where 'n' is the number of pattern and the ith element of the vector stores the frequency of the ith pattern found in the episode. If there are 'm' episodes, then we can construct an $m \times n$ episode pattern matrix $X = [x_1, x_2, \ldots, x_n]$ whose (j, i)th element indicates the number of times the ith pattern appeared in the jth episode. For instance, Pearson correlation (e.g. relationship between two values), Kullback-Leibler divergence (e.g. to measure the information gain between pattern vector and outcome vector), p-value (to measure the significance of correlation) and odds ratio to be computed for statistical relation between the episodic elements.

4.2.3 Interactive Visualization

Recently, the process of interactive visualization with public health data has become increasingly important for understanding huge amounts of data accumulation. The geography of health and disease has great potential to benefit from such interactive visualization. GeoComputation has already become a significant visual tool to support medical practitioners, epidemiologists, public health planners and environmental specialists for understanding geographical distribution of disease-related data. Pattern diagram is an important element of visualization. GeoComputation can not only perceive associations across space but also observe associations between numerous variables and over time. Patterns, colour and sizes are animated to represent statistical shifts at different locations.

4.3 Exploration of Public Health Data on 2-D, 3-D and 4-D

Multiple linked views are used to identify and test different hypotheses about specific datasets in imaginative or exploratory data analyses (Godinho et al. 2007). Combining 3-D data views with conventional 2-D view displays gives you the advantages of providing 3-D visualization when it has been considered useful or successful for seeking new insights into a dataset. Essentially, by integrating different views and interactively connecting them, we can use it or play with various forms of visualization because we do not need to rely on a single data exploration representation. Various studies employ combinations of 2-D (Fig. 4.4) and 3-D displays (Fig. 4.5). Several Geo-visualization and exploratory visualization tools employ a number of different displays that linked together 2-D and 3-D representations. 3-D maps can be viewed as a conceptual bridge between very simplistic 2-D maps and reality. Although 2-D cartographies may be adequate for simple things to give geometrically correct topography, they dramatically simplify reality and therefore do not pay tribute to the highly complex capabilities of human spatial cognition (Resch et al. 2013) (Fig. 4.5).

3-D visualization of multi-temporal geo-data is progressively in the focus of academic research as it has substantial recompenses associated with the 2-D methods in efficiently assigning geographical content (International Organization for Standardization 2006). Since the last few years, there have been improved applications of 3-D illustration of spatial structures and procedures. Nevertheless, cartographic philosophies and variables were researched extensively, and their effect was researched for 2-D (Garlandini and Fabrikant 2009). In various research projects, specific technical methods have been explored in terms of principles for web-based geo-applications like 3-D data representation. The 3-D WebGIS based on the Virtual Reality Modelling Language (VRML) has been created by Shan (1998). Ming (2008) extended this concept by the use of the X3-D language. X3-D has several interfaces to control other tools, which are successors to VRML. Nonetheless, its

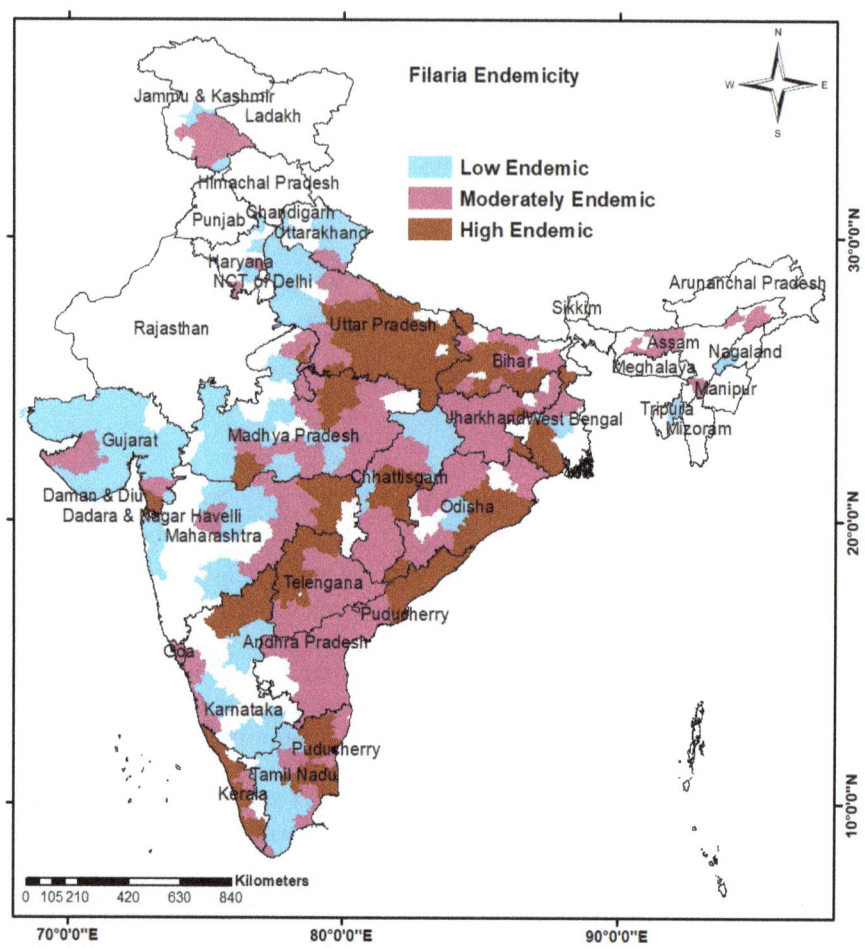

Fig. 4.4 2-D representation of filaria endemicity in India at district level. The 'brown colour' represents the 'high endemic' zone; 'pink colour' shows the moderate endemic zone; and 'light blue' colour shows the low endemic zone of filariasis

simulation abilities for massive data sizes are scarce because there is no graphics processing unit (GPU) (Ming, 2008). The use of 3-D is a significant limitation in viewing real-work scenes which have been skewed by the use of various spatial reference systems by spherical perspective, topographical irregularities, occurrence objects, scale-dependent depiction of features or geometric inconsistencies. In spite of these deficiencies, 3-D maps are better than 2-D mapping for the mapping of a spatio-temporal phenomenon.

There have been three spatial and one temporal dimensions, e.g. four-dimensional (4-D) mapping of geospatial data on the web can suggestively subsidize to solve the challenges of apprising decision-makers in addition to the public about expansion in

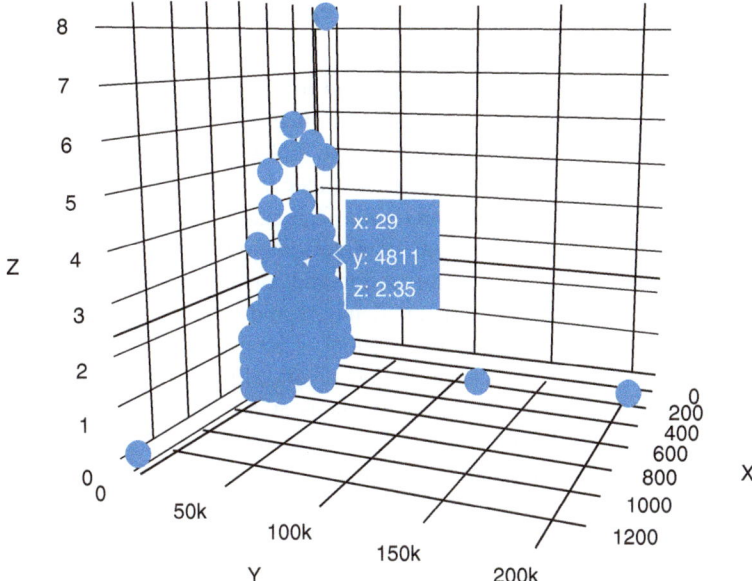

Fig. 4.5 Shows the 3-D scatter plot of a vector layer (scatter plot prepared through Q-GIS software). The 'x' represents the 'village code'; 'y' represents the 'population of the village'; and 'z' represents 'incidence rate'

public health areas. The essential for the 4-D Web-based Geographic Information System (WebGIS) is therefore growing, and web-based geo-data visualization approaches are being developed (Feng et al. 2011). However, 4-D representation of time-varying phenomena in geo-visualization techniques has not yet been studied systematically. Web-based geo-visualization application users often face difficulties in interrelating with structures and data evidence that is far from ideal. The viewing, understanding and interaction with a 4-D map, however, requires considerable cognitive and perceptive capabilities. These comprise the variation of graphical variables such as colour, shape or intensity and more. In addition, understanding spatio-temporal processes can be fostered by conveying information such as the occurrence of disease, geographic distribution or spatial association with variables of environment.

It has not been widely considered the interrelated issues of the public health influence of visualization intricacy on the user, ranging from simple abstraction to photorealistic representation of the spatio-temporal phenomenon. Most 4-D interface users find a pseudo-photorealistic illustration of the natural phenomenon that is more helpful because it conveys a more realistic and understandable view of a real-world situation. This makes it easier to understand and interpret the actual real-world processes of communication. Despite these shortcomings, 3-D cartographic representations are better for representing spatio-temporal phenomenon than 2-D representation.

4.4   Exploratory Analysis of Clustering of Public Health Data

Searching for accurate diagrams of disease and risk trends and identifying whether there is a greater risk in those areas are performed by public health officials and researchers (Mclafferty 2015). In the 1970s and 1980s, local space disease clusters were studied and identified in order to raise understanding of the health effects of environmental pollutants and raise residents' fears about their neighbourhood disease clusters. With the development of GeoComputational tools for performing large-scale spatial analytical computation and for layering spatial information on disease incidence with data on social, demographic and environmental phenomenon (Fig. 4.6). Recognizing the local spatial clusters typically depends on two basic components – (i) local concentration of disease cases which is searching 'field-based', 'site-side' and 'case-side' methods at recurrently spaced interval to check for local clusters and object based approaches that search by building clusters around disease cases or areal units containing case concentrations (Shi 2010) and (ii) hotspot detection system that requires a statistical model for deciding if the local disease concentration is unusual. In the geographical search and statistical hotspot detection process, factors such as uneven population distribution, transportation networks and topography can be incorporated.

Spatial clusters of diseases and health-related activities have long been of concern to researchers and policymakers in public health. Described as an unusual number of cases within a population, location and timeframe, a disease hotspot is a geographic structure that can be defined, visualized and explored using the GeoComputation and Geostatistical method (Table 4.1). Clusters may provide information about disease aetiology and risk behaviours, indicating local environmental or social factors causing increased risk. In addition, the cluster may provide clues on disease aetiology and risk behaviours, indicating local environmental or social characteristics which promote increased risk. GeoVisualization application for data exploration and analysis addresses the issues like (i) filters and selection, (ii) making comparisons, (iii) seeking statistical feedback, (iv) analysing number of

Fig. 4.6 Integration between Geo-visualization and GeoComputation in Public Health Surveillance

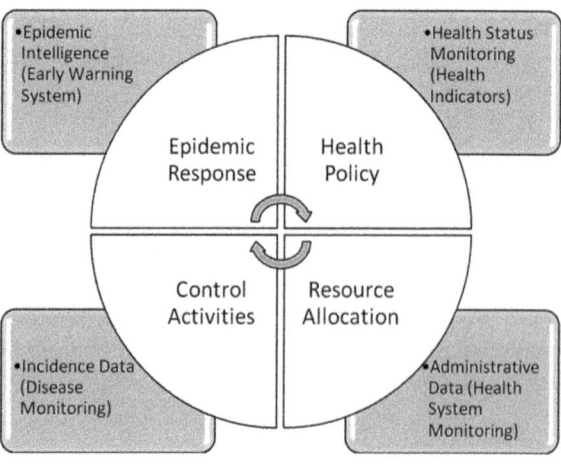

Table 4.1 Spatial and spatio-temporal cluster detection methods

Type	Methods	Applications	References
Spatial search	Local spatial autocorrelation	Detects irregularly shaped cluster	Aldstadt and Getis (2006)
	Spatial filters	Determines minimum population threshold	Cai et al. (2012)
	Multi-objective optimization method	Detecting irregularly shaped cluster	Duczmal et al. (2008)
	SaTScan model	Detects irregularly shaped clusters	Tango and Takahashi (2005)
	Adaptive spatial filters	Minimum population threshold	Tiwari and Rushton (2004)
Network-based	Kernel estimation	Density distribution of point events over space and evaluate network performance in terms of centrality	Okabe et al. (2009)
	Network implementation	To detect bridge nodes between single-hop destination clusters and to find path towards a disjoint destination cluster	Shiode (2011)
	Space-time STAC	Analysing the distribution of disease incidences in time and space; for instance, all features within a given year as a 'layer' and to all features from the same geographical location as a 'stack'	Shiode and Shiode (2013)
	Bayesian method	For the identification of spatial and space-time clusters with significantly higher/lower risk of incidence of the disease. Model is used to partition the study region into a set of areas which are either 'null' or 'non-null', the latter corresponding to clusters (excess risk) or anti-clusters (reduced risk)	Vandenbulcke et al. (2014)
	Local k function	Considers those pairs of points having a given point i as one of its members; number of permutations for creating the confidence envelope	Yamada and Thill (2007) and Getis (1984)
	Ripley's K-function	Global spatial autocorrelation; to evaluate spatial clustering, the scale of clustering and the temporal stability of these clusters	Hinman et al. (2006)
	Local Moran's I	To identify spatial cluster patterns and estimated degree of autocorrelation may vary significantly across geo-space; allowing assessment of dependency relationships in different areas	Yamada and Thill (2010) and Tsai et al. (2009)

(continued)

Table 4.1 (continued)

Type	Methods	Applications	References
Space-time		Longitudinal data with residential histories	Cook et al. (2009)
	Q-method	Case control data with residential histories	Jacquez et al. (2006)
	Bayesian space-time model	Estimating the spatial pattern in disease risk across nth areal units	Li et al. (2014)
	Autologistic regression model	Space-time kernel estimation; distribution of facility-based deliveries identification	Nakaya and Yano (2010)
Statistical modelling	Bayesian method	This algorithm is applied to data prior to the study period and produces n potential cluster structures for the disease data. Moreover, a separate Poisson log-linear model to be applied for each cluster structure, which allows for step-changes in risk where two clusters meet	Lawson (2006)
	Monte Carlo method	Downscale case location	Shi (2010)
	Dirichlet process mixture model	For clustering longitudinal gene expression data; utilized for the means of the regression coefficients to induce clustering; potential convergence to the actual clusters in the data	Verity et al. (2014)

Modified after Mclafferty (2015)

variables at a time and (v) exporting information. The exploratory analysis is being carried out in two stages – (i) based on the numeric attributes (e.g. normalization, principal component analysis, outlier detection and removal, clustering, etc.) and (ii) iterative process have been repeated with the nominal/categorical attributes added. Geo-visualization has effectively been used to sustain the investigation of health-related information, though this triumph is inadequate in accomplishment-beleaguered researchers to use such Geo-visualization applications. Robinson et al. (2005) used a succession of usability practices with GIS and epidemiological connoisseurs to iteratively expand the Exploratory Spatio-Temporal Analysis Toolkit (ESTAT). Geographic investigation of the public health data is reliant upon an amalgamation of indicators generally establishing a number of complexes of attributes on health, environment and/or demography. In such multi-dimensional datasets, the extraction of patterns and the sighting of novel acquaintances may be problematic, as forms remain veiled (Koua and Kraak 2004). Because of the quantities of data collected, the identification of trends and relationships in broad health statistics and survey data is often a challenging activity for data analysis.

New approaches to data analysis and viewing are required to provide a visual representation of data, which can better stimulate the recognition of patterns and generate hypothesis and enhance understanding and knowledge building. Knowledge Discovery in Databases (KDD) has provided a geographical knowledge discovery window for several years now. Moreover, density-based DABSCAN clustering depends upon two input parameters, i.e. number of neighbours essential to flinch a new cluster-*K* and the proximity identifying the neighbourhood of a point-*epsilon*. Data mining, knowledge finding and visualization approaches are often united stab to recognize structures and forms in multifaceted spatial data. Using cartographic methods and information visualization techniques, some graphical illustrations are used to depict consequential structures and forms in a graphical form that can permit for better understanding of the assemblies and the topographical progressions. The usage of these graphical illustrations plays a role by offering pictorial depictions of information that convey the possessions of human insight to bear.

4.4.1 Spatial Search Processes

These inventions mainly focus on the search window (spatial filter or kernel) by detecting local areas in which to test for spatial clusters. Most conventional approaches, for example, the spatial scan statistic, Disease Mapping Analysis Programme (DMAP) and Geographical Analysis Machine (GAM), depend on simple geometric forms – circles or ellipses – of uniform size in probing for clusters; nevertheless, these systems bound our capability to recognize unevenly formed clusters that imitate irregular environmental and habitat patterns. For instance, AMOEBA technique pursuits for cluster successively linking areas or disease cases apparent from case location (Aldstadt and Getis 2006). Tango and Takahashi (2012) have improved the search process and statistical model considering new technique to make it more computationally viable. In identifying clusters that are intermittently created by treading ideas related to cluster type and local relative disease risk, multiple target optimization strategies were explored (Duczmal et al. 2008). Further invention through variable-size spatial filters is included in the spatial search process. Hotspot methods depending on fixed scanning windows typically influence widely different levels of population sustenance like windows in rural areas with a small proportion of the population (Shi 2010). This issue was addressed by a few early cluster detection methods with a minimum limit for spatial filters (Tiwari and Rushton 2004). However, this inconsistency is not well deduced and is different between physical sceneries.

4.4.2 Network-Based Cluster Detection

Network-based methods' pursuit for distance clusters encompasses a transportation network that channels people's places, migrations and connections, and distances between locations are calculated through this network space. Black (1992) assumed Moran's *I* to perceive geographical clustering of flows along a transport link. The estimate of core density has changed to include the recognition of pedestrian accident clusters (Ha and Thill 2011) and network spaces for the spatial scan (Shiode 2011). Vandenbulcke et al., (2014) projected a Bayesian, case-control technique for hotspot identification along network. SANET (Okabe et al. 2006) and GeoDANet (Hwang and Winslow 2012) have developed network solutions for real-world problems. Moreover, topological characteristics of network areas are computational issues requiring consideration in science.

4.4.3   Statistical Analysis and Modelling of Local Clusters

Advances have occurred in local disease risk statistical modelling over the past few decades. Lawson (2006) projected a Bayesian method for classifying disease clusters and hotspot mapping, and spatial extent of disease clusters depends upon case event data. The Bayesian model was developed by Wakefiled and Kim (2013) to test the meaning of the space clustering within the area, which was established by a space growth and trim process. Spatial cluster identification approaches that include multifactorial safety and disease structures have been prolonged (Fig. 4.7). Any present attempts are aimed at modelling the area migration of suburban residents effectively by rationalizing property values into smaller, more organized geographical units for large regions (Shi et al. 2013). Feigning large numbers of probable disease outcomes in complex environmental sceneries with changing statistical assumption necessitates massive GeoComputational resources, producing prospects to yield benefit of current developments in high-performance calculating and CyberGIS (Wang et al. 2013a, b).

4.4.4   Space-Time Cluster Detection Methods

Space scanning statistics and kernel figures have existed for cluster identification in space and time for recent decades, but new efforts have made a move towards bringing them into network perspective. Rogerson and Yamada (2004) have devised approaches for perceiving spatio-temporal clusters depending upon clustering of disease incidence data for interrelated regional systems. The multi-distance spatial cluster analysis tool analyses the geographical pattern of disease incident data. It also summarizes geographical clustering or spatial distribution over a range of distances. Ripley's K-function demonstrates how the geographical clustering or diffusion of feature centroids changes when the neighbourhood size is altered (Fig. 4.8). However, the bulk of these approaches often requires applying time as 3-D to 2-D spatial techniques, without thoroughly analysing people's behaviour during time and space that disrupt the possibility of disease. Nordsborg et al. (2014) used Q-statistics for identifying space-time clusters of breast cancer in Denmark.

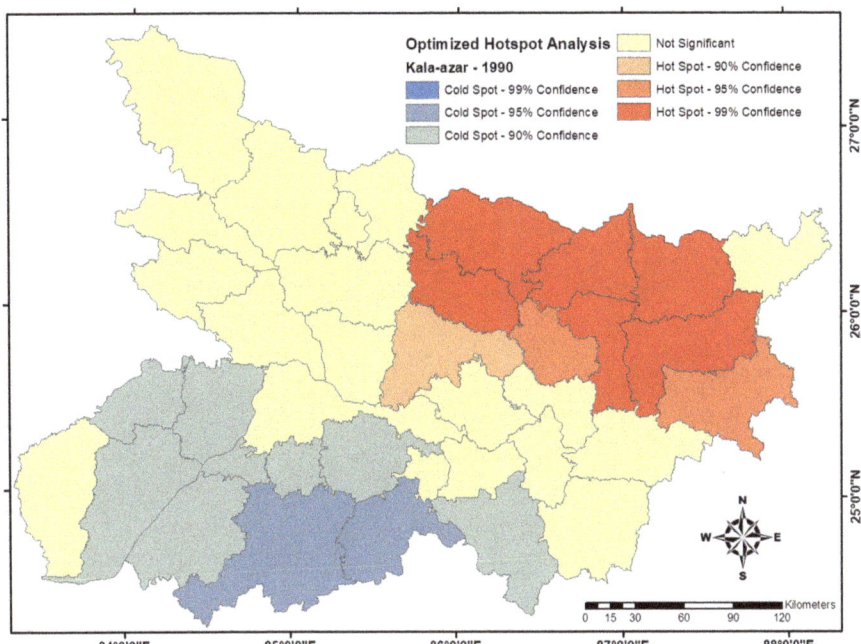

Fig. 4.7 Shows the hotspot and cold spot areas of kala-azar disease, generates a map using the Getis-Ord G_i^* statistic. This tool categorizes statistically significant geographical clusters of high values (hotspots) and low values (cold spots). Finally, this tool creates identifies significant clusters, using the False Discovery Rate (FDR) correction method. The *Getis-Ord* local statistics (Mitchell 2005) is given as:

$$G_i^* = \frac{\sum_{j=1}^{n} w_{i,j}\, x_j - \bar{x} \sum_{j=1}^{n} w_{i,j}}{S \sqrt{\dfrac{\left[n \sum_{j=1}^{n} w_{i,j}^2 - \left(\sum_{j=1}^{n} w_{i,j} \right)^2 \right]}{n-1}}}$$

where x_j is the attribute value for feature j, $w_{i,j}$ is the spatial weight between feature i and j, n is equal to the total number of features and

$$\bar{X} = \frac{\sum_{j=1}^{n} x_j}{n}$$

$$S = \sqrt{\frac{\sum_{j=1}^{n} x_j^2}{n} - \left(\bar{x} \right)^2}$$

4.5 Mashups in Epidemiology

Generating knowledge by merging evidence and amenities from various sources is called 'Mashup'. First of all, it was used to describe the combination of musical tracks to produce a new piece of music. It was first launched in 2004 through the use of social media and web technology (Zhang et al. 2019). Mashups are becoming progressively prevalent, particularly in the context of coalescing spatial data and demonstrating such unified data on maps and web-based mapping applications like

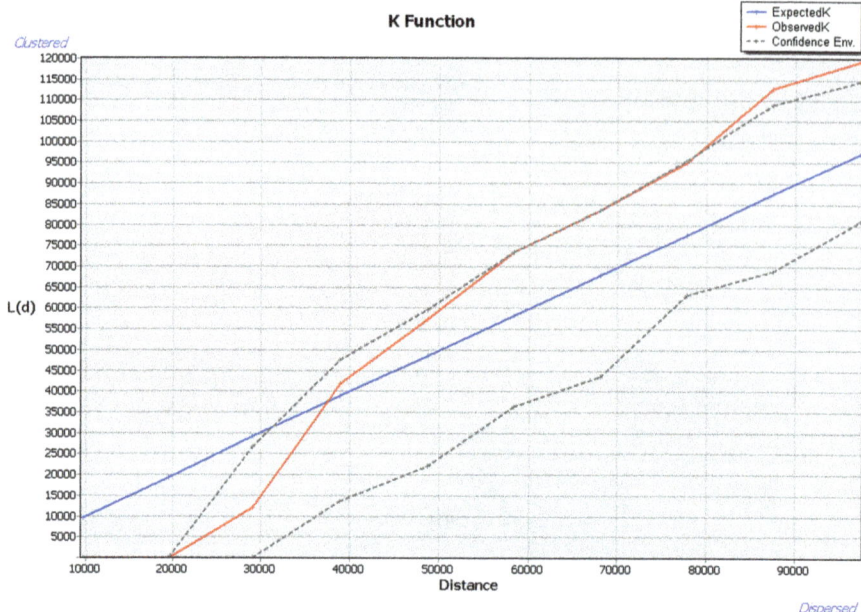

Fig. 4.8 Shows the multi-distance spatial cluster analysis using Ripley's K-function

Mathematically, the multi-distance spatial cluster analysis tool uses a common transformation of Ripley's K-function where the expected result with a random set of points is equal to the input distance (Bailey and Gatrell 1995). The transformation $L(d)$ is calculated as:

$$L(d) = \sqrt{\frac{A\sum_{i=1}^{N}\sum_{j=1,j\neq i}^{N}k(i,j)}{\pi N(N-1)}}$$

where A is the area, N is the number of points, d is the distance and $k(i,j)$ is the weight, which is 1 when the proximity between i and j is less than or when the proximity between i and j is more than d. When edge correction is applied, the weight of $k(i,j)$ is modified at some extent. The K-function continuously appraises feature geographical distribution in association with complete spatial randomness (CSR). When the observed K-value is larger than the expected K-value for a particular distance, the distribution is more clustered than a random distribution at that scale of analysis. When the observed K-value is smaller than the expected K-value, the distribution is more dispersed than a random distribution at that distance. When the observed K-value is larger than the upper confidence envelope value, spatial clustering for that distance is statistically significant. When the observed K-value is smaller than the lower confidence envelope value, spatial dispersion for that distance is statistically significant.

Google Maps, Bing Maps and Open Street Maps which allow to create datasets programmed using the Keyhole Markup Language (KML) format to be displayed via a two-dimensional or three-dimensional map (Boulos et al. 2008). Map mashups incorporate multiple data sets (or 'mash up') that are visually illustrated (Liu and Palen, 2010). With the appearance of Web mapping, geo-browsing activities are amplified rapidly, leading to the quick progress of map mashups (Fig. 4.9). Using this up-to-date, self-motivated and collaborating demonstration and propagation of numerous geospatial information can be executed excellently.

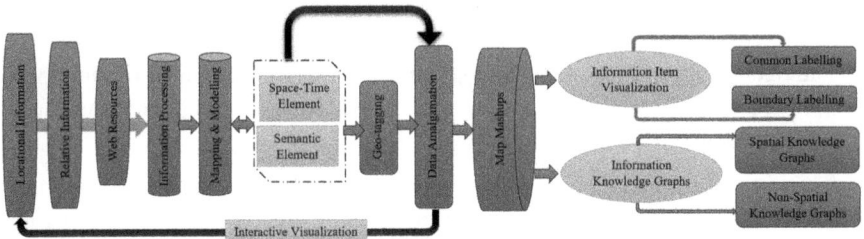

Fig. 4.9 Map mashup model illustration and visualization. (Modified after Zhang et al. 2019)

Global climate change presents a number of fears of health and wellness at a broad sequence of scales, including changes in the incidence and circulation of vector-borne diseases, rising water- and foodborne diseases, intensifying malnutrition and a range of health and well-being effects, linked to extreme conditions (Costello et al. 2009). An unbiased decision support mechanism offers an enhanced connection of data between weather, atmosphere, environmental data and wellbeing and goodwill repositories and expanded access to these data. Data mashups will contribute to creative and original research activities that are helpful to human health and climate change's wellness impacts, offering both evidence and decision support mechanisms through:

- Perceiving and attributing health problems by using cohesive model's new research into ecological knowledge and health
- Quickly detecting hotspots to susceptible populations
- Information sharing of public health planners and environmental management with appropriate monitoring of location information and population risk classifications
- Introducing and appraising interferences to endorse revision by plummeting the exposures
- Circulating and providing admittance to data as a part of outreach and assignation with the research group, policymakers and civil world
- Encouraging the adaptation strategies with recognized efficacy and sustainability for individuals, families, communities and regions, as well as the impact of climate change on health and wellbeing

The Medical and Environmental Data Mashup Infrastructure (MEDMI) is a project to promote the research-grade mashup of temperature, economic, demographic and public health data. MEDMI has been established with the assistance of Medical Research Council and Natural and Environmental Research Council, a coalition of four main collaborators (Leonelli and Tempini 2018). This project aimed to combine data relation to:

- Territory: levels of urbanization, population density as well as socio-economic data, for example, educational levels, average income.
- Climate, i.e. temperature, rainfall, altitude that will provide insight into the weather conditions at various sites

- Patients-related data like symptoms, rate of incidence, time and severity of disease
- Pathogen biology such as physiology, life cycle and nutrition of the microbes which aids to evaluate the circumstances under which pathogen is probable to prosper and most detrimental to hosts

Through collecting all of these data through interdisciplinary studies, the atmosphere and public safety have been connected to a broader extent and the construction of more munitions.

Several research works have been carried out by various scientist and researchers in the world. For instance, *Nature* has formed its individual geo-mashup with Google Earth for pursuing avian flu outbreaks, and HealthMap recognized by the Children's Hospital Informatics Programme in Boston carries together incongruent data sources within Google Maps to attain an amalgamated and complete view of the current global state of infectious diseases and their consequence on human and animal health. The benefit of such easy-to-use GeoComputational analytics is evident in the growing variety and reach and the increasing interest in using GIS and other Web-based technologies to integrate data with public health.

4.5.1 Basic Components of Mashups in Public Health Data

- *Data Formatter* – Information obtained on numerous websites is accumulated and illustrated in various ways. To address this issue, Dapper may be used with the capability to visually map the web content to a specific structure. Such tools allow the content taken to be generated in several formats, including Really Simple Syndication (RSS). These tools help to quickly delete material and customize the site (Cheung et al. 2008).
- *Data Visualization* – When many databases are combined in a standard format, automated data visualization tools are available. For instance, Yahoo! Pipes can be used to integrate and format geo-referenced data into KML format for visualization by Google Maps or Google Earth or Bhuvan.
- *Data Sharing* – One important aspect of Web data sharing and community collaboration. For example, Dapper and Yahoo! Pipes both contain collaboration forums in which users can view and utilize the work of others. In the context of GIS, Websites, for instance, GeoCommons, permit geo-referenced data (e.g. KML files) to be ticketed, shared, reused and mixed.
- *Web API* – Another key part of the web trend features growing use of various Web application programming interfaces (APIs) for developers to build rich client applications that can programmatically access online services such as Google Maps and GeoCommons. Such Web APIs allow existing functionalities to be reused. For example, using the GeoIQ Javascript API provided by GeoCommons, one can develop client applications that include content such as heat maps, concentration indices or intersection indices in custom data. EpiCollect has been

intended to permit the gathering of epidemiological data through two-way communication (e.g. between mobile phone and central database could be utilized)
and could meaningfully increase the assemblage and collection of data for these
types of public projects, particularly among the young where the mobile phone
in the natural way of relocating information.

The network is a tool used by field workers from many places in biology in order
to capture and interpret data. The methods for uploading, visualizing and analysing
data obtained by multiple users can be found in a centralized integrated database
accessible through a website. The database can be efficiently evaluated and explored
by connecting these repositories to Google Maps, which allows for the spatial
spread of genotype pathogens. Centralized repositories for data gathering by field
staff may often be of use to ecologists as well as monitoring habitats of endangered
species. The time period between analyses of data gathered across various fields
may dramatically be shortened through a centralized online interface. For example,
EpiSurveyor (www.episurveyor.org) allows the use of standard Nokia mobile
phones for the collection of text-based data and has been utilized in resource-poor
areas of Africa for many kinds of data collection. The emergence of a new wave of
cell phones, backed up by the word 'smartphones', recently provides new approaches
by technological creation, which allows submission and bidirectional retrieval of
data from the field to the central cellular database (Morris 2009).

Eco-studies like the influenza analysis carried out by Google rely on multitudes
of aggregated knowledge and are therefore vulnerable to issues like ecological mis-
understanding and lack of effective control. Online profiles can generate misleading
information and selective reporting. The Internal Protocol address defines most
Internet computers and can be connected to geographical areas in a specific manner.
Knowledge techniques like page crawling, mashups, natural language processing
and machine learning can be useful for web content translation at different points in
the workflow.

4.5.2 Geotagging

The use of geotagging technology and the map machining framework provides a
new perspective for navigation and location-based knowledge perception. In terms
of many reasons, however, the latest findings are incomplete. Firstly, the dominant
work focuses primarily on material achievement and display and the exhaustive
tenacity and knowledge to be considered radical, and somaticized material has not
been properly examined. Furthermore, the cartographic administration of maps is
important for the methods to simulation of such work, and related cartographic
philosophies and enplanements have not been exploited. The eminence of the visual
diagrams and of the map mashups interface cannot be guaranteed because of the
efficient characteristics of referred position data. Additionally, because the multidimensional content of the evidence has not been explored, conception is

incomplete to roughly extent the location of information; accordingly, the further level acquaintance of the material set has not been systematically pulled out. The specific processes of geotagging are as follows:

(i) *Assignment* – Geographic locations are assigned based on the gazetteer.
(ii) *Disambiguation* – Locational ambiguity due to multiple geographic locations with similar names. The local ambiguity is used, according to hierarchical relations and the assurance score algorithm, by selecting the perhaps most possible set of tasks for individual comparison positions.
(iii) *Focus determination* – Based on the incidence counts and the corresponding confidence scores of the locations, numerous sites mostly related to the information are taken out to epitomize the geographic focus of the page.

4.6   Visual Approaches to Data Exploration and Knowledge Construction

4.6.1   *Population Mapping*

Mapping of populaces of epi-unit (e.g. spatial unit of analysis) is vital as it is controlling the data gathering procedure and it outlines the level upon which analysis and inferences will be strained. Pathogens exist and feast in the populaces, not just in the discrete epi-units. Depending upon the scale of analysis, the geographical unit may be categorized from objects denoted as points (e.g. host location, reservoir location, breeding sites, etc.) to any administrative units or eco-epidemiological regions defined by polygons. From the epidemiological lookout, a clear difference should be demarcated between many kinds of these units depending upon their extent, species alignment, population turnover, biosecurity level and other physiognomies (seasonality, suppleness, etc.). While the large-scale data would unceasingly be selected to the accrued topographies, conservancy of such comprehensive sign might not be continually feasible and less cost explanation for disease mitigation measures.

Depending upon the disease epidemiology, a map of some species to be mapped that may play a key role in sustaining its transmission cycle (reservoir), and other susceptible populations (accidental hosts) separately. The epidemiological reservoir is a species ecological system capable of maintaining the life cycle of a pathogen over a sufficiently extended period of time and involving multiple generations of host species. Hence, careful consideration of all available disease data and epidemiological indications, including results of special studies, is sometimes needed to clearly identify the reservoir (Haydon et al. 2008). However, there are four main ways to describe distribution of host populations:

- *Point units*: representing disease location at household level or village level
- *Polygons*: areal units equal to administrative division and considering the spatial range of villages, districts, provinces nations
- *Grids*: systematic homogeneous rectangular polygons not connected with the administrative divisions portraying populations as incessant spatial variable
- *Eco-epidemiological zone*: encircling areas employed by parallel production schemes and atmospheres

4.6.2 Point Epi-units

Epidemiological production units may be represented as points in two ways: (a) intensive system and (b) extensive system. Creating and maintaining such a database up to date is a very imperative part of epidemiological mapping (Fig. 4.10). It is also strappingly simplifying any disease inhibition and control activities, vacillating from surveillance to inoculation. Data gathered in the databases should not be implicated with data on extensive production structure in the similar area and classically should comprise evidence reliable of biosecurity, species, definite and strategic scale of production, specialty, immunization status, import/export alignment, etc. Information on typical species compositions, production system, seasonal migration, etc. is usually collected during household surveys, which are carried out rarely and are quite expensive. Such investigations are typically used for more calculations and accumulation of data at micro level or across higher administrative units.

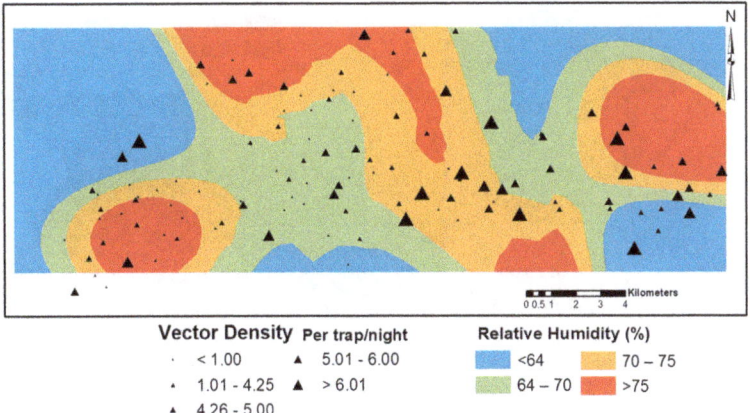

Fig. 4.10 Shows the vector density distribution and its association with indoor relative humidity at micro level. Such investigation helps to demarcate threshold value of relative humidity for determining vector abundance

4.6.3 Aggregated Mapping

Aggregated representation of epidemiological information is most commonly used for statistical, economical and administration purposes. Most of the public health data are gathered, arranged and preserved based on the administrative unit. Higher-resolution administrative division can be a more appropriate technique to explain epidemiological units (Fig. 4.11). However, administrative division better corresponds to geographical distribution patterns of disease. The main drawbacks of this type of analysis are that:

(a) The administrative units are varying in size and shape and do not capture necessary level of details,
(b) Non-analogous between neighbouring units which obscures regional and transboundary extents of analysis.
(c) These are often reformed, accumulated or disaggregated, making it difficult to use chronological data along with probable prophecies and classification.

District-level division for aggregated kala-azar case counts. From a technical perspective, the outlines of administrative units are produced at various scales and degrees of generalization in a topologically correct manner. Here, visualization of aggregated quantitative disease incidence data is normalized in order to account for variation in polygonal areas. However, the administrative polygons are changed,

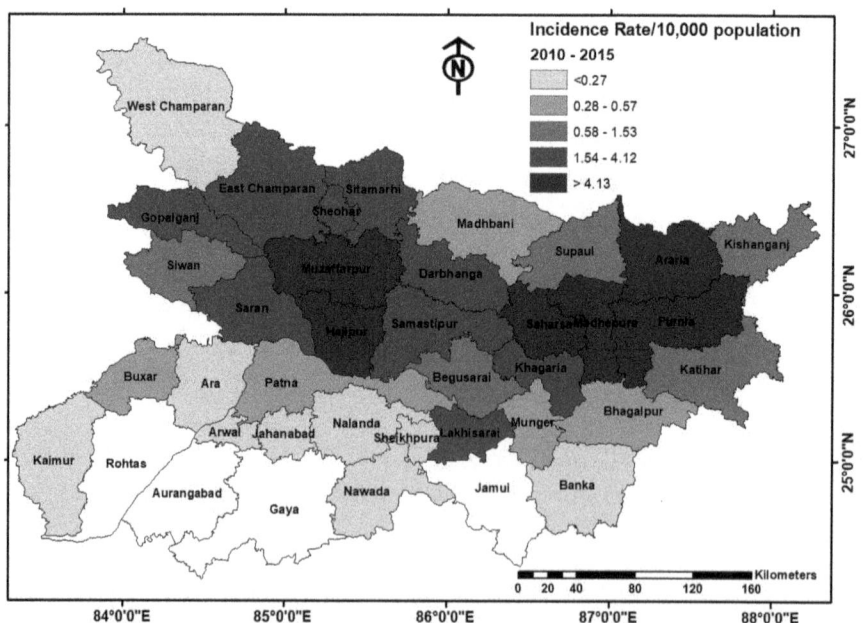

Fig. 4.11 Shows the disease incidence rate/10,000 population at district level of Bihar state in India

aggregated or disaggregated as well as do prospective predictions and characterization of *kala-azar* incidence within the state.

From technical side, the outlines of administrative units are produced at various scales and degrees of generalization, which makes it difficult to align them together in a topologically correct manner. However, when using polygon units for describing disease distribution and estimating aggregated disease frequency matrices for this kind of geographical objects, it is important to understand the 'true epidemiological units'. Geographical segregation of disease in a quasi-epidemiological unit, nevertheless putative for statistical tenacities, expediency of illustration and practical management. Visualization of collected numerical population data always necessitates standardization in order to account for distinction in polygon areas.

4.6.4   Gridded Maps

A good solution to overcome some of the existing data gaps and also better reflect continuous, rather than discrete, patterns in distribution of disease is gridded maps. Grid cells are regular standardized rectangular polygons with epidemiological data expressed as a number of diseases in a grid cell or as a density. This method is also a promising way to plot infectious and vector-borne diseases which better reproduces the role of geographically adjustable ecological factors in their circulation and immigration patterns across time and space. Overall, gridded maps are a much more flexible way to deal with large spatial scales and particularly suitable for disease having regional to global distribution. It also helps to harmonize data collection across international borders, evade discrepancies and breaches in the information, develop statistically rigorous models and forecast and circumvent difficulties associated with recurrent variations in administrative boundaries. The main weaknesses of the approach are lower precision of disease prediction due to use of obsolete historical data for modelling.

4.7   Challenges of Geo-visualization and GeoComputational Analysis in Public Health

The visualization of geographic information has become progressively significant because of the intensifying necessity for better and more accurate viewing, more supple data models and more assorted data integration, in a wide diversity of applications. Geo-visualization analytics deals with the issues regarding terrestrial space and many objects, events, phenomenon and progressions peopling it. Special attention must be given to temporal variability and connotations between space and time. *Daniel Keim* fears about the shortage of interdisciplinary support between the researchers specifying in visual and cooperating practices. Scientists focused on computational analysis tactics are now established with several application areas, which promotes opinions about the probable destiny of the Geo-visualization as an independent study field. *Stefan Wrobel* offers a belief of a computer scientist concentrating in computational analysis approaches. He has confidence in a great possibility presented by coalescing these approaches with visual and interactive manners. In another way of computer science, Geo-visualization has not yet documented standards appraising outcomes of research and development. Evolving such standards is crucial for the GeoComputational process. Recently, researchers on Geo-visualization have focused little consideration to secrecy issues even though most of the tools have established discretion insinuations. This circumstances prerequisite to be rehabilitated and handy co-operation between Geo-visualization and secrecy fortification research is mandatory.

4.7.1   Data Containers

CSV files are among the easiest types of files to interpret in reflection of the public health data (e.g. columns are divided by viruses, rows are divided by a newline). XLSX is a Microsoft Office Open XML (OOXML) format, and it is a specification consisting of zipped XML files and other embedded data that is common among practitioners of public health. However, there are a number of challenges in these formats. Because of the complexity of this file type, a third-party library will be needed to read and write XLSX archives almost in all circumstances (for instance, xlrd and xlwt for Python, POI for Java for Apache). JSON is simple and quick to use for programming languages and is another popular data format. JSON is a simple, but not just tabular, format, as opposed to CSV. JSON can be more complicated and conceptually similar to the XML relation between records. Due to its ubiquity and structure, GeoJSON (http://geojson.org/) and TopoJSON (https://github.com/topojson/topojson) enable sharing geographic data. In fact, a number of photographs and information on public health can be downloaded in PDF format. Information extraction, however, is the key challenge in mixed formats, such as text (e.g. photographs of descriptive text in a report), graphic (e.g. bar and line charts) and tabular. Because of the variety of ways a table can be epitomized in a PDF format, the graphical data cannot be continuously mechanized and even the tabular data are archetypally problematic to abstract.

4.7.2   Data Privacy

Technologies empowering gathering and investigation of many classes of private data have developed speedily. The drawback of these developments is increasing risk to personal secrecy comprising locations of people. Investigation of these data may counter the rights of individuals in preventing the disclosure of their home, workplace, activities, etc. However, in the field of geovisualization, privacy concerns have not yet been closely considered, and some researchers may be unaware of their particular significance for the systematic process. The European research project on Geographic Privacy-Aware-Knowledge Discovery and Delivery (GeoPKDD) has focused on data mobility and created new methods for anonymization and privacy-preserving analysis of geographic data (Andrienko et al. 2011). In addition, GeoPKDD is continuing to work towards mobility, data mining and privacy (MODMP), by synchronizing and improving research activities in the areas of agility, data extraction and privacy conservation. Hence, integrating GeoComputation and Geo-Visualization can enhance inherent capabilities while complementing and amplifying the capabilities to establish various associations among public health information.

4.7.3   Evolution Data Standards

There are several uses of epidemiological data. It may also be used from a public health perspective to better explain decision-making and resource management for population-level production of diseases. High-quality location data are needed to allow prophecy and prediction from a modelling point of view. Depending on this model, the performance criteria of the relevant elements in these two disciplines must be empirically established. The Internet has been the primary way for epidemiological knowledge to be distributed, shared and compiled. However, no such standard exists for the publication of public health-related epidemiological data (Sane and Edelstein 2015). However, these data are numerous machine-readable formats like Comma-Separated Values (CSV), Microsoft Excel, Extensible Markup Language (XML) and Adobe Portable Document Format (PDF). Data containers (e.g. CSV, JavaScript Object Notation (JSON)) and element formats (e.g. timestamp format, location name format) may vary (Fairchild et al. 2018). Character encodings (e.g. ASCII, UTF-8) and line endings (e.g., \r\n, \n) may also change. In addition, the introduction of a homogenized epidemiological data context faces many legal, scientific, political and cultural challenges (Pisani and AbouZahr 2010). These challenges are closely linked to the scheme, data model and standard fluency. Moreover, the goals to be set up for array of methods, one wants to rely, and follow a standard that allow us to recognize when a certain piece of research or a certain piece of software has actually reached the goals that we are setting forward for ourselves.

4.7.4   Interface Challenges

Some data is freely accessible via simple, modern web interfaces and well-documented programming application interfaces (APIs), while others are presented as static web pages, which require unreliable and vulnerable web disclosure. Such an interface provides the simplest and most powerful method of data collection based on US Centers for Disease Control and Prevention (CDC) (https://data.cdc. gov/) and the World Health Organization (WHO) Global Health Observatory (GHO) (http://www.who.int/gho/en/). Many departments of public health are subject to resource constraints and depend on older websites that appear to be published in some standard format (e.g. CSV, Microsoft Excel, PDF) and presented in tables of Hypertext Markup Language (HTML).Web scraping is based on the trends in a website's HTML/CSS source code. Although HTML scraping is often straightforward, it can be very difficult in some instances. When an organization changes its layouts, scrapers can exhibit unpredictable behaviour. One example of a difficult-to-scrape data source is the Robert Koch Institute SurvStat 2.0 website (Robert Koch Institute 2016). In certain instances, direct communication must be made with a person, who then prepares and sends the requested data. These physically implored and ordered results, however, are also plagued by various limitations. Finally, the outcomes of epidemiological data interfaces or system knowledge are always a burdensome and error-prone chore. Having multiple interfaces boosts the possibility of man-made errors when collecting epidemiological data.

4.7.5 Geography

Subtle differences of political boundaries and names may lead to incorrect result during an analysis. The ISO 3166 standard outlines country and principle subdivision names, but it does not knob finer-than-subdivision sections. Subsequently, political boundaries are modified over time. Hence, the incidence rate for a disease must take into account these changes.

4.7.6 Data Reporting

Bureaucratic report process of public health data is another major challenge for geo-visualization and GeoComputational analysis. Modern disease surveillance system relies on multifaceted reporting hierarchies. Raw data are initially captured and then anonymized and then aggregate data sent to the next level in the hierarchy. As such, most of the developed regions and developing countries of the world pass the data by hand. Accordingly, most disease surveillance systems across the world experience reporting lags of at least 1 to 2 weeks that affect a spontaneous understanding of the state of affairs in addition to computational forecasting models.

Another aspect is heterogeneous case definitions across the jurisdictions. Due to undefined epidemiological data, it is often tough to circumnavigate websites to recognize the definitions. Moreover, some public health data are available in the local language of the region of the world in which they originate. Although the online verbal conversion facilities do exist, these are not continually reliable, and they cannot simply decode text in the images. Moreover, the historical data is not published at regular intervals. These are updated retroactively, i.e. maybe updated next week, or the following week, as new data appear. Hence, the analyst and modeller must be aware of these potential issues.

4.8 Conclusion

Given considerable progress in the detection and modelling of disease hotspot over the past decades, there are many methods for incorporating process-based identifications that are relevant in determining spatial and spatio-temporal trends of public health-related issues. Although identifying the cluster of diseases in the absence of these empathies can be helpful for policymakers in determining where geographically to target interferences in public health, predicting the physiognomies of these intercessions needs understanding why disease risk is increasing in the target area. Incorporating time variations into the visualization scheme based on maps on the Internet remains a methodological and technical challenge. There are also no well-documented collaborative functionality and/or non-optimal data presentation for the users of web-based geovisualizing systems. Over the past decade, spatial cluster detection methods were however significantly advanced, driven by advances over spatial search and statistical and computational methodological growth, network-based approaches and time analysis. WebGIS has gradually transformed and analysed geographic information and changed the audience of this information. Various communities like 'Consumer Health Geoinformatics' or 'Wikification of GIS' or 'Participatory GIS' opening up the opportunity of several new applications and comprehending the visions. Presently, geotagging and other web-based applications exist where people can add their own individual data to a share in real time over the web. GPS enabled cell phones and cameras to allow crowds of people to cooperatively view the earth in ways that have never been done before and to open up various mobile location-based service opportunities and openings. Furthermore, the incorporation of environmental factors makes sense to adopt a step-by-step process similar to that used to incorporate individual-level covariates in cluster detection, which helps to observe changes in cluster locations, sizes and spatial extent, suggesting whether the environmental factor has an impact on the clustering of local diseases.

References

Aldstadt J, Getis A (2006) Using AMOEBA to create a spatial weights matrix and identify spatial clusters. Geogr Anal 38(4):327–343

Andrienko GL, Andrienko N, Keim D, MacEachren AM, Wrobel S (2011) Challenging problems of geospatial visual analytics. J Vis Lang Comput 22(4):251–256

Ayres J, Flannick J, Gehrke J, Yiu T (2002) Sequential pattern mining using a bitmap representation. In: Proceedings of the eighth ACM SIGKDD international conference on Knowledge discovery and data mining, pp 429–435

Bailey TC, Gatrell AC (1995) Interactive spatial data analysis, vol 413, no 8. Longman Scientific & Technical, Harlow

Black WR (1992) Network autocorrelation in transport network and flow systems. Geogr Anal 24(3):207–222

Boulos MNK, Scotch M, Cheung KH, Burden D (2008) Web GIS in practice VI: a demo playlist of geo-mashups for public health neogeographers. Int J Health Geogr 7:38

Cai Q, Rushton G, Bhaduri B (2012) Validation tests of an improved kernel density estimation method for identifying disease clusters. J Geogr Syst 14(3):243–264

Cheung KH, Yip KY, Townsend JP, Scotch M (2008) HCLS 2.0/3.0: Health care and life sciences data mashup using Web 2.0/3.0. J Biomed Inform 41(5):694–705

Cook KA, Thomas JJ (2005) Illuminating the path: the research and development agenda for visual analytics (No. PNNL-SA-45230). Pacific Northwest National Lab. (PNNL), Richland

Cook AJ, Gold DR, Li Y (2009) Spatial cluster detection for repeatedly measured outcomes while accounting for residential history. Biometr J 51(5):801–818

Coppock JT (1995) GIS and natural hazards: an overview from a GIS perspective. In: Carrara A, Guzzetti F (eds) Geographical information systems in assessing natural hazards. Kluwer, Dordrecht, pp 21–34

Costello A, Abbas M, Allen A, Ball S, Bell S, Bellamy R, Friel S, Groce N, Johnson A, Kett M, Lee M (2009) Managing the health effects of climate change: lancet and University College London Institute for Global Health Commission. Lancet 373(9676):1693–1733

Duczmal L, Cançado ALF, Takahashi RHC (2008) Delineation of irregularly shaped disease clusters through multiobjective optimization. J Comput Graph Stat 17(1):243–262

Fairchild G, Tasseff B, Khalsa H, Generous N, Daughton AR, Velappan N, Priedhorsky R, Deshpande A (2018) Epidemiological data challenges: planning for a more robust future through data standards. Front Public Health 6:336

Feng L, Wang C, Li C, Li Z (2011) A research for 3D WebGIS based on WebGL. In: Proceedings of 2011 international conference on computer science and network technology, vol 1. IEEE, pp 348–351

Gahegan M (2000) Visualization as a tool for geocomputation. In: GeoComputation. Taylor and Francis, London/New York, pp 253–274

Garlandini S, Fabrikant SI (2009) Evaluating the effectiveness and efficiency of visual variables for geographic information visualization. In: International conference on spatial information theory. Springer, Berlin/Heidelberg, pp 195–211

Getis A (1984) Interaction modeling using second-order analysis. Environ Plan A 16(2):173–183

Godinho PIA, Meiguins BS, Meiguins ASG, do Carmo RMC, de Brito Garcia M, Almeida LH, Lourenco R (2007) PRISMA-a multidimensional information visualization tool using multiple coordinated views. In: 2007 11th international conference information visualization (IV'07). IEEE, pp 23–32

Gotz D, Stavropoulos H, Sun J, Wang F (2012) ICDA: a platform for intelligent care delivery analytics. In: AMIA annual symposium proceedings, vol 2012. American Medical Informatics Association, p 264

Gotz D, Wang F, Perer A (2014) A methodology for interactive mining and visual analysis of clinical event patterns using electronic health record data. J Biomed Inform 48:148–159

Ha HH, Thill JC (2011) Analysis of traffic hazard intensity: a spatial epidemiology case study of urban pedestrians. Comput Environ Urban Syst 35(3):230–240

Haydon DT, Morales JM, Yott A, Jenkins DA, Rosatte R, Fryxell JM (2008) Socially informed random walks: incorporating group dynamics into models of population spread and growth. Proc R Soc B Biol Sci 275(1638):1101–1109

Hinman SE, Blackburn JK, Curtis A (2006) Spatial and temporal structure of typhoid outbreaks in Washington, DC, 1906–1909: evaluating local clustering with the G i* statistic. Int J Health Geogr 5(1):13

Hwang MH, Winslow A (2012) User manual for GeoDaNet: spatial analysis on undirected networks. 2012-03-22. https://geodacenter.asu.edu/drupal_files/Geodanet_Manual_03_2012.pdf

ISO, I (2006) 9241: ergonomics of human–system interaction – Part 151: Guidance on world wide web interfaces. International Organisation for Standardisation

Jacquez GM, Meliker JR, AvRuskin GA, Goovaerts P, Kaufmann A, Wilson ML, Nriagu J (2006) Case-control geographic clustering for residential histories accounting for risk factors and covariates. Int J Health Geogr 5(1):32

Koperski KJ, Han J (1999) Adhikary: mining knowledge in geographic data. Communications of the ACM

Koua EL, Kraak MJ (2004) Geovisualization to support the exploration of large health and demographic survey data. Int J Health Geogr 3(1):1–13

Lawson AB (2006) Disease cluster detection: a critique and a Bayesian proposal. Statist Med 25:897–916. https://doi.org/10.1002/sim.2417

Leonelli S, Tempini N (2018) Where health and environment meet: the use of invariant parameters in big data analysis. Synthese:1–20

Li G, Haining R, Richardson S, Best N (2014) Space–time variability in burglary risk: a Bayesian spatio-temporal modelling approach. Spat Stat 9:180–191

Liu SB, Palen L (2010) The new cartographers: Crisis map mashups and the emergence of neogeographic practice. Cartogr Geogr Inf Sci 37(1):69–90

MacEachren AM (1995) How maps work: representation, visualization, and design. Guilford Press, New York

MacEachren AM, Kraak M-J (2001) Research challenges in Geovisualization. Cartogr Geogr Inf Sci 28(1):3–12

Mclafferty S (2015) Disease cluster detection methods: Recent developments and public health implications. Ann GIS 21(2):127–133

Ming W (2008) A 3D web GIS system based on VRML and X3D. In: 2008 second international conference on genetic and evolutionary computing. IEEE, pp 197–200

Mitchell A (2005) The ESRI guide to GIS analysis, vol 2. Redlands

Morris K (2009) Mobile phones connecting efforts to tackle infectious disease. Lancet Infect Dis 9(5):274

Nakaya T, Yano K (2010) Visualising crime clusters in a space-time cube: an exploratory data-analysis approach using space-time kernel density estimation and scan statistics. Trans GIS 14(3):223–239

Nordsborg RB, Meliker JR, Ersbøll AK, Jacquez GM, Poulsen AH, Raaschou-Nielsen O (2014) Space-time clusters of breast cancer using residential histories: a Danish case–control study. BMC Cancer 14(1):255

Okabe A, Okunuki KI, Shiode S (2006) SANET: a toolbox for spatial analysis on a network. Geogr Anal 38(1):57–66

Okabe A, Satoh T, Sugihara K (2009) A kernel density estimation method for networks, its computational method and a GIS-based tool. Int J Geogr Inf Sci 23(1):7–32

Pisani E, AbouZahr C (2010) Sharing health data: good intentions are not enough. Bull World Health Organ 88:462–466

Resch B, Hillen F, Reimer A, Spitzer W (2013) Towards 4D cartography – four-dimensional dynamic maps for understanding Spatio-temporal correlations in lightning events. Cartogr J 50(3):266–275

Robert Koch Institute (2016) SurvStat@ RKI 2.0. Web-based query on data reported under the German 'Protection Against Infection Act'

Robinson AC, Chen J, Lengerich EJ, Meyer HG, MacEachren AM (2005) Combining usability techniques to design geovisualization tools for epidemiology. Cartogr Geogr Inf Sci 32(4):243–255

Rogerson PA, Yamada I (2004) Monitoring change in spatial patterns of disease: comparing uni-variate and multivariate cumulative sum approaches. Stat Med 23(14):2195–2214. https://doi.org/10.1002/sim.1806

Sane J, Edelstein M (2015) Overcoming barriers to data sharing in public health. In: A global perspective. Chatham House

Shan J (1998) Visualizing 3-D geographical data with VRML, computer graphics international, proceedings, pp 108–110

Shi X (2010) Selection of bandwidth type and adjustment side in kernel density estimation over inhomogeneous backgrounds. Int J Geogr Inf Sci 24(5):643–660

Shi Q, Zong QG, Fu S et al (2013) Solar wind entry into the high-latitude terrestrial magne-tosphere during geomagnetically quiet times. Nat Commun 4:1466. https://doi.org/10.1038/ncomms2476

Shiode S (2011) Street-level spatial scan statistic and STAC for analysing street crime concentra-tions. Trans GIS 15(3):365–383

Shiode S, Shiode N (2013) Network-based space-time search-window technique for hotspot detec-tion of street-level crime incidents. Int J Geogr Inf Sci 27(5):866–882

Tango T, Takahashi K (2005) A flexibly shaped spatial scan statistic for detecting clusters. Int J Health Geogr 4(1):11

Tango T, Takahashi K (2012) A flexible spatial scan statistic with a restricted likelihood ratio for detecting disease clusters. Stat Med 31(30):4207–4218

Tiwari C, Rushton G (2004) Using spatially adaptive filters to map late stage colorectal cancer incidence in Iowa. In: Developments in spatial data handling. Springer, Berlin/Heidelberg, pp 665–676

Tsai PJ, Lin ML, Chu CM, Perng CH (2009) Spatial autocorrelation analysis of health care hotspots in Taiwan in 2006. BMC Public Health 9(1):464

Vandenbulcke G, Thomas I, Panis LI (2014) Predicting cycling accident risk in Brussels: a spatial case–control approach. Accid Anal Prev 62:341–357

Verity R, Stevenson MD, Rossmo DK, Nichols RA, Le Comber SC (2014) Spatial targeting of infectious disease control: identifying multiple, unknown sources. Methods Ecol Evol 5(7):647–655

Wakefiled J, Kim AY (2013) A Bayesian model for cluster detection. Biostatistics 14(4). https://doi.org/10.1093/biostatistics/kxt001

Wang F, Lee N, Hu J, Sun J, Ebadollahi S, Laine AF (2013a) A framework for mining signatures from event sequences and its applications in healthcare data. IEEE Trans Pattern Anal Mach Intell 35(2):272–285

Wang S, Anselin L, Bhaduri B, Crosby C, Goodchild MF, Liu Y, Nyerges TL (2013b) CyberGIS software: a synthetic review and integration roadmap. Int J Geogr Inf Sci 27(11):2122–2145

Wongsuphasawat K, Gotz D (2011) Outflow: visualizing patient flow by symptoms and outcome. In: IEEE VisWeek workshop on Visual Analytics in Healthcare, Providence, Rhode Island, USA. American Medical Informatics Association, pp 25–28

Wongsuphasawat K, Guerra Gómez JA, Plaisant C, Wang TD, Taieb-Maimon M, Shneiderman B (2011) LifeFlow: visualizing an overview of event sequences. In: Proceedings of the SIGCHI conference on human factors in computing systems, pp 1747–1756

Yamada I, Thill JC (2007) Local indicators of network-constrained clusters in spatial point pat-terns. Geogr Anal 39(3):268–292

Yamada I, Thill JC (2010) Local indicators of network-constrained clusters in spatial patterns represented by a link attribute. Ann Assoc Am Geogr 100(2):269–285

Zhang H, Li L, Hu W, Yao W, Zhu H (2019) Visualization of location-referenced web textual infor-mation based on map mashups. IEEE Access 7:40475–40487

Chapter 5
GeoComputation and Disease Exploration

Abstract In public health, GeoComputation application faces additional challenges: legal issues related to personal data collection, social and cultural areas and
human mobility. The use of space data opens up the prospect of increased productivity in the private and public sectors as well. The introduction of specific and
accurate paraphernalia in the health and GeoComputational analysis can enhance
these approaches. The ability to transform and visualize space is defined here, but
has been enhanced by using GeoComputational analyses that are still dependent on
the validity and lawfulness of the requested sources. This chapter demonstrates how
biological applications affect the core organizational concepts and techniques of
GeoComputations and how computationally complex biological process simulation
is at the cutting edge of biological science.

Keywords Data quality · Data exploration · Decision modelling · Ecological
fallacy · Spatial scale · Spatio-temporal trends

5.1 Introduction

The incidence and form of health measures in a population are concerned with the
exploration of diseases. Epidemiologists should equate the occurrence of a disease
in various populations with the consequential pace. Pattern shows by time, location
and person the incidence of health measures. The disease maps are used mostly for
informative purposes to rapidly review the data in the assessment of high-risk areas
and to assist in the implementation of policies and services in such areas. Health
planners use spatial methods to identify ways and methods of viewing the crowds of
people and resources across space. In a certain order, a time process may then be
observed. The 'World Atlas of Epidemic Diseases' (Pollitzer 1954) and the
sequences of map generated by May (1959) for the American Geographical Society
are particularly valuable for those involved in the environmental ailment of worldwide disease. Learmonth (1978) provides a useful illustration for the spatial

treatment of 20 years of health figures for ex Britain's India, which shows how much evidence can be conveyed in one single map.

Geography as a discipline is primarily concerned with researching the causal linkages of disease emergence in different geographic areas. Statistical data mapping is now widely recognized as the way of solving many problems in ecology, geology, geography, meteorology, climatology, economics, government, etc., and its apparent significance in recognizing the visual relationships of diseases is such that it is somewhat shocking that the medical profession is not doing more. Comparing infectious areas with disease-free areas can well throw useful sidelights at source. If the cause is unclear, a map showing the geographical variations of the specific disease can well explain the complicated early stages of the epidemic. The underlying influences of the distribution will then be investigated, and it may well be that the observed distribution dynamics are similar to those of some other known phenomena.

In public health, GeoComputation was also identified as the technology side of research on the generic issues surrounding information technology, impeding their successful execution or emerging from understanding of the potential of geospatial technology (Fig. 5.1). In the analysis of the relationship between location, environment and disease, epidemiologists have traditionally used maps (Gesler 1986). GC may translate the spatial data into the structure of geography or coordination. The GC splits into four major processes: research (statistics and spatial information, data conversion and categories), production (database architecture and data processing and spatial data editing), visualization (creating perspective and geovisualization and using maps to better understand) and application development (software development, Web site establishment and improvement, web design and support). GC programs may illustrate different details in the form of charts, tables and graphs, in an environmental context. GC may use satellite maps to compare vector details such as temperature, soil type, environmental patterns with spread of diseases and occurrence in different areas. Its program is capable of integrating data from various regional regions and will put together experts from various epidemiological and healthcare institutions to create sophisticated, detailed spatial models. Therefore, GC is regarded as a decision-making method for different health-related issues.

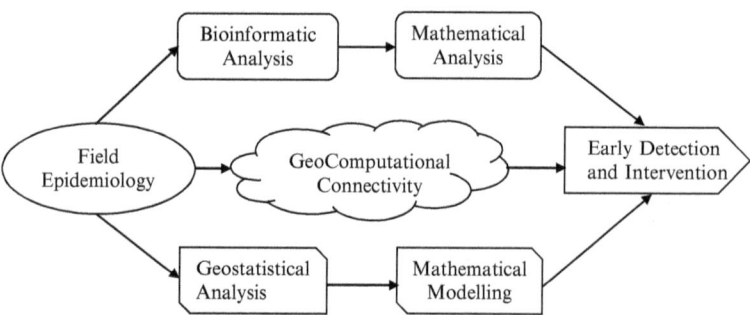

Fig. 5.1 Integrated approach for early epidemic detection and intervention as evidenced by the GeoComputational modelling

The data analysis focuses on the quest for data attributes such as trends, patterns and outliers. It is especially important where low-quality proof or plausible theories are missing. Numerous of these approaches highlight graphical views of the information to display different features and allow the analyst to recognize trends, relations, outlines, etc. Exploratory spatial point pattern research is concerned with investigating the properties of the spatial point pattern processes in the first and second order. The results of the first order contribute to variance in the mean value of the phase (a large-scale trend), whereas the results of the second order derive from the statistical correlation mechanism or statistical dependence. Exploratory field data analysis involves recognizing and explaining different types of spatial variability in the data. Spatial interaction can be rigorously described using accurate spatial statistics on self-correlation in several ways (Cliff and Ord 1981), or more informally, for example, using a scatter plot and displaying each value against the mean of neighbouring areas (Haining 1993). The overall trend of dependency in the data is outlined in a single predictor in the systematic approach to spatial autocorrelation, such as *Moran's I* and *Geary's c*. These measurements of the spatial association can be used to determine spatial behaviour of the data, which can be easily visualized by the spatial variogram, a collection of measurements of autocorrelation in various contiguity commanded. No doubt, ESDA support valuable tools of spatial data sets to produce insights into global and local patterns and connections. Nevertheless, the application of ESDA technologies is typically restricted to specialist users engaging with the displays of data and mathematical diagnostics to analyse spatial knowledge and fairly basic low-dimensional data sets.

5.2 Issues of Exploration in Public Health Through GeoComputation

As in many other sectors, public health can now rely on nearly infinite data streams ('Big Data'), collected from multiple outlets, ranging from satellite imagery to mobile telematics to voluntary geographic information. Parallel to this wealth of data is accessible and inexpensive computing power which allows fast processing. Therefore, GeoComputation's problems have moved to research, including definitions, description, simulation and dependability. GeoComputation application in public health faces more challenges: legal concerns surrounding personal data analysis; social and cultural spaces; and human mobility. During the latter half of the twentieth century, massive developments in computing technology brought about an immense transformation of the medical geography and made modern medical GIS possible. Some electronic mapping systems revolutionized the field at the end of the 1960s, thereby allowing modern automated mapping technology to emerge that made updating information simpler and more accurate. Maps could now be produced much faster and more precisely than before, and the dissemination of diseases could be measured and assessed more conveniently. Throughout the early 1970s, a new generation of theoretical methods was developed, making hypothesis testing more important to the industry. Scientists have also been involved in not only identifying pathogens and their propagation but also in how human activity and environmental conditions impact one another. By the mid-1980s, computer mapping technology flourished to make it possible to integrate more non-space data known as 'reference data' into space point data in digital maps, resulting in the birth of modern GIS on the desktop. In addition to measuring point densities, one may also use GIS to systematically check whether such points appeared to overlap at some sites and whether clusters of disease were correlated substantially with certain forms of human experiences. In determining the proximity, aggregation and clustering, spatial smoothing, interpolation or regression of data, the epidemiologists utilize spatial information. Identification of disease clusters relating to non-random geographical distributions of disease occurrences, frequency or prevalence remains the most common use of GeoComputation in these areas. To recognize multiple ways of clustering, multiple analytical and statistical techniques were developed.

In the coming decades, communicable diseases are expected to remain a major public health problem that poses a challenge to national and international health security. Digital records do exist in many situations, but there are confidentiality problems, social security, etc. that have limited their use by infectious disease and health-related agencies. It continues to be a challenge to find the resources to collect new data and turn paper maps and data into digital format. In addition to infectious diseases like immune deficiency syndrome (HIV/AIDS), tuberculosis (TB), malaria and neglected tropical diseases, transmittable diseases would also pose a public health threat and require a high degree of early detection preparedness and fast response. Communicable diseases are among the key challenges for human disease. The epidemiological transition is also motivated by the social and economic influences of health and certain old and new risk factors, like globalisation, unplanned and unregulated housing growth, the shift in behaviour, environmental issues (e.g. climate change and air pollution) and media and advertisement impact. The development in biomedical and emerging research and technology, which along with population patterns and people's aspirations are accused of working to deepen the distance between available funding and healthcare requirements, is undoubtedly a significant threat to sustainable healthcare.

5.3   Data Availability and Quality

Spatial data is a tool on an equal basis with employees, funds, etc. The use of spatial information opens up the prospect of rising productivity in both the private and public sectors. Unlike other tools, there is no wear and tear on spatial data from repetitive use. In the opposite, the reuse of data improves incentives for increasing the consistency of data processing information. Since, the exploration of diseases is based entirely on the analysis of information and its various visual images, data collection, methods, accuracy and public access. Furthermore, many data sources were placed without regard to the geographical epidemiologist's advantage (Staines et al. 2000); precise and accurate data on health points or areas in spatial epidemiology should be completely ensured (Jarup 2004). These approaches can be upgraded by introducing specific and accurate paraphernalia in health and GeoComputational analysis (Fig. 5.2). The following causes of GIS data error were established by Oppong (1999): errors in object orientation, errors in object-related parameters and errors in spatial variance modelling (e.g. suggesting spatial uniformity between objects). In addition, the correct tools and processes can be effected efficiently and practically by assembling and normalizing data. The geographic factor such as environmental (quality of surrounding areas), demographic statistics (income and race) and the possessions thereof should be investigated in relation to geographic health consequences. According to Gervais (2004), a computer-based guidebook is important to mitigate violence by offering geospatial data information that is more easily accessible to users. Several authors emphasized the need to develop such a tool, often referred to as 'Quality-aware GIS', 'Quality GIS' or 'Error-conscious GIS', which will automatically take into account quality details through data processing (visualization, requests and updates) to deter users from 'illogical operations'.

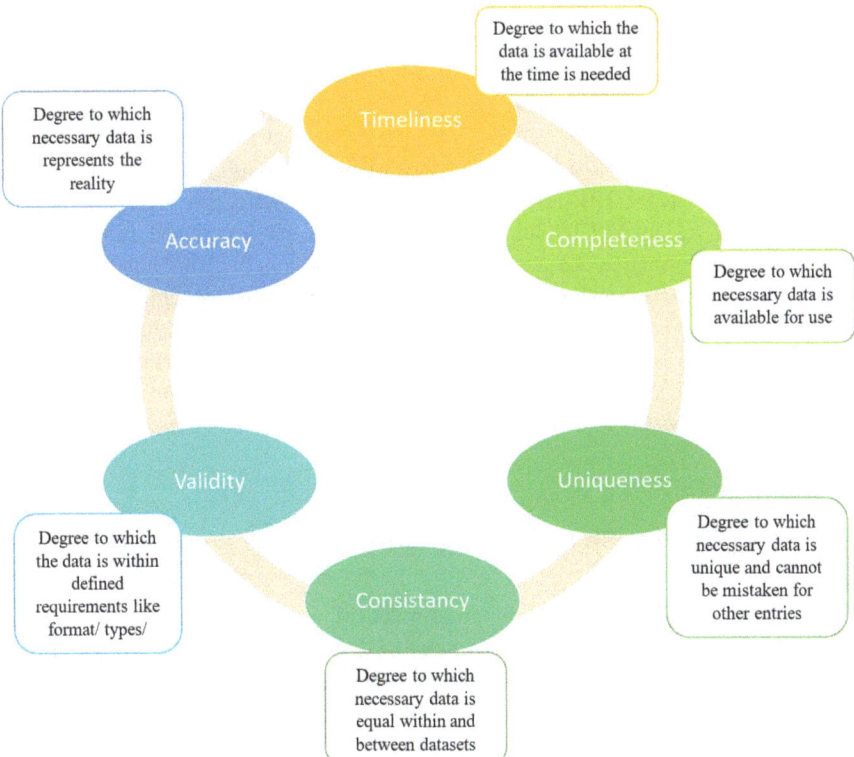

Fig. 5.2 Data quality dimension in GeoComputation

According to Armstrong et al., the ability of researchers to diagnose clusters of diseases or to examine possible associations between environmental conditions and disease incidents is compromised in four ways as data are spatially aggregated over broad areas:

1. Absolute and relative positions within the spatial framework of each region make it difficult to carry out clustering experiments, even those expressly intended to work on data aggregated to areas.
2. The geographical effect of the aggregation in the geographical size of the clusters means that the aggregation level employed in the analysis restricts the size of the clusters to be calculated.
3. Structure and location of areas of aggregation in connection with the distribution of disease in the world or clubs under study, for instance, where the disease cluster is two or more areas of aggregation, which show uncertain or negative results.
4. Precise analyses are only possible if health data is spatially encrypted on the edge of regions with typical threshold levels of environmental exposure, probably because exposure assessment data for several areas are generally obtained, other than health and population data. (Armstrong et al. 1999; Boulos 2004)

Many GIS authorities may mention errors arising from GIS procedures on spatial data (transformation and interpolation), the consequences of generalization operations (aggregation), errors induced by temporal variations and conceptual errors. For the problems being explored in a study, the scale level should be sufficient, or the findings may not be relevant and could even be deceptive (Freier 2000). Different diseases have patterns that are interesting on different spatial scales, and the optimal size is the one with the most fascinating pattern (Rushton 2003).

Devillers et al. (2005) assess the quality of geospatial data, thus reducing the risk of data misuse. Geospatial data have usually been generated and implemented within the same entity in the past. Knowledge of the mechanisms and features of data processing, including consistency, was more implicit (i.e. organizational memory) than explicit (e.g. metadata). The manner in which organisations or entities share information relating to geospatial data underwent a transformation through which transmitted information became more available or relevant to a wider community of data consumers in the geospacer (Fig. 5.3).

5.4 Data Protection and Confidentiality

The privacy of spatial data is a dynamic matter. Data protection restrictions also prohibit the release of individual disaggregated data which limits the types and precision of the analytical results that could be achieved (Fig. 5.4). Throughout public health globally, with a few exceptions, the public recognition of an individual's health status and residency is generally forbidden, irrespective of the extent of contagion or threat. Safe and protected data are a key factor for successful research on epidemiology. Information can be obtained in different formats or with varying

Information richness for
accessing data quality

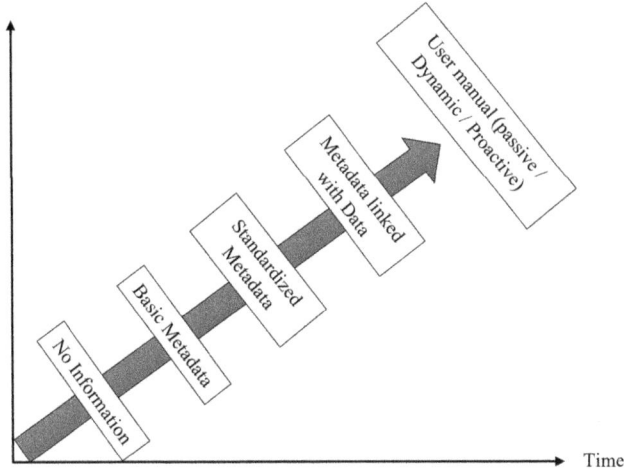

Fig. 5.3 Evolution of the usefulness of the information communicated to data users for assessing geospatial data quality. (Source: Devillers et al. 2005). No content information situation is also very prevalent, and it is not unusual to see users expressly asking that the metadata not be submitted, even though it exists. Basic quality information includes some information as the geospatial datasets are transmitted, such as dataset comparison structures, geographic precision, or date of development. This material, though, is unique from one entity to another, not consistent with the norm, explaining various aspects of the datasets at various levels of income. Normalized metadata specifications to homogenize the knowledge exchange among the organisations. Various criteria may also be used by one entity to another. These requirements, however, are more producer-oriented than user-oriented and more formalized than realistic information which data-recovery experts understand to a general audience and can be used for decision-making processes. Metadata related to data presented with datasets are often stored routinely in a text file separate from their data file, without any clear connection between the two; several academic and industry work projects are now being carried out to improve the relation between the metadata and the data mentioned, up to the instance and attribute rates. These specifications would be easier for customers or automated systems to access, but would also be more difficult to generalize if the granularity of quality information is very strong. Nonetheless, in terms of the types of metadata that can be stored, and the extent of metadata information, these tools are also limited

degrees of precision in certain situations, but still generally follow the same pattern in the same geographic region. On the basis of choropleth mapping and analysis of aggregate data in administrative areas, traditional ecological analysis has been strongly criticized. In the field of public health, it is becoming abundantly clear that individual patient data amassed to pre-existing political or other administrative areas for the preservation of individual privacy frequently removes knowledge required for spatial research, rendering it difficult to resolve certain critical public health concerns, e.g. the threat of injuries in different settings, risks of living near toxic waste facilities, the danger of lead pollution associated with urban highways, etc. Just micro-data may overcome these issues. Even if a single database may seem to have adequate secrecy protection, the 'total' may be less well-secured than the

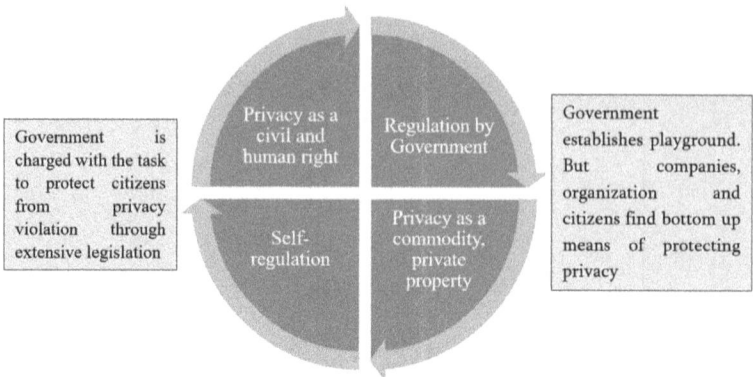

Fig. 5.4 Different understandings of privacy and the roles of governments. (Source: Modified after Resch et al. 2012)

'bits' when several databases are connected within GIS. A false identity can be as harmful to a person as a true identity which is not held confidential (Richards et al. 1999). It creates what Jacquez calls 'spatial confusion' by using aggregated data instead of address-level data (when required). In comparison, centres of the region can produce misleading results rather than exactly locations (Jacquez 1998).

5.5   Exposure Assessment and Mapping

In recent years, the proliferation of Internet access has significantly enhanced the potential utility of GIS, as it is now possible to find and leverage data from multiple different GIS repositories to create very rich empirical knowledge on virtually every subject correlated with physical locations. Notwithstanding the advent of an increasing number of special purpose GIS software packages, the standard standardized software packages, regardless of their state of the art, still basically only require fairly straightforward and easy applications. An analytical limitation, data quality or specified precise spatial grasp of epidemics are identified in GeoComputational analyses. The capabilities for spatial interpolation and visualization, however, have been greatly improved through the use of GeoComputational analyses, which still rely on the validity and legality of the source data requested. However, existing GIS packages do have significant limitations to tackle problems of a more complicated nature that are characteristic of many types of applications, including especially those in the health sector. Usually, these kinds of problems need more than description; they are far more concerned with modelling and scenario-building and potential innovation.

5.6 Application Services and Decision Modelling

GIS will in principle act as an effective evidence-based analysis method to identify and address issues early and as a community health solution. When appropriately used, GIS can: advise and educate (professionals and the public); facilitate decision-making at all levels; assist in the preparation and modification of therapeutic and cost-effective actions; forecast outcomes before taking any financial commitments and prioritizing in an environment of finite resources; change practices; and continually track and evaluate adjustments and sentinel incidents. The failure of the existing GIS to manage these more complicated scenarios effectively is a significant factor for inhibiting the full capacity of GIS systems in the health and medical fields, as well as in many other regions. These are particularly relevant to tackle issues of optimum resource distribution in space, optimum geographical distribution or regionalization for administrative purposes, for forecasting and in a wide range of epidemiological studies. Lying with maps is not only easy but also essential, Monmonier says. The cartographer's fallacy is that the map must give a narrow, distorted perception of nature in order to prevent concealing vital details in a cloud of complexity (Monmonier 1996). Public health practitioners must be vigilant to 'errors' that differ from fair and appropriate failure to selectively warn tourists of what is to be perceived as more serious discrepancies, where the visual picture includes findings that are not accompanied by robust epidemiological studies. For instance, if there are low denominators in any regional unit of measurement, reported disease rates may be extraordinarily high if there are cases in these areas (the 'low numbers' problem). If the concentrations are shown on a map for such geographic areas, readers may mistakenly infer that they are 'hot spots', high-priority places for targeted interventions. Such regions should be marked more correctly to suggest that concentrations are statistically unpredictable due to low numbers and thus not seen (Richards et al. 1999).

Jacquez described the 'gee whiz' effect as 'the development of theories to justify an obvious (visual) phenomenon whose presence has not been verified' and stressed the importance of using sufficient and reliable statistical methods to help the thematic data layers to be viewed and evaluated in order to avoid the effects of visual bias in GIS processes where spatial trends are powerful (Jacquez 1998). Such approaches play a part in confirmatory and exploratory experiments, according to Jacquez. He continues: 'In the exploratory context in particular, one needs to be able to distinguish real trends from obvious trends that may be interpreted by chance'. Without this capability, both confirmatory and exploratory analyses spin their wheels as they lack an analytical method to classify and measure interactions in the results.

5.7   The Ecologic Fallacy and the Atomistic Fallacy

The basic issue with ecological inference is that the aggregation process eliminates information, and this lack of information typically avoids the detection of value parameters in the overarching individual-level model. If there is no variation of exposures and explanatory variables within the environment, so there would be no ecological discrimination; thus, ecological bias exists according to the heterogeneity in exposures and confounders within the region. The fundamental issue with ecological analyses is the lack of knowledge due to accumulation – the mean function, which is mostly based on regression, is not generally detectable from ecological data alone. This lack of identification can contribute to the ecological error in which human and ecological relations vary between the outcome and an explanatory parameter and can also reverse the course (Fig. 5.5). There are two main problems that need to be addressed – First, hierarchical models cannot compensate for knowledge loss and, in particular, the use of spatial models does not overcome the ecological fallacy. Second, the only approach to the ecological flaw, and therefore to make an accurate inference, is to replace data at the ecological level with evidence at the individual level. Users, like decision-makers, can be able to conclude from association causation and draw inferences from demographic evidence (the ecological fallacy) regarding individuals. While decisions based on an experiment at the aggregate level are likely to be constrained by accumulation bias and ecological fallacy (failure to identify the true nature of causal-effect relationships at the individual level), atomic fallacy may also restrict decisions based on a study at the individual level (failure to consider the greater context in which individual behaviour). GIS technologies can be used to combine human (human) data with contextual knowledge and ecological predictors aggregated at a variety of regional

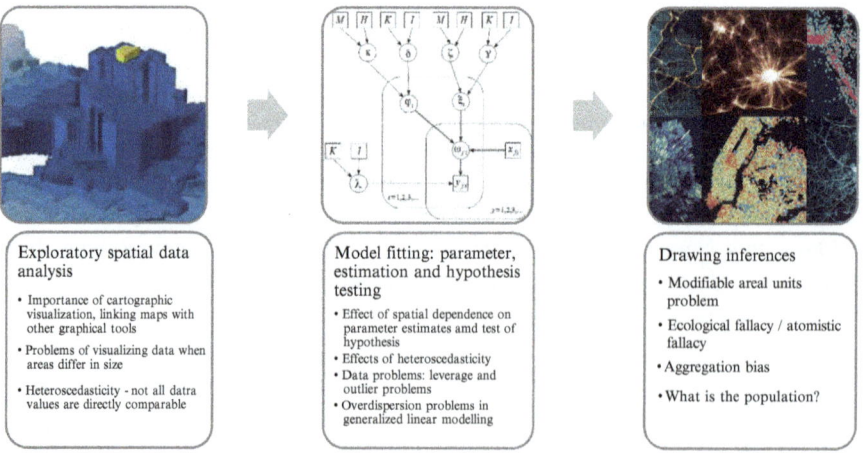

Fig. 5.5 Illustration of ecological fallacy

(community) levels to assist in the assessment and discernment of biological, contextual and ecological results.

5.8 Spatial Scale of Disease Exploration

Scale has been described as 'the fundamental philosophical problem in ecology, if not in all science' (Levin 1992), debated in numerous volumes and numerous articles but hardly resolved (O'Neill and King 1998). Scale appears to be a 'conceptual puzzlement' case where 'the different contexts from which the meaning of a word is expanded should not be exclusively congruent; they may have contradictory meanings' (Pitkin 1972). Lee (1993) postulates that 'since human obligation will not fit the geographical, temporal or functional size of natural events, it is possible the wasteful usage of resources may occur until the mismatch between scales is fixed. The suitable scales for social and natural process analysis may often be somewhat different, but the two forms of systems are not philosophically separate, and the epistemological problems they present are practically similar.

5.8.1 Definitions of Spatial Scale

Spatial scale refers to the order of magnitude or size of a studied or identified land area or geographic space. Exploring spatial scale in geography offers a qualitative and realistic guide to spatial-scale problems in both physical and social science fields (Fig. 5.6). For reference, the word scale was estimated to have three dimensions, including size (e.g. district census, state, continent), level (e.g. local regional, national) and relationship (as a complex mix of volume, location and environment). Throughout the disciplines of GeoComputation, scale concepts are well established as a crucial element when researching patterns in nature and the processes that cause them. In an effort to boost the efficiency of results from spatial planning models, a range of different types of scale have been established (Sheppard and McMaster 2008; Wu and Qi 2000).

The advent of scalable parallel hardware significantly expands prospects for large-scale spatial analysis within GIS, using modern methods that seek to address some of the conventional challenges by moving to a more computationally efficient framework. GeoComputation itself is a fairly recent concept, described as the adoption of a computationally demanding strategy to physical and human geography and geosciences in general. This is a concept which is obviously applicable to GIS but which goes well beyond it as well. Scale is a crucial influence in the analysis of patterns, mechanisms and changes in physical bodies and was generally accepted as a central aspect in the analysis of relationships between the person and environment. In GeoComputational analysis, the spatial scale has been analysed as follows.

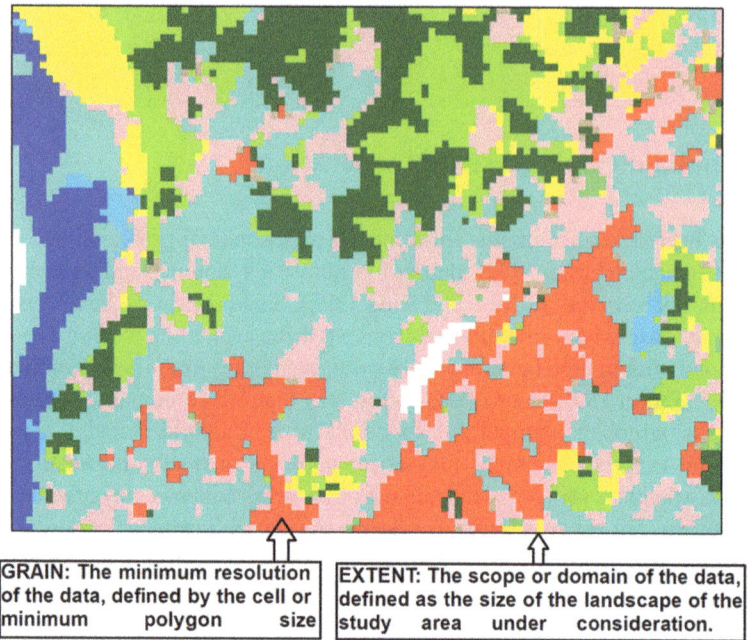

| GRAIN: The minimum resolution of the data, defined by the cell or minimum polygon size | EXTENT: The scope or domain of the data, defined as the size of the landscape of the study area under consideration. |

Fig. 5.6 Component of spatial scale (Grain: minimum smallest entities that can be distinguished – in raster lattice data – the cell size; in the field sample data, the quadrant size; in imager,- the pixel size; in vector data, minimum mapping unit. Extent: size of the landscape or spatial domain encompassed by an investigation or the area included within the landscape boundary)

5.8.2   *Cartographic Scale*

This is the earliest geospatial scale of any form. It is known as the relation between a map and what it represents in geometry. A cartographic scale is a statistical term showing the relationship between the region being modelled and the area's real-world size (Fig. 5.7). Maps are smaller than the part of the atmosphere of the planet they represent. Therefore, a 'large-scale map' shows a relatively small region of the earth, like a county or city, while a 'medium-scale map' shows a fairly wide region, like a globe or an earth hemisphere. Comprehension of conventional cartographic scale is important, as developments in geospatial technology quickly apply carto-graphic scale in computer programs. Now the user can easily switch from one car-tographic scale to another via a computer screen view, rather than drawing a different map on a piece of paper.

'Cartographic scale' was a solely geometric spatial term that limited the capacity to classify geographic objects along with the resolution of databases. It is part of the 'modelling scale' which applies to show map representations which are expressed as a ratio. This concept on a cartographic scale is frequently thought to be counter-intuitive when applied to the size of analyses or phenomena, where small- and large-scale entities are generally referred to, respectively, by small and large

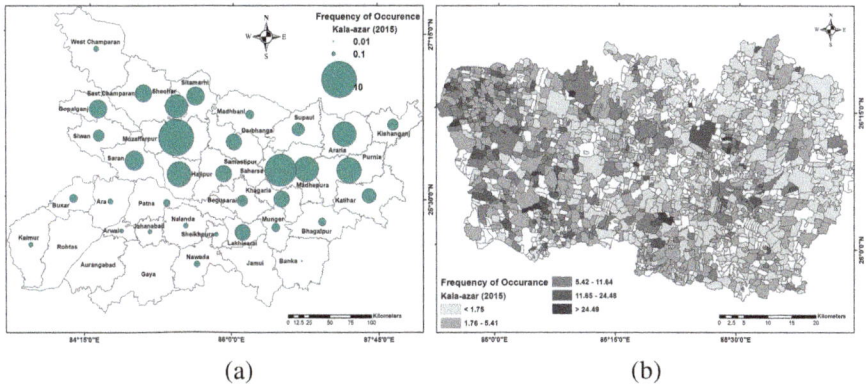

Fig. 5.7 Two different scales (1:5000 and 1:30,000) were measured at these cartographic scales with a separate minimum mapping unit. This type of scale can be expressed as a ratio or fraction (e.g. 1:30,000) meaning that 1 cm is equal to 30,000 cm on the ground (**a**). Now, this can shift from one cartographic scale to another effortlessly through a computer screen view rather than having to draw a different map on a piece of paper (**b**)

entities. An essential problem about the size of the cartography is that flat maps inevitably distort spatial relationships on the surface of the earth: distance, direction, form and/or area. Whether these interactions are skewed is a part of the analysis of map forecasts. For certain projections, particularly small-scale maps of large areas of the world, this bias is severe such that on one part of the globe, the linear or areal scale is very unique than for other areas.

5.8.3   Geographic Scale

'Geographic scale' is most commonly interpreted in terms of dimensional characteristics such as magnitude (or size). This is known as the geographic extent of the objects being studied on Earth. The geographic scale offers a view of how to treat a given sample. This is expressed mathematically in square units. To expand on a regional scale (geographic scale) the analysis of disease distribution, the representation of disease groups may be provided on a cartographic scale of 1:50,000, 1:750,000 or 1:500,000 (Fig. 5.8). This represented the amount of detail about how to view at various cartographic scales. It was proposed that the definition would provide an explanation of what the local, national and global scales mean and how to differentiate them from each other.

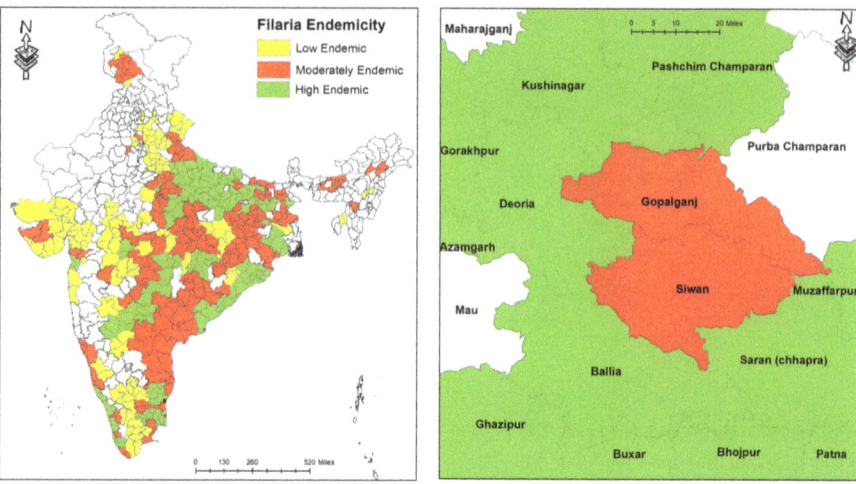

Fig. 5.8 Illustration of geographical scale

5.8.4   Ecological Scale

To our understanding of the dynamics of biological ecosystems and the relationship of individual species with the environment, the definition of scale in ecology is seen as increasingly important. Krebs (1985) stated: 'The value of ecological scale is a key concern in the 1980s as we seek to get a deeper understanding of how ecological processes function in the real world'. Throughout spatial ecology, scale refers to the magnitude of biological processes spatially and data temporally perceived. An individual or species' response to the environment is unique to a given scale and may respond differently on a larger or smaller scale (Fig. 5.9).

The ecological system can be divided into temporal and spatial scales, contributing to the most relevant issues and perspectives. Temporary scale is central to the broader conversation, often defined as a 'paradigm shift,' between equilibrium and non-equilibrium models and ecological expectations (Briske et al. 2003). Spatial size poses related issues. When analysing the findings on the smallest scale (i.e. the smallest plot), they showed the greater variance on the ungrazed areas; on the intermediate scale, the variance on all three treatments was identical; and on moderately grazed areas, the lowest variability was observed on a larger scale.

5.8.5   Operational Scale

Operational scale describes a metric (e.g. length metric, volume measurement, etc.) as does all other meanings except 'Cartographic scale', which is a ratio instead of a size. 'Operational scale' was too general because the relation between the procedures and the corresponding geographical extent cannot be considered. This applies

Fig. 5.9 Ecological spatial scale considered in terms of landscape size and extent. (**a**) Landsat Thematic Mapper data with 30 m pixel size in 1989, (**b**) Indian Remote Sensing – Linear Imaging Self Scanning with 5.8 m pixel size in 2009, (**c**) Google Earth image with <1 m pixel size in 2020. It describes the size of individual landscape features. The extents describe volume within which all the samples are taken. Temporal scale can be thought of as the frequency of sampling

to an object being traced at a given spatial resolution. The spatial resolution is a sensor's ability to recognize the smallest detail in a scale of an object being analysed on an image. Spatial precision on a digital image is constrained by pixel size. The smallest tracked object can't be smaller than the size of the pixel. This results in high and low resolution of the areas. Low spatial resolution refers to observed coarse features on an image. For instance, individual buildings can be observed with remote sensing imagery from Seninel-2 (with 10 m spatial resolution), whereas only urban area can be identified with MODIS (Moderate Resolution Imaging Spectroradiometer with 250 and 1000 m spatial resolution) and not individual buildings (Fig. 5.10).

5.8.6   Modelling Scale

A mathematical expression describes the dimension of the data being collected or generated. The scope of the study includes the unit size in which anomalies are measured and the unit size in which observations are aggregated for data processing and visualization. Essentially, it's the degree of spatial phenomenon understanding. This has long been understood that the size of research must fit the real scale of the phenomena in order to analyse and study a phenomenon more accurately. This refers to all three scale dimensions – spatial, temporal and thematic.

Fig. 5.10 Example of
Operational scale (**a**)
Landsat Thematic mapper
with 30m pixel resolution,
(**b**) Sentinel-2 with 10m
pixel resolution

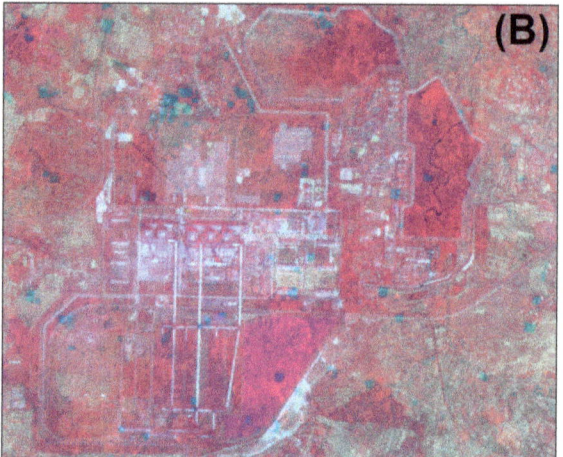

5.8.7   *Policy Scale*

The policy scale reflects the administrative level (or territorial boundaries) at which
decisions (e.g. local, regional or national) are made. This refers to the scale at which
spatial systems occur and in which spatial mechanisms in the world. It's the 'real'
geographic range of phenomena. Terminological continua such as 'local-global' or
'micro-, meso-, macroscale' are also used to describe the spectrum of scales of concern to geographers and epidemiologist. It is generally accepted that different sizes
of spatial anomalies communicate, or that one-dimensional anomalies arise from
phenomena of smaller or greater size. It is expressed by the concept of a 'size hierarchy', in which smaller phenomena nest inside bigger phenomena. Conceptualizing
and modelling these hierarchies of size can be very complicated, and the conventional method of concentrating on a single size persists mostly throughout the
geography.

5.9   GeoComputation in Epidemiological Analysis

The epidemiologist and public health specialists' ability to work with spatial data has significantly changed with GeoComputation. There are many advantages to GeoComputation and the ability to perform recurrent tasks, compare spatial information quickly and handle large amounts of data via overlay operation (e.g. superimposes multiple datasets), Boolean searches (e.g. finding places that fulfil two or more criteria), buffer analysis (e.g. zone around map features measured in a unit of distance or time) and statistical analysis (e.g. clustering and pattern analysis). The demographic, environmental, behavioural, socio-economic as well as genetic and infection risk factors are taken into account for spatial epidemiology. Different areas of GeoComputational epidemiology can be identified, which reflects various needs of health experts and epidemiologist in the evaluation of epidemiological conditions (Rezaeian et al. 2007). The definition of the geographical trends of morbidity and death can be described as part of epidemiologic descriptive studies to establish aetiology-related disease hypotheses. GeoComputational epidemiological analysis is scale-dependent and descriptive. However, when regional aggregate data is combined with individual data, an overall spatial problem can be solved (Dooley et al. 1989). The phrases defining one of the various statistical methods in which this mixture is enabled are multiple-level modelling, hierarchical regression and background analysis. Multilevel modelling is an effective, relatively recent approach used to assess how much the ecological impact can be explained by changes in the distribution of risk factors at different rates (Kwok et al. 2008). GeoComputation has been developed by incorporating time changes with spatial variations and ecological impact integration into epidemiological analyses.

GeoComputation succeeds in capitalizing on the first law of geography, which states: 'Everything is more connected to everything else, but close to things more important than far-off things' (Tobler 1970). The geostatistical spatio-temporal models provide a probabilistic structure for data processing and forecasts based on the mutual spatial and temporal observational dependency (Kyriakidis and Journel 1999). Some of the first examples of the power of visualization and interpretation of health data were the research carried out by Dr. John Snow on the cholera outbreak that devastated London in 1854. Snow has been able to deduce that the public pump is the source of the cholera outbreak by means of a map showing where the pumps and homes of those who died of cholera have occurred (McLeod 2000). In fact, one of the key characteristics of the above-mentioned data types is their structured spatial and time distribution, reflecting the impact of several factors operating at different time and spatial levels (e.g. geology, climate, human activities, land coverage). Descriptive disease speculation has also given rise to the scientific analysis of disease spatial patterns, including assumption testing, multilevel modelling, regression and multivariate analysis.

The epidemiology of public health includes information about the onset of diseases, the infection rate, age group, sex, transmission of the disease, patient site specifications, parasite and virus load availability host, etc. This was used to denote the horizontal and vertical structure of the diseases, history and so on, with respect to space and time. GIS is used to map the geographic distribution of disease incidence (e.g. transmissible and non-communicable), disease transmission pattern and spatial analysis of the disease's environmental aspects. GeoComputation is designed with thematic mapping and/or personalized mapping designs, symbols and colour and likely integrated mapping facilities, overlay analysis, cluster analysis, nearest neighbourhood checks, pattern recognition, time analysis, point data interpolation, spatial correlation, fuzzy analysis, linear determinant analysis and minimum and maximum likelihood analysis. The current GIS World Source booklists hundreds of system providers and system sources and catalogues (GIS 1995). There are surprising comparisons in the field of forest, ecology, archaeology and epidemiology, for example, which may provide significant benefits in sharing of experience and pooling of resources. There is a strong need for GeoComputation. Good epidemiological science is in common with good science of geographical information. In order to set their own goals, epidemiologists will seize the opportunity to direct science and technology against public health objectives.

Today, GeoComputation's research group indicates that increasingly sophisticated metaphors are being created for the organization of the machine with the 'desktop metaphor' being more widely recognized. Most operating systems provide context-sensitive provide, documentation, installation and automatic updates. Such software was called an intelligent agent and used over a network. Most future GeoComputational analyses will use these methods to search for new data over a network linked to your problem, to alert you about errors in your data management and analysis and maybe to automatically compose maps and reports after completing a project. Multimedia and hypermedia become quickly a GIS software component. Simultaneous text, sound, animation and graphics can be used by multimedia.

Disease mapping is one of the geographical epidemiological branches that meet the requirement that accurate morbidity and mortality maps are created (Mala and Jat 2019). For instance, dot-density maps are used for displaying data points, while for areal data choropleth maps are used, and continuous surface data are made possible by contour or isopleth maps. Over recent decades, the use of visualization has grown quickly in the medical context. The presentation of maps is now a fundamental tool for analysing information about public health. Closing of diseases is one of the branches of disease exploration that involves determining local and worldwide disease accretion (Lawson et al. 1999). Different types, including general and specific clusters, exist. Clustering means evaluating the current pattern of regional diseases clustering and compares it to an evaluation of the current spatial autocorrelation, in which clusters are located exactly (Besag and Newell 1991). The detection clusters in points data are nevertheless more extensive than for data in a single format. Spatial immediacy classification and outbreak-related, which is generally involves an excess of cases above a certain time-space background rate (Paul and Daniel 2004). Methods for point-format detection clusters are numerous more than for data in areal formats and typically are divided into three groups: globally, locally and oriented. There are numerous tests available to evaluate various kinds of epidemiological data in point and areal format, as described below.

5.10   GeoComputation and Spatio-temporal Trends

At least one object with a spatial and a temporal property may be defined as a spatio-temporal object. The location and orientation of the object are the spatial properties. The time-stamp or time-interval of the object is the temporal property. For most instances, this geographical, temporal and thematic object includes geographical or non-spatial attributes. Some of these things occur in real world and display spatial and temporal variations that can help to grasp their physical manifestations. The spatial and time properties of these events and their relations are also very important for the efficient classification of broad spatio-temporal data settings on a given application domain. With the growing availability and knowledge of vast quantities of spatial and spatio-temporal datasets in many key fields of use such as epidemiology, spatio-temporal data processing and mining is increasing importance.

Since the early 1980s, systematic research can be conducted on the temporal dimensions of both non-spatial and spatial structures (Allen 1983). Spatio-temporal databases contain data gathered in both space and time that explains a process at a given place and time period. It is an evolving field of research due to the emergence and implementation of modern analytical techniques that enable vast spatio-temporal datasets to be analysed. The occurrence in a spatio-temporal dataset describes a spatial and temporal process where 't' and 'x' position or space happens at a given time (Fig. 5.11). Epidemiology has a long tradition of examining conditions that affect patients with severe diseases incidence or mortality variability. Spatial patterns in population health have played a key role in determining the delivery and efficiency in health services by those variables. Spatial variability in health outcomes has also shown dependency patterns and noise rates in the results. Most specifically, analyses of the time series were used to investigate the way health indicators differ over time. Spatio-temporal analysis benefits from strict spatial or time-series experiments as they allow the investigator to both study the consistency of patterns over time and display unusual patterns. The introduction of terms for space-time interaction can also pinpoint data clusters that might be emblematic of evolving infectious disease or recurrent data recording errors.

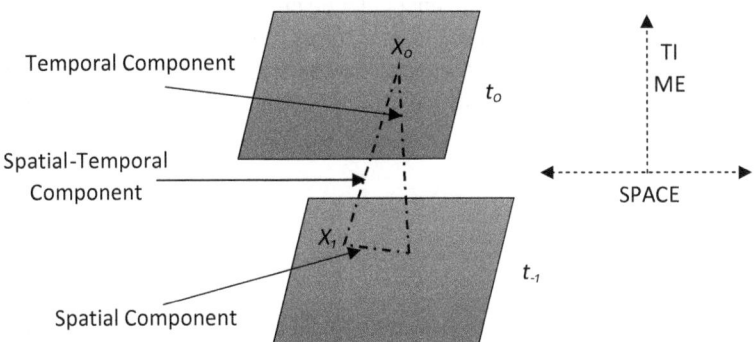

Fig. 5.11 Spatio-temporal component through GeoComputation

Spatial pattern analysis is related to the use of spatial data within GeoComputation to analyse sequence changes over time. In these analyses, the temporal aspect is also critical for measuring progress over time and presenting time-relevant strategies for environmental management. Spatial patterns may generally be described as having one of three features: being uniformly distributed (or often referred to as dispersed), so the distance between points is unusually large; being randomly dispersed, so the distance between points is equivalent with that of a group of points with randomly generated or clustered coordinates, such that the range between points is unusually high. Tuberculosis, human immunodeficiency virus (HIV), pneumonia, malaria, dengue, Zika and other diseases transmitted by mosquitoes are a chronic problem for health agencies, physicians and patients. Its spatio-temporal variations pervade a number of geographical, climatic and socio-economic influences. Factors such as climate change, behaviours or land use often disturb and hinder the perception of such systems. To determine both the time and space dimensions of the results, the computer processing model has major difficulties for two crucial reasons:(1) continuous and discrete changes in the spatial and non-spatial attributes of entities in space-time and (2) the effect of adjacent spatio-temporal entities collected on each other.

5.11   Spatio-temporal Analysis of Epidemic Phenomenon

Epidemic information are epiphanies of spatio-temporal systems with action that is autoregressive or 'self-exciting'. Epidemic data are different in relation to at least three facets which impede the implementation of conventional statistical solutions: (i) the data are typically the result of organized observations, (ii) the measurements (cases, events) are not consistent and (iii) the mechanism is often only marginally measurable. Although extensive work is undertaken in the literature on disease management, work into the application of spatial, temporal and spatio-temporal data analysis techniques into the clustering of diseases is very limited and has recently begun to attract interest. A clearer understanding of spatio-temporal trends may provide critical knowledge for efficient epidemic prevention, such as preparation for disease containment and response efforts in terms of risk detection, resource selection and routing of reservoir and agents, distribution of preventive measures and policy evaluation of disease-related mortality, transmission propagation and epidemic mitigation strategies.

5.11.1   Scan Statistics

The spatial variations in temporal trends (SVTT) scan statistics is developed to identify clusters of areas with irregular temporal biases. The linear SVTT (Kulldorff 2013) procedure is based on measurements of scans and is used to predict the pattern in Poisson over time as an independent variable. The pattern prediction is carried out here using a Poisson equation with time as an independent variable, time as a contingency and the number of events as a dependent variable. Such patterns are then used to change the estimated number of cases for any position and period where, because of the various forecasts, the modification would be various inside and outside the window. The chance of this opening is determined for the currently measured estimates, and the average value is observed for all windows. The maximum probability from a large range of random datasets is then compared. That is achieved by randomization and not by randomization of the periods found.

5.11.2   Lattice Statistics

Lattice statistics refers to a divisible array within a spatial structure of normal or irregular cells. A neighbourhood relationship represents the spectrum of spatial dependence among cells, which may be expressed by a contiguity matrix called a W-matrix. The behaviour of the space neighbourhood may be defined by means of a space (e.g. rook, queen) or Euclidean distance, or by means of more precise structures, cliques and hypergraphs (Warrender and Augusteijn 1999). Spatial autocorrelation statistics can be described on the basis of a W-matrix to quantify the correlation of a non-spatial feature across neighbouring locations. Common spatial autocorrelation statistics include *Moran's I*, *Getis-Ord G_i^**, *Geary's C*, *Gamma index*, etc. as well as their local versions called local indicators of spatial association (LISA) (Fig. 5.12). Numerous mathematical spatial models, including the SAR, the conditional autoregressive (CAR), Markov random fields and other Bayesian hierarchical methods are used for analysing lattice data. Data from lattice simulations can also be used (Banerjee et al. 2004).

5.11.3   Topological Relationship Patterns

The topological relationship between two spatial objects varies if the structure or position of one of the spatial objects' changes. Spatial objects are typically collected and stored in spatio-temporal libraries of geometry and translation changes over time. The evolving topological relation between spatial objects and time is demonstrated by spatio-temporal patterns of topological contact (Rao et al. 2012). The change in topology between two O_1 and O_2 space entities from t_1 to t_4 as seen in Fig. 5.13. This illustration may be epitomized as D-O-C-T in the topological relationship pattern where D, O, C, T corresponds to disjoints, overlaps, contains and touches.

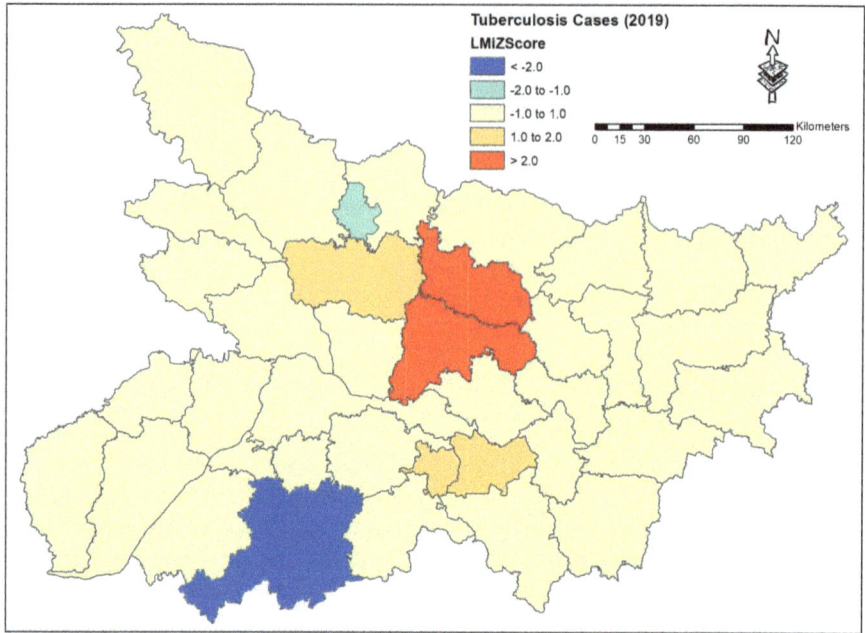

Fig. 5.12 Cluster-outlier analysis of tuberculosis distribution in Bihar (India) using Anselin Moran's Index (Data source: Bihar State Health Society, Bihar, India). A high positive z-score for a feature indicates that the surrounding features have similar values (either high values or low values). A low negative z-score (e.g. less than −3.96) for a feature indicates a statistically significant spatial data outlier

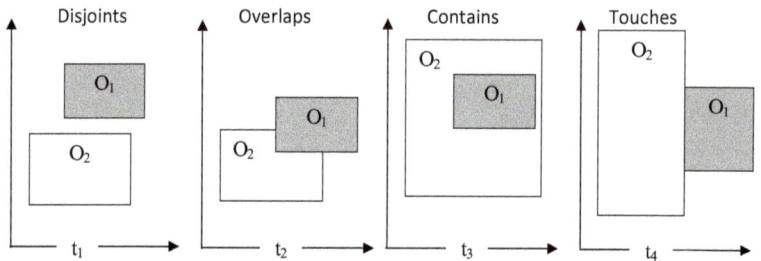

Fig. 5.13 Topological relationship in an interval ($t_1...t_4$). (Modified after Rao et al. 2012)

5.11.4 Spatio-temporal Neighbourhood

Any temporal object in space and a specific time stamp(s) related to some position(s) (x, y). O_1, O_2 are the spatial neighbour if the spatial difference between the two spacious objects is less than the defined threshold value of the neighbourhood. SQRT ((o1.x−o2.x)2 + (o1.y−o2.y). 2) can be used for spatial distance between O_1 and O_2. Similarly, O_1 and O_2 are temporary neighbours if their temporal difference is greater than the time period defined. As a module (o1.ts–o2.ts), the temporal difference can be measured. The O_1 and O_2 are spatio-temporal neighbours; both spatial and temporal spatio-temporal regions are intended to provide places with information where activities like the clustering and outward identification of details can be centralized. A set of 'N' is known as a set of objects to capture the idea of 'nearby', so that the spatio-temporal neighbours are all these objects in the group. Neighbourhood configuration estimation can be used as a pre-processing method for the exploration of clustering, outlining and collocating trends and also for online research.

5.11.5   Collocation Pattern

The concept of collocation describes two or more types of objects, whose instances in the vicinity of space and time are always located. A series of time-sized drawings of certain specific object forms through successive time slot episodes is a collocation episode. Sets of different object types, for example, that change directions, velocity, positions and travel closely for a time. The discovery of spatio-temporal collocation episodes takes the regularity of movement between varying object groups (Cao et al. 2006). When a vector travels past a reservoir, for example, the source would always travel with a high probability to the same reservoir. A specific element (e.g. the reservoir) is called central function in a collocation segment, engaging in a collocation sequence (e.g. reservoir-vector, reservoir-host).

5.12   Spatio-temporal Point Process

By adding the time factor, a spatio-temporal point method extrapolates the spatial point method. Like spatial point systems, the Poisson process, Cox process and cluster process are spatio-temporal. There are also statistical tests which include a spatio-temporal K feature and spatio-temporal scanning statistics. The lattice statistics are similar to spatial and temporal autocorrelation, spatio-temporal autoregressive regression (STAR) pattern and Bayesian hierarchical models. A significant range of spatio-temporal statistics include the study of empirical geophysical orthogonal structure, canonical-correlation (CCA) and spatio-temporal computational model assimilation (Kalman filter) (Cressie and Wikle 2011).

5.12.1   Gaussian Kernel Density Analysis (GKDA)

In statistics, the calculation of kernel density is a non-parametric method of calculating a random variable's probability density function. Estimation of kernel density is a central problem in data smoothing, where population inferences are made dependent on a finite data set. A density map is developed in Gaussian Kernel Density Analysis (GKDA) which represents the density of the infection (Fig. 5.14). In kernel density measurements, the quest parameter does not impact the density estimation (Cai et al. 2012). In the case of a zoonotic reservoir and anthroponotic reservoir, a search radius for the GKDA may be assumed to imitate the range of infected activity before bites and human transmission (Yin et al., 2012). In this analysis kernel function is used as:

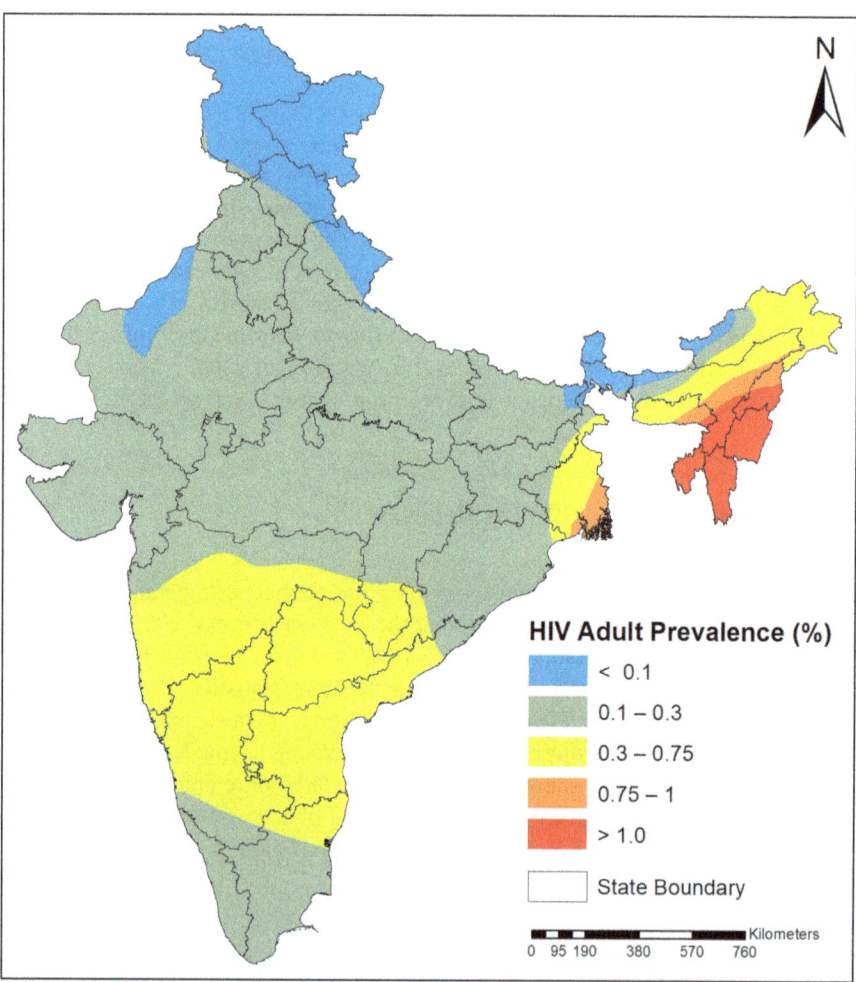

Fig. 5.14 Estimation of HIV adult prevalence rate of India based on Gaussian Kernel Density Analysis using 2017 data (Data source: India HIV estimations 2017 Technical Report). In this analysis, radius is used as 5 km. The result is a smooth approximation of density that is derived from the data, which acts as a strong non-parametric model of point distribution

$$e^{-3\left(\frac{r}{h}\right)^2}$$

where r is the radius centre at point and s and h is the bandwidth.

5.12.2 Average Nearest Neighbour (ANN) Distance

This approach was first introduced by J.G. Skellam in 1952 and expanded by P.J. Clark and F.C. Evans in 1954. The method is used to set the disease clustering (Clark and Evans 1954) and can be calculated as:

$$\bar{d} = \frac{\sum_{i=1}^{N} d_i}{N}$$

where N is the number of points and d_i is the nearest neighbour distance for point i. In a random pattern, the predicted value of the nearest distance is

$$E(d_i) = 0.5\sqrt{\frac{A}{N}} + \left(0.0514 + \frac{0.041}{\sqrt{N}}\right)\frac{B}{N}$$

where A is the region and B is the perimeter length.

The distance observed is the average of all neighbouring distances. A hypothetical random distribution with the same number of cases estimates the estimated average distance in the same geographical area. The ANN is determined by segregating the average distance observed with the expected average distance (Fig. 5.15). The case pattern will be clustered and the index will be higher than 1, if the ANN ratio is less than 1; the tendency is to dispersal. The case data were analysed annually based on the time of infection for comparing yearly variations in case distribution patterns.

5.12.3 ST-DBSCAN

DBSCAN (Density-Based Spatial Clustering of Applications with Noise) is a prominent unsupervised method used to build model and machine learning algorithms to identify a number of clusters from the estimated density distribution of the appropriate nodes (Ester et al. 1996). In 1972, Robert F Ling published a related algorithm in 'The Theory Construction of k-clusters' in the *Computer Journal* with an estimated runtime complexity of $O(n)^3$. DBSCAN is primarily designed to find high-density regions that are separated by low density regions. It follows two important steps:

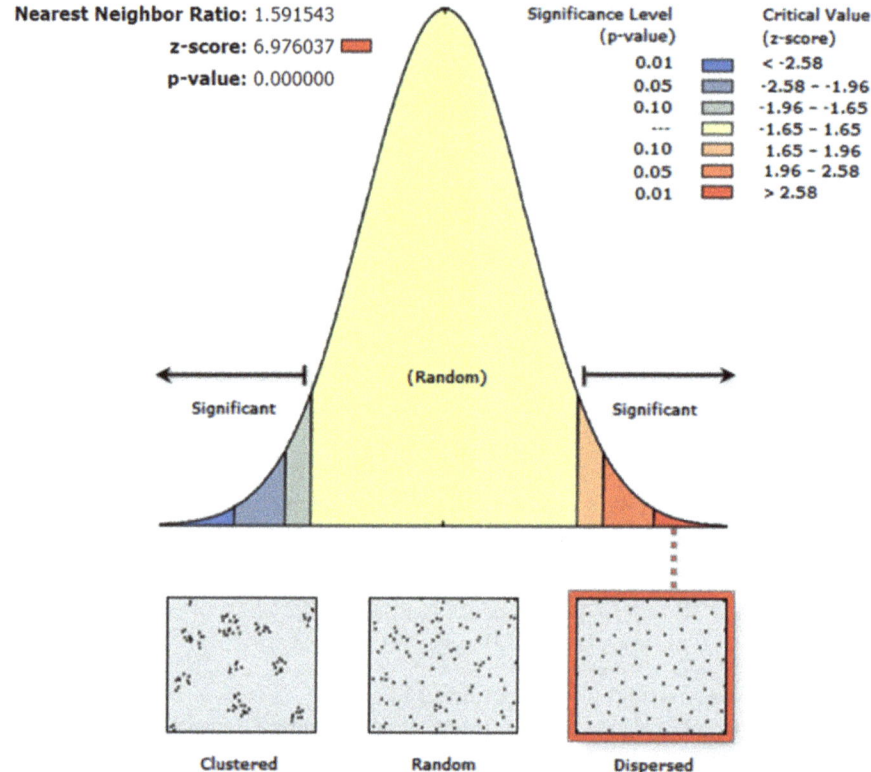

Fig. 5.15 Average nearest neighbour analysis of HIV in Bihar district of 2019. Results also showed the z-score of 6.976; there is a less than 1% likelihood that this dispersed pattern could be the result of random chance. The NN ratio is calculated as 1.59, $P < 0.0000$

- Point Density (P): Number of points in the Radius *Eps* (ϵ) circle from point P.
- Density Region: The circle of radius ϵ contains a least number of points of every Cluster point (*MinPts*).
- The key points are usually located inside a cluster here. A Border Point has fewer that MinPts within its ϵ – neighbourhood (N), but there's another core point in the neighbourhood. Noise is a data point which is neither border nor centre (Fig. 5.16).

For example, a case q is *directly density-reachable* from p if there is a sequence $P_1...P_n$ of cases with $p_1 = p$ and $p_n = q$ where each $p_i + 1$ is *directly density-reachable* from p_i. Compared to DBSCAN, spatial distance can be extended to composite spatio-temporal distance and is defined as:

$$d_{st} = D_s + \delta * D_t$$

Fig. 5.16 Process of
DBSSAN algorithm

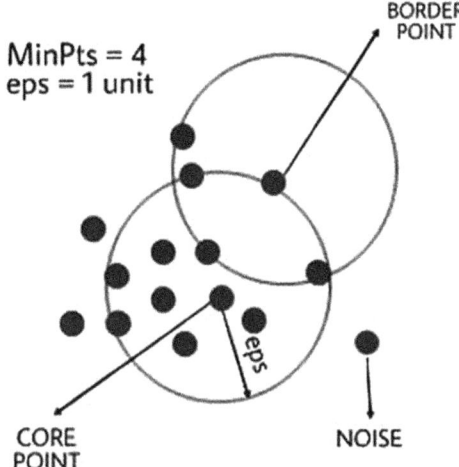

where

D_s is the spatial distance.
D_t is the temporal distance.
δ is a temporal to spatial ratio.

DBSCAN works as follows:

- Divides the dataset into n dimensions.
- This counts the number of points within this form for each point in the data set.
- DBSCAN considers this type as a cluster and spreads the cluster over numerous nearby data points by crossing each individual cluster point.

5.13 Spatio-temporal Outlier

Exterior analysis shows unusual objects that are incompatible with the other data objects. The spatio-temporal outlier can be defined as the reference space point whose non-spatial values differ significantly from those of other objects in spatial and temporal contexts (Birant and Kut 2006). Spatio-temporal outlier has been privately described as occurrences in a data set which tend to be incompatible with the rest of the data set, or which diverge so far from other measurements as to give rise to concerns that a separate process has produced them. On the other hand, a spatio-temporal outlier is an object whose non-spatio-temporal characteristic values differ significantly from those of other objects in its spatio-temporal group. Spatio-temporal outlier identification represents 'discontinuity' on non-spatio-temporal features within a spatio-temporal environment. Strategies should be outlined as per the categories of input data.

5.13.1 Outliers in Spatial Time Series

With a description of spatio-temporal neighbourhoods, simple outer spatial identification methods such as visualization-based techniques and neighbourhood-based strategies can be extended for spatial time series (e.g. data for the reference point, raster data and graph data). The *approach to visualization* plots spatial positions on a graph to recognize outliers in the room. This methods are followed clouds with variogram and Moran scatterplot, as previously implemented. The *neighbourhood method* describes a spatial or spatio-temporal neighbourhood, and the transformation between the non-spatial dimension of the existing position and that of the neighbourhood median is calculated as a spatial statistic (McGuire et al. 2014). This work has been expanded in several ways to enable numerous non-spatial attributes (Chen et al. 2008), average and mean value of attributes (Lu et al. 2003), weighted spatial outliers (Kou et al. 2006), numerical spatial outliers (Liu et al. 2014), local spatial outliers (Schubert et al. 2014) and rapid identification (Wu et al. 2009).

5.13.2 Flow Anomalies

Flow anomaly exploration, given a series of measurements across several geographical positions on a distributed network flow, intends to locate prevalent time periods where the proportion of time instance of substantially incorrectly matched sensor readings beats the percentage threshold assigned. Flow anomalies can be defined as the detection of discontinuities or incoherencies within an area detected by the node flow of a non-spatio-temporal function, where these discontinuities will continue over time. A time-scalable practice known as SWEET (Smart Window Enumeration and Evaluation of persistent-Thresholds) has been suggested using many algebraic characteristics in the flow anomaly issue to effectively explore these phenomena.

5.13.3 Anomalous Moving Object Trajectories

Due to its high feature space and fluid nature, spatio-temporal outliers are difficult to differentiate from moving body courses. A context-aware stochastic model was suggested to identify anomalous motion pattern in trajectories of indoor devices (Liu et al. 2012). For anomaly detection over running object trajectory source, another spatial deviation (distance)-based approach was introduced. A controlled solution called Motion-Alert was also suggested to detect disturbances in large moveable structures. This approach eliminates opportunities from motion pathways first, clusters features and utilizes a controlled model to assess whether a pathway is an artefact (Li et al. 2006).

5.14   Spatio-temporal Couplings and Tele-couplings

Spatio-temporal coupling patterns reflect forms of spatio-temporal artefacts whose appearances frequently occur in near proximity to geography and time. Such patterns can be classified on which temporal sorting of object types takes place: spatio-temporal (mixed drove) co-occurrences are worn for automatically generated patterns, spatio-temporal cascades for partly organized patterns and spatio-temporal serial patterns for completely organized trends. Spatio-temporal tele-coupling is the phenomenon of substantially rational or irrational temporal similarity between long distance spatial time series data (Zhang et al. 2003a, b). Recognizing spatio-temporal cascade trends from databases of disease incidents, for example, will help the health department identify the epidemic hotspot in a region and therefore take appropriate steps to minimize disease. The explanatory statistic for spatio-temporal patterns is the space-time K function, which extends the space-time K of Ripley to several variables. Mixed Drove Spatio-temporal Co-occurrence Patterns (MDCOPs) are subtypes of two or more related object types whose appearances are often placed in close contact to space and time (Lynch and Moorcroft 2008). Nonetheless, mining MDCOPs are very costly computer-based because interest measurements are computationally complex and because of their archival background, and because of the number of object types, the set of subject patterns is proportional to their number (Guting and Schneider 2005).

5.14.1   Spatio-temporal Sequential Pattern

A spatio-temporal successive pattern is a series of forms of spatio-temporal events in form of $f_1 \rightarrow f_2 \rightarrow \ldots \rightarrow F_k$. It is a 'chain reaction' from event type f_1 to event type f_2 and then to event type f_3 before the event type f_k is achieved. A sequential spatio-temporal model varies from a sequence of colocation in that it has a complete series of case forms. These trends are essential in areas such as epidemiology where some propagation of the disease can follow pathways via spatial interactions between several organisms (Cao et al. 2005; Verhein 2009). Mining spatio-temporal serial trends are difficult due to the absence of statistically relevant steps and the high cost of computing. A sequence index measure using K-function to represent spatio-temporal and sequential pattern.

5.14.2   Cascading Spatio-temporal Patterns

Partially arranged subsets of event types are termed spatio-temporal cascading patterns (CSTPs), whose occurrences are clustered together as well as appear in stages. The discovery of CSTP plays an important role in public health, such as monitoring the growth, spread and re-emergence of multiple infectious diseases (Morens et al. 2004).

5.14.3   Spatial Time Series and Tele-connection

The goal of tele-connection research, given a set of spatial time series at various locations, is to distinguish spatial time series pairs whose correlation exceeds a given threshold. The nature of the time series and the number of claimant couples and the time series are statistically troublesome. The technique uses a 'wild boot-strap' to catch spatio-temporal correlations and takes into account over a span of time the spatial autocorrelation, variability and pattern in the time sequence (Zhang et al. 2003a, b). He also proposed an effective index system, called the cone-tree, and a filtration and complexity method, used to flush out redundant peer-wise correlation calculations using spatial autocorrelation of adjacent spatial time series.

5.15   Spatio-temporal Prediction

The issue of spatio-temporal prediction is aimed at developing a model which can forecast the dependent variable from the explanatory variables. The phenomenon is called the spatio-temporal classification, if the dependent variable is discreet. If the dependent variable is continuous, the dilemma is called spatio-temporal regression. For example, the complexities of digital image classifications over time are made up by various spectral bands or channels (e.g. blue, green, red, infrared or thermal), and the dependent variable is a thematic category such as forest, village, water and cultivation (Almeida et al. 2007). This classification problem consists of specific spectral channels. For example, the annual crop yield and daily temperature prediction in various locations are space-temporal regression (Little et al. 2008). The techniques of spatio-temporal prediction comprise classical statistics, spatial and time-self-relation, spatial and temporal non-stationary variability and a multi-scale function.

5.15.1   Spatio-temporal Autoregressive Regression (STAR)

The spatial dependency of the dependent variable is specifically determined by the regression equation in the spatial auto-regression model. The regression equation can be modified by means of the $y = \rho Wy + X\beta + \epsilon$, where '$W$' is the neighbourhood relationship contiguity matrix and ρ is a parameter that represents the degree of spatial dependency between the dependent variable elements through the logistic function for binary dependent variables (Anselin 2013). Spatio-temporal Autoregressive Regression (STAR) extends the SAR specifically by modelling time and space-time dependency between parameters at multiple locations.

5.15.2   Spatio-temporal Kriging

A typical activity in geostatistics is an interpolation of spatial random fields. Easy methods such as weighted inverse distance projections or well-known kriging techniques have been used on a regular basis for many years. Modern sensors nowadays require various variables to be tracked at a growing time resolution, generating rich spatio-temporal data sets. Kriging is a geostatistical technique for forecasting where predictions are uncertain dependent on observations. Spatio-temporal kriging is computationally challenging because in each position in a spatio-temporal prediction grid, it requires a spatio-temporal semivariance calculation. We restrict the temporal measurement positions used for forecasts, i.e. perform local kriging on the temporal portion, to increase performance and to reduce the computing time. Spatial dependency can be calculated by spatial variograms by spatial covariance matrix. Spatio-temporal kriging extrapolates spatial kriging with a matrix and variograms (Cressie and Wikle 2011). It can be used to forecast spatio-temporal data from inadequate and noise. It is assumed that the Gaussian spatio-temporal random field Z is defined for the S and the T-spaces. Typically a sample $z = (z(s_1, t_1)\ldots, z(s_n, t_n))$ has been observed at a set of distinct spatio-temporal locations $(s_1, t_1),\ldots, (s_n, t_n) \in S \times T \subseteq R^2 \times R)$ that can require repetitive observations at the same location or several spatial locations at the same time. Here, the random zone Z around the ST area of concern is stationary and isotropic spatially.

5.15.3   Hierarchical Dynamic Spatio-temporal Models

Hierarchical spatio-temporal dynamic models (DSMs) have the goal of dynamically modelling spatio-temporal processes using a hierarchical Bayesian structure. DSM is conditionally based on (actual or potential) assumptions of latent inputs on the underpinning hidden mechanism. In the middle is a process model that encompasses the reliance spatio-temporally on the business strategy. The bottom of the parameter model contains the previous parameter distributions. DSMs are commonly used for simulating population change or atmospheric and oceanic cycles, in environmental and climatology science. Kalman filter can be used for pattern inference under the repression of Gaussian and linear simulations.

5.16   Spatio-temporal Clustering

Spatio-temporal partitioning, refers to the identical spatio-temporal data objects which are clustered, and hence the corresponding time and space are annexed. For many societal applications, this is significant. This should be borne in mind the close correlation that exists but does not exist with space time partitioning or clustering. Hotspots may be identified as special clusters which are slightly larger than outside when there is an occurrence or event within a cluster.

5.16.1   Spatio-temporal Event Partitioning

Such methods can be categorized as global, distance, hierarchical and graphical partitioning. Spatial objects are partitioned geographically to optimize internal group resemblance. The following are the examples: K-means, K-Medoids, EM algorithm, CLIQUE (Agrawal et al. 1998), BIRCH and CLARANS (Ng and Han 2002). Density-based methods first define 'dense' points and link them to form neighbours. Examples include the spatio-temporal extension ST-DBSCAN (Birant and Kut 2007) to its spatial variant, DBSCAN (Ester et al. 1996), and the ST-GRID (Wang et al. 2006), which splits time and space into 3-D cells and fuses compact cells into clusters. Hierarchical methods such as agglomerative, dendrogram, and BIRCH, spatio-temporal divisions or classes at the various levels of hierarchy.

K-Means Algorithm
K-means is a common numerical advantage algorithm, which restricts the dynamic data mining method to large-size data. It is recognized in real-time datasets for its usefulness of grouping spatial and temporal data points. The algorithm K-means follows:

- Numbers of 'k' cluster centres are identified, and 'k' data points are allocated arbitrarily.
- Assign the nearest cluster location to all data points.
- Calculate a new location for a cluster centre by measuring the median of all cluster centre data points.
- Continue the cycle before equilibrium and the position of the centre.

Using Euclidean distance, the K-means algorithm historically tests the difference between data points:

$$dist\ p_1 p_2 = \sum_{i=0}^{n} \ln\left(p1_i - p2_i\right)^2$$

5.16.2   Spatial Time-Series Partitioning

The spatial time series is separated by regions so that the correlation or similarity between time series in the same regions is maximized. Global approaches such as K means, K medoids and EM can be applied to this effect. But density-based methods and graphical solutions are also not efficient due to the high dimensionality of the time series.

5.16.3 Trajectory Data Partitioning

The classification of trajectory data aims at separating trajectories into classes by resemblance. There are two types of algorithms, namely, those based on density and frequency. The density-dependent strategies first split trajectories in small parts to link large segment areas by applying clustering algorithms similar to DB-SCAN. The frequency-based approach (Lee et al. 2009) uses algorithms to classify sub-sections of high-frequency trajectories (also termed high 'support') through interaction rule mining.

5.16.4 Spatio-temporal Summarization

The aim of the data description is to compactly represent a data set for compression as well as to allow pattern analysis more conveniently (Chandola and Kumar 2007). Analysis can be achieved with spatial time series data by eliminating spatial and temporal redundancies due to autocorrelation effects. Likewise, the spatial time series can also be represented by the centroids from K-means. For trajectory results, in particular trajectories of the space network, the description is more difficult because of the huge expense of estimating similarities (Evans et al., 2012).

5.17 Spatio-temporal Hotspots

Spatio-temporal hotspots are a special cluster type, whose internal strength is considerably higher than external. In epidemiology, hotspot disease may be identified and services provided by authorities to reduce the spread of the outbreak. Space-time scan statistics are used to find statistically important space-time datasets.

5.17.1 Clustering-Based Approaches

Methods for clustering may be used to classify regions for further measurement of spatio-temporal hotspots. Those include regional partitioning, clustering dependent on the density and hierarchical clustering. These techniques can be used to create potential hotspot areas as a pre-processing step, and statistical tools can also be used to check statistical significance. CrimeStat, a statistical mapping program of crime incidents, provides a variety of clustering approaches in order to classify the crime hotspots. The CrimeStat kit consists of the k-means tool, NNH clustering of the closest nearby individuals, NNH Risk Modified (RANNH) tool, the STAC Hot Spot Area Tool and a Spatial Interaction Geographic Predictor (LISA) tool to determine possible hotspot areas. Although numerous of the above-mentioned clustering methods are usually intended for dual-dimensional Euclidean space and are mainly employed for pure spatial data, space-time candidate hotspots may be identified by taking the time part of the data as the 3-D dimension.

5.17.1.1   Diagnostics for Spatio-temporal Clustering

Estimated K-functions are commonly accepted as valuable methods to evaluate trends in space points (Sarıkaya 2001). In this article, basic diagnostic techniques are followed to analyse potential dependency between the spatial and temporal components of the spatial-temporal point structure that underlies it. Firstly, it considers the following functionalities.

5.17.1.2   Cylindrical Space-Time Analysis

The space-time measurement takes place using a spherical spatial base and time-related height cylindrical frame. In both space and time, this cylindrical window is shifted. The number of cases, i.e. μ_{zd}, is measured for each cylinder, while C represents the cumulative number of cases observed. The number of cases inside the cylinder detected is C_A, and the predicted number of cases is μ_A for one particular cylinder A. The approximate number of cases μ_{zd} is measured using the following equation for each location and day on the observed marginal. The generalized likelihood ratio is calculated for every cylinder (if $C_A > \mu_A$) as given in the following equation:

$$\mu zd = \frac{1}{C}\left(\sum_z c_{zd}\right)\left(\sum_d c_{zd}\right)$$

$$T_A = \left\{ \begin{array}{l} \dfrac{C_A}{\mu_A}^{c_A} \dfrac{C-C_A}{C-\mu_A}^{C-C_A} \\ 1, \qquad \text{otherwise} \end{array} \right\}$$

The cylinder with the largest T_A is the space-time cluster in cases and is least likely to happen at random of the cylinders evaluated. It is also the central cluster for a possible epidemic. Generate random data replicas by permutation of space positions and time pairs. Monte Carlo hypothesis checking (*p*-value) is used to research statistical significance. The *p*-value is determined by combining the maximum widespread probability ratio from 999 simulation data sets with the maximum widespread probability ratio from real data. The *p*-value is calculated as:

$$p-\text{value} = \frac{1+\sum_{i=1}^{k} I\left(LLR_i \geq LLR(z)\right)}{k+1}$$

The number of random data sets generated is k in this equation, LLR_i is the maximum random data set logarithm probability ratio, $LLR(z)$ is the total chance for the detected areas and I is the indicator function. The cluster is extremely important if the p-value <0.05 is for all of the identified clusters.

5.17.2 Spatio-temporal Scan Statistics-Based Approaches

Spatio-temporal hotspot detection can be used as a special case of pure spatial hotspot detection, by adding time as a third dimension. There are two major forms of spatio-temporal hotspots: the 'persistent' temporal hotspot and the spatio-temporal hotspot 'emerging'. A 'persistent' spatio-temporal hotspot is a zone that continuously exaggerates the rate of development of observations over time. Continuous measurement of the hotspot then assumes that the probability of a hot spot is stable over time and checks for a hotspot for space and time, merely adding up the number of observations in each cycle. The area where observed concentrations increase monotonically over time is an 'emerging' spatio-temporal hotspot. The kind of transient geographic hotspot happens when an epidemic happens and the number of observations increases unexpectedly (Tango et al. 2011). Epidemiology may detect these anomalies, with the number of disease cases rapidly increasing at the beginning of an epidemic. The spatial scan statistics are used to identify emerging spatio-temporal hotspots though perceptions have shifted over time (Neill et al. 2005).

5.18 Spatio-temporal Analysis Tools

Present spatio-temporal research methods include GIS, spatial-temporal modelling methods, spatial database management systems and geographic big data frameworks.

5.18.1 Softwares

ArcGIS is the commercial GIS platform used the most in the field for maps and spatial information as well as for spatio-temporal data visualization and interpretation. A very common open-source GIS software is QGIS (formerly Quantum GIS). R packages useful for time series analysis analysis, spatial data, spatial and temporal statistics, spatial modelling. With comprehensive space-time research, computation and simulation tools, GRASS GIS is the first temporary open source. It allows the monitoring, study and simulation of climatic data, time series of vegetation indices, harvest data or land-use adjustments over the years. TerraLib extends object-related DBMS technology to handle spatio-temporal data such as data models, geographic ontology, spatial statistics, spatial econometrics, dynamic modelling, cellular automation and modelling of the environment.

5.18.2   Spatial Statistical Tools

R packages provide specific spatial and spatial statistical analysis, such as spatstat for analysing dot patterns, gstat and geoR for geostatistics and spdep for analysing spatial data. MATLAB provides toolbox and other toolboxes for mathematical space mapping. SAS historically assisted spatial statistics such as KRIGE2D, SIM2D for Gaussian random field and Space Point Pattern and VARIOGRAM for variograms method. OpenStreetMap Statistics, OSM stats and OSM Tag History are Web applications which present global and/or country-aggregated statistics (Raifer et al. 2019). STTOP and STTCOP (topology-based map algebra) to specify spatio-temporal operations of different time series data between topological-based map layers (Gebbert et al. 2019).

5.18.3   Spatial Database Management Systems

Due to the complexities of data systems involving a thorough study of the structuring of measurements and the description and interpretation of data involved, space-time data management was not straightforward. Most commercial databases provide spatial data extensions, such as Oracle Space and DB2 Data Extender. PostGIS is an extension of Postgres, an object-related DBMS, and a commonly used open-source spatial database management systems. VLSI (Very Large Scale Integration) stores millions of blueprints, drawings and performance data on electronic chip components.

5.18.4   Spatial Big Data Platform

The forthcoming space large-scale data from GPS, smartphone position data and remote sensing pictures surpass conventional space DBMS capacity and need new tools to enable flexible space analysis. ESRI GIS on Hadoop, Hadoop GIS and Space Hadoop are present spatially big data systems. OpenStreetMap (OSM) is a worldwide, open and free geographical knowledge resource that offers Volunteered Geographic Information (VGI). It is designed to work internationally with OSM background data and enables one to analyse system evolution and user input in a scalable manner. For the broad storage of point, line and polygon data, GeoMesa (https://www.geomesa.org/) provides spatio-temporal indexing over Accumulo, HBase, Google Bigtable and Cassandra databases. GeoMesa also supports spatio-temporal data analysis by laying spatial semantics on top of Apache Kafka in almost real time.

5.19   Conclusion

Disease outbreaks are phenomena that arise spatio-temporally. Applications of a computational simulation approach to spatial propagation to analyse spread and management of infectious diseases have expanded in the last two decades. Fast response and decision-making around control, treatment, eradication and disease prevention are progressively critical aspects of public health officers' and medical providers' jobs. Diseases and the pathogens that cause them are spatially and temporally dispersed, and successful countermeasures are focused on strategies that can identify the infection's focal points in a timely manner, forecast the spread of diseases and their causes and determine the risk of epidemics. Such approaches include the use of broad data sets from genetics, microbiology, ecology and geography of the environment. The epidemiological analysis which combines data from multiple data sets is managed within the GeoComputation system. Many GIS software packages can be used to query, map, interpret and visualize the data with limited preparation. The predictive and spatio-temporal modelling may be done in conjunction with other statistical or simulation applications. Models built for epidemiology often usually use a state-based approach, where a person or group may be at any one-time stage in one of four states – vulnerable, latent (infected), resistant and recovered/removed. The dynamic behaviour and evolution of epidemics over time is replicated by the repetitive implementation of the laws governing disease propagation and the temporal change between sequential individual regions. The topics discussed in this chapter demonstrate how the key organizational principles and strategies of GC are influenced by biological applications and how computationally complex simulation of biological processes is at the cutting edge of biology science. A lot of biologically focused GC work has already begun, but there is still a plethora of possibilities for GeoComputational researchers to use and develop biological data and applications.

References

Agrawal R, Gehrke J, Gunopulos D, Raghavan P (1998) Automatic subspace clustering of high dimensional data for data mining applications. In: Proceedings of the 1998 ACM SIGMOD international conference on Management of data, pp 94–105

Allen JF (1983) Maintaining knowledge about temporal intervals. Commun ACM 26(11):832–843

Almeida CM, Souza IM, Alves CD, Pinho CM, Pereira MN, Feitosa RQ (2007) Multilevel object-oriented classification of quickbird images for urban population estimates. In: Proceedings of the 15th annual ACM international symposium on Advances in geographic information systems, pp 1–8

Anselin L (2013) Spatial econometrics: methods and models, vol 4. Springer

Armstrong MP, Rushton G, Zimmerman DL (1999) Geographically masking health data to preserve confidentiality. Stat Med 18(5):497–525

Banerjee S, Carlin BP, Gelfand AE (2004) Hierarchical modeling and analysis for spatial data. Chapman & Hall/CRC

Besag J, Newell J (1991) The detection of clusters in rare diseases. J R Stat Soc A Stat Soc 154(1):143–155

Birant D, Kut A (2006) Spatio-temporal outlier detection in large databases. J Comput Inf Technol 14(4):291–297

Birant D, Kut A (2007) ST-DBSCAN: an algorithm for clustering spatial–temporal data. Data Knowl Eng 60(1):208–221

Boulos MNK (2004) Towards evidence-based, GIS-driven national spatial health information infrastructure and surveillance services in the United Kingdom. Int J Health Geogr 3(1):1

Briske DD, Fuhlendorf SD, Smeins FE (2003) Vegetation dynamics on rangelands: a critique of the current paradigms. J Appl Ecol:601–614

Cai Q, Rushton G, Bhaduri B (2012) Validation tests of an improved kernel density estimation method for identifying disease clusters. J Geogr Syst 14(3):243–264

Cao H, Mamoulis N, Cheung DW (2005) Mining frequent spatio-temporal sequential patterns. In: Fifth IEEE international conference on data mining (ICDM'05). IEEE, p 8

Cao H, Mamoulis N, Cheung DW (2006) Discovery of collocation episodes in spatiotemporal data. In: Sixth international conference on data mining (ICDM'06). IEEE, pp 823–827

Chandola V, Kumar V (2007) Summarization–compressing data into an informative representation. Knowl Inf Syst 12(3):355–378

Chen D, Lu CT, Kou Y, Chen F (2008) On detecting spatial outliers. GeoInformatica 12:455–475

Clark PJ, Evans FC (1954) Distance to nearest neighbour as a measure of spatial relationships in populations. Ecology 35(4):445–453

Cliff, A. D., & Ord, J. K. (1981). Spatial processes: models and applications,(pion limited)

Cressie N, Wikle CK (2011) Statistics for spatio-temporal data. Wiley, Hoboken

Devillers R, Bédard Y, Jeansoulin R (2005) Multidimensional management of geospatial data quality information for its dynamic use within GIS. Photogramm Eng Remote Sens 71(2):205–215

Dooley D, Catalano R, Rook K, Serxner S (1989) Economic stress and suicide: multilevel analyses: Part 1: Aggregate time-series analyses of economic stress and suicide. Suicide Life Threat Behav 19(4):321–332

Ester M, Kriegel HP, Sander J, Xu X (1996) A density-based algorithm for discovering clusters in large spatial databases with noise. Kdd 96(34):226–231

Evans MR, Oliver D, Shekhar S, Harvey F (2012) Summarizing trajectories into k-primary corridors: a summary of results. In: Proceedings of the 20th international conference on advances in geographic information systems, pp 454–457

Freier J (2000) Mapping outbreaks using GIS. In: Proceedings of outbreak symposium at AVMA (American Veterinary Medical Association), 23 July 2000, Schaumburg, IL

Gebbert S, Leppelt T, Pebesma E (2019) A topology based spatio-temporal map algebra for big data analysis. Data 4(2):86

Geographic Information Systems (GIS). GIS world sourcebook (1995). GIS World, Inc., Fort Collins, CO

Gervais M (2004) La pertinence d'un manueld'instruction au seind'unestratégie de gestion de risquejuridiquedécoulant de lafourniture de donnéesesgéographiquesnumériques. Ph.D. thesis, Université Laval, Québec, Canada, 344 p (in French)

Gesler W (1986) The uses of spatial analysis in medical geography: a review. Soc Sci Med 23(10):963–973

Guting R, Schneider M (2005) Moving object databases. Morgan Kaufmann, Burlington, MA, USA

Haining R (1993) Spatial data analysis in the social and environmental sciences. Cambridge University Press

Jacquez GM (1998) GIS as an enabling technology. GIS Health 6:17–28

Jarup L (2004) Health and environment information Systems for Exposure and Disease Mapping, and risk assessment. Environ Health Perspect 112:9 CID. https://doi.org/10.1289/ehp.6736

Kou Y, Lu CT, Chen D (2006) Spatial weighted outlier detection. In: Proceedings of the 2006 SIAM international conference on data mining. Society for Industrial and Applied Mathematics, pp 614–618

Krebs CJ (1985) Ecology: the experimental analysis of distribution and abundance. Harper & Row, New York

Kulldorff M (2013) SaTScan user guide for version 9.0. 2010. http://www.satscan.org

Kwok OM, Underhill AT, Berry JW, Luo W, Elliott TR, Yoon M (2008) Analyzing longitudinal data with multilevel models: an example with individuals living with lower extremity intra-articular fractures. Rehabil Psychol 53(3):370

Kyriakidis PC, Journel AG (1999) Geostatistical space–time models: a review. Math Geol 31(6):651–684

Lawson AB, Böhning D, Biggeri A, Lesaffre E, Viel JF, John W (1999) Disease mapping and its uses. In: Disease mapping and risk assessment for public health, pp 3–13

Learmonth ATA (1978) Patterns of disease and hunger. Newton Abbot: David & Charles

Lee KN (1993) Greed, scale mismatch and learning. Eco Appl 3:560–564

Lee AJ, Chen YA, Ip WC (2009) Mining frequent trajectory patterns in spatial–temporal databases. Inf Sci 179(13):2218–2231

Levin SA (1992) The problem of pattern and scale in ecology: the Robert H. MacArthur award lecture. Ecology 73(6):1943–1967

Li X, Han J, Kim S (2006) Motion-alert: automatic anomaly detection in massive moving objects. In: International conference on intelligence and security informatics. Springer, Berlin/Heidelberg, pp 166–177

Little B, Schucking M, Gartrell B, Chen B, Ross K, McKellip R (2008) High granularity remote sensing and crop production over space and time: NDVI over the growing season and prediction of cotton yields at the farm field level in Texas. In: 2008 IEEE international conference on data mining workshops. IEEE, pp 426–435

Liu C, Xiong H, Ge Y, Geng W, Perkins M (2012) A stochastic model for context-aware anomaly detection in indoor location traces. In: 2012 IEEE 12th international conference on data mining. IEEE, pp 449–458

Liu X, Chen F, Lu CT (2014) On detecting spatial categorical outliers. GeoInformatica 18(3):501–536

Lu CT, Chen D, Kou Y (2003) Algorithms for spatial outlier detection. In: Third IEEE international conference on data mining. IEEE, pp 597–600

Lynch HJ, Moorcroft PR (2008) A spatiotemporal Ripley's K-function to analyze interactions between spruce budworm and fire in British Columbia, Canada. Can J For Res 38(12):3112–3119

Mala S, Jat MK (2019) Geographic information system based spatio-temporal dengue fever cluster analysis and mapping. Egypt J Remote Sens Space Sci 22(3):297–304

MAY, JM (1959) Ecology of human disease. New York: 7.tI Publications, 327 pp. $7.50. LC 5813432

McGuire MP, Janeja VP, Gangopadhyay A (2014) Mining trajectories of moving dynamic spatio-temporal regions in sensor datasets. Data Min Knowl Disc 28(4):961–1003

McLeod KS (2000) Our sense of Snow: the myth of John Snow in medical geography. Soc Sci Med 50(7–8):923–935

Monmonier M (1996) How to lie with maps. University of Chicago Press, Chicago

Morens DM, Folkers GK, Fauci AS (2004) The challenge of emerging and re-emerging infectious diseases. Nature 430(6996):242–249

Neill DB, Moore AW, Sabhnani M, Daniel K (2005) Detection of emerging space-time clusters. In: Proceedings of the eleventh ACM SIGKDD international conference on Knowledge discovery in data mining, pp 218–227

Ng RT, Han J (2002) CLARANS: a method for clustering objects for spatial data mining. IEEE Trans Knowl Data Eng 14(5):1003–1016

O'neill RV, King AW (1998) Homage to St. Michael; or, why are there so many books on scale? Ecological scale: theory and application. Columbia University Press, New York

Oppong JR (1999) Data problems in GIS and health: setting an agenda for research on health and the environment. Workshop

Paul E, Daniel W (2004) Spatial epidemiology: current approaches and future challenges. Environ Health Perspect 112(9):998–1006

Pitkin HF (1972) Wittgenstein and justice: on the significance of Ludwig Wittgenstein for social and political thought. University of California Press, Berkeley

Pollitzer R (1954) World-Atlas of epidemic diseases. Edited by Ernst Rodenwaldt and Helmut J. Jusatz, under the Sponsorship of the Heidelberger Akademie der Wissenschaften. First of four issues in the third part of the series. 36 pp., 10 maps. Falk-Verlag, Hamburg. 7(4): 469–470

Raifer M, Troilo R, Kowatsch F, Auer M, Loos L, Marx S, Przybill K, Fendrich S, Mocnik FB, Zipf A (2019) OSHDB: a framework for spatio-temporal analysis of OpenStreetMap history data. Open Geospat Data Softw Stand 4(1):3

Rao KV, Govardhan A, Rao KC (2012) Spatiotemporal data mining: issues, tasks and applications. Int J Comput Sci Eng Surv 3(1):39

Resch B, Zipf A, Beinat E, Breuss-Schneeweis P, Boher M (2012) Towards the live city-paving the way to real-time urbanism. Int J Adv Intell Syst 5(3 and 4):470–482

Rezaeian M, Dunn G, St Leger S, Appleby L (2007) Geographical epidemiology, spatial analysis and geographical information systems: a multidisciplinary glossary. J Epidemiol Community Health 61(2):98–102

Richards TB, Croner CM, Rushton G, Brown CK, Fowler L (1999) Information technology: geographic information systems and public health: mapping the future. Public Health Rep 114(4):359

Rushton G (2003) Public health, GIS, and spatial analytic tools. Annu Rev Public Health 24(1):43–56

Sarıkaya Y (2001) Defining urban fire risk: exploratory analyses of fire incidents with socioeconomic characteristics for Altındağ and Cankaya districts, Ankara. Unpublished master thesis, Department of City and Regional Planning, Middle East Technical University, Ankara, Turkey

Schubert E, Zimek A, Kriegel HP (2014) Local outlier detection reconsidered: a generalized view on locality with applications to spatial, video, and network outlier detection. Data Min Knowl Disc 28(1):190–237

Sheppard E, McMaster RB (eds) (2008) Scale and geographic inquiry: nature, society, and method. Wiley

Staines A, Elliot P, Wakefield JC, Best NG, Briggs DJ (2000) Spatial epidemiology: methods and applications. Oxford University Press

Tango T, Takahashi K, Kohriyama K (2011) A space–time scan statistic for detecting emerging outbreaks. Biometrics 67(1):106–115

Tobler WR (1970) A computer movie simulating urban growth in the Detroit region. Econ Geogr 46(sup1):234–240

Verhein F (2009) Mining complex spatio-temporal sequence patterns. In: Proceedings of the 2009 SIAM international conference on data mining. Society for Industrial and Applied Mathematics, pp 605–616

Wang M, Wang A, Li A (2006, August) Mining spatial-temporal clusters from geo-databases. In: International conference on advanced data mining and applications. Springer, Berlin/Heidelberg, pp 263–270

Warrender CE, Augusteijn MF (1999) Fusion of image classifications using Bayesian techniques with Markov random fields. Int J Remote Sens 20(10):1987–2002

Wu J, Qi Y (2000) Dealing with scale in landscape analysis: an overview. Geogr Inf Sci 6(1):1–5

Wu M, Song X, Jermaine C, Ranka S, Gums J (2009) A LRT framework for fast spatial anomaly detection. In: Proceedings of the 15th ACM SIGKDD international conference on Knowledge discovery and data mining, pp 887–896

Yin CP, Zhou H, Wu H, Tao XY, Rayner S, Wang SM, Tang Q, Liang GD (2012) Analysis on factors related to rabies epidemic in China from 2007–2011. Virol Sin 27(2):132–143

Zhang P, Huang Y, Shekhar S, Kumar V (2003a) Correlation analysis of spatial time series datasets: a filter-and-refine approach. In: Pacific-Asia conference on knowledge discovery and data mining. Springer, Berlin/Heidelberg, pp 532–544

Zhang P, Huang Y, Shekhar S, Kumar V (2003b) Exploiting spatial autocorrelation to efficiently process correlation-based similarity queries. In: International symposium on spatial and temporal databases. Springer, Berlin/Heidelberg, pp 449–468

Chapter 6
GeoComputation and Disease Ecology

Abstract There are subtle variations in the ecological requirements of a number of disease vectors, and complex patterns of transmission occur in different parts of the world. This section discusses numerous infectious and vector-borne diseases and their interaction with weather and environmental variables. The spatial analysis will be performed through GeoComputaion technique. GeoComputation is used to recognize event/disease clusters and to generate maps that statistically show an excess of concentrations. GeoComputing will be a useful method for identifying target areas for public health initiatives that would not attempt to further understand the anomalies that emerge. GeoComputation can lead to systems for monitoring by integrating a spatial dimension that enables a finer assessment of the risk in grains. The nature of this chapter setting creates study-dependent solutions and prevents deeper incorporation of landscape approaches into operational models and instruments.

Keywords Disease ecology · Earth observation data · Landscape ecology · Climate · Topography

6.1 Introduction

Disease ecology is an ecological subdiscipline that deals with the causes, dynamics, and consequences of host-pathogen interactions, in particular those of infectious diseases, in the sense of environmental factors. It is an evolutionary division that explores the occurrence and distribution of species linked to the disease. This research focus seeks to extend work into the evolving field of disease ecology to encompass actively transmitted diseases, vector-borne diseases, and environmental, public health, and habitat conservation issues related to zoonotic and non-zoonotic diseases. Disease ecology investigates the spatio-temporal processes and the evolution of pathogens in natural and regulated environments (e.g. viral, bacterial, fungal). Infectious diseases have long been acknowledged to cause destructive diseases in humans, crops, and cattle, but pathogens were thought to have meagre effect on populations of wild plants and animals until recently, except in occasional and

© The Author(s), under exclusive license to Springer Nature
Switzerland AG 2021
G. S. Bhunia, P. K. Shit, *GeoComputation and Public Health*,
Springer Geography, https://doi.org/10.1007/978-3-030-71198-6_6

sometimes dramatic die-offs. It has become increasingly evident over the past two decades that parasitic species not only are a widespread and important part of ecosystems but also affect the productivity of wild populations, may cause extinction of their hosts, and act as chauffeurs of evolution (Hudson et al. 2002). Of specific concern is the study of impacts of environmental change on the ecology of infectious organisms or diseases across broad areas (land cover change, climate change, social movement). From this understanding of the ubiquitous role of pathogens in environments, the pitch of disease ecology is described as the ecological study of host-pathogen communications within the framework of their climate and progression (Fig. 6.1).

A reuse cycle accompanies the transfer of pathogens from present to potential host. This process may be simple, with a straightforward transmission from existing to future host, or dynamic, where transmission persists through transitional (multiple) hosts or vectors. There are different elements to the transmission cycle:

- The pathogen: the infection-causing organism
- The host: the infected human or animal 'carrying' the pathogen
- The exit: the form used by the pathogen to leave the host's body

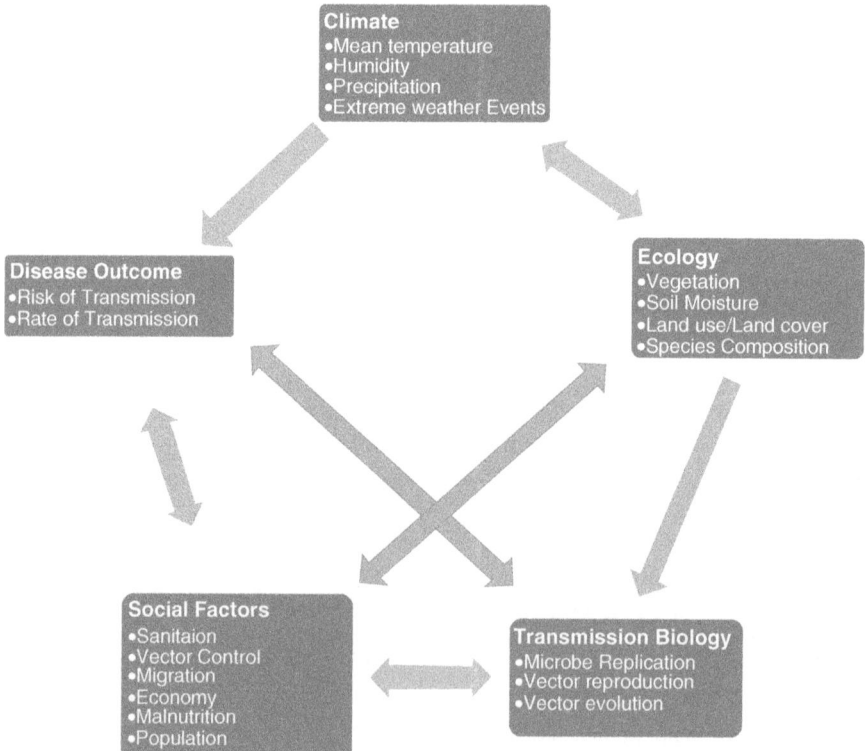

Fig. 6.1 Linkages between climate, ecosystems, and infectious diseases

- Transmission: how the pathogen is passed from the host to the sensitive person or animal, which may include developmental stages in the ecosystem, in secondary hosts, or in vectors
- Climate: the climate in which the pathogen is transmitted
- The entry: the form used by the pathogen to reach the susceptible human or animal's body
- The contagious person or animal: the likely future host, responsive to the pathogen

Nonetheless, the propagation processes of such diseases need to be studied to explain when diseases arise in a given situation and to learn how to avoid them. The ecological solution has its definitions and specificities, based on the question one wants to solve. Studies will thus concentrate on mapping the geographic spread of diseases, defining geographical clusters of outbreaks, and examining correlations between the occurrence of disease/events and collective-related environmental or social exposures (Albuquerque 2001). The use of GeoComputation is capable of recognizing clusters of events/diseases and generating maps where excess concentrations are identified as statistically significant. It will, for example, be a helpful method for identifying target areas for public health initiatives, which would not try to further understand the emergence of anomalies, whereas 'spatial autocorrelation' identification methods, calculated by the Moran coefficient or by semi-variograms, will detect an association between geographically proximate occurrences 'explicitly recognizing the possible importance of their spatial structure in the study or understanding of the data' (Bailey and Gatrell 1995).Using GeoComputational technologies, an in data-rich contexts, the artificial neural network (ANN) can be used as an exploratory method and is able to incorporate various forms of nature into a single spatial database. Meanwhile, in the case of integrating complex elements into models, cellular automata push even faster. These models 'would free us from static images of space' and may reflect the transition in space as the result of human behaviour over time.

6.2 Basics of Host-Parasite Ecology

In disease ecology, the terms worms, disease pathogen, and infectious diseases are often used interchangeably to describe organisms in or from which a host receives nutrients, solely for the benefit of the host. Nonetheless, illness is sometimes a host's pathogenic condition caused by a pathogen or parasite, and therefore diseases are not transmitted between hosts, but pathogens and parasites that cause the disease. In order to reflect variations in their population biology, parasites are often classified into microparasites and macro-parasites (Anderson and May 1979). During ancient times in the hosts, sometimes smaller than that of their hosts, microparasites (including viruses, bacteria, fungi, and most protozoans, including malaria) replicate. For microparasites, disease ecologists group hosts that indicate their exposure stage including susceptible (S), exposed (E, infected but not yet conta-gious), contagious (I, infected and capable of spreading the pathogen), and recov-ered (R, or immune) for certain pathogens. In contrast, the macro-parasites (generally parasite worms called helminths and parasite arthropods like lice) are larger, lived longer, and never complete their life cycle inside a single host. Alternatively, adult macro-parasites usually leak infectious stages such as eggs or larvae into the atmosphere, and these may or may not invade the same host in which adult macro-parasites live. Because the impacts of macro-parasites on their hosts, and sometimes parasite survival and fecundity, are highly dependent on the number of adult parasites in each host, disease ecologists control the number of macro-parasites in each host and measure their spread across hosts mathematically.

Pathogens may be transmitted in a number of ways, including through direct contact with hosts; through air, water, soil, or other surfaces; or by biting arthropods which may be hosts where the pathogen reproduces. External factors such as tem-perature and humidity may determine pathogenic survival outside a host, with sig-nificant implications for transmission. External factors may also influence pathogen replication within arthropod vectors, which in turn determines if the pathogen is subsequently transmitted by the vector. Disease ecologists use a population ecology-developed quantitative methodology and commonly use statistical models to estab-lish and evaluate theories, analyse evidence, and directly attempt to manage disease. A central cornerstone of this strategy is the pathogen's specific reproductive ratio (R_0), which defines the initial development of a pathogen in a formerly unexposed host population. R_0 determines, in a deterministic scenery, if the pathogen can con-quer and feast (e.g. if R_0 exceeds 1). The length of transmission chains in the actual stochastic universe is longer, and pathogen invasion is more likely (but not certain) for values of $R_0 > 1$. In many instances, R_0 can be instinctively interpreted as the number of new-fangled infections (βSI) induced by one infectious person (I = 1) when the population becomes entirely vulnerable (S = N), separated by the rate of loss of infectious individuals (sum of the three terms of mortality: natural death, disease-caused death, and regeneration $- \mu + \alpha + \ddot{y}$).

Experts are trying to work out how to change the world – with, for example, a new field or bridge – where the next viruses are expected to spill into humans and how they can be detected before they appear so they can blow out. They obtain blood, saliva, and other samples from high-risk species of animals to build a virus database so that it can be detected more easily if one infects humans. And they are researching ways to manage trees, animals, and pets to prevent disease from enter-ing the wilderness and transforming into the next pandemic. The International Livestock Research Institute found that diseases which infect humans from wild and domestic animals kill more than two million people a year. Peter Daszak, a disease ecologist and EcoHealth president, says 'Every new epidemic has arisen over the last 30–40 years as a result of destruction of forests and population changes'.

6.3 Emerging Infectious Diseases

Emerging infectious diseases, as the flu does every year, are either new forms of viruses or existing ones that have evolved to become original. Scientists researching the novel coronavirus have been able to recognize infectious disease and note that the time between cases in a transmission chain is less than a week and that more than 10% of people are contacted by someone who has the virus but has no symptoms yet. The pace of an outbreak depends on two factors – how many people ingest each case and how long it takes to spread infection among people. The first quantity is considered the number of reproductions; the second is the length of sequences. COVID-19's short serial interval means that new outbreaks would expand rapidly and will be difficult to stop. Previously, researchers had some confusion with the coronavirus about asymptomatic transmission. This new evidence could give public health officials advice on how to control the disease's spread (Zhanwei et al. 2020).

Of example, HIV spread into humans from chimpanzees in the 1920s when they were slaughtered and butchered by bushmeat hunters in Africa. Diseases have always come from forests and wildlife and have become part of human populations – the plague and the malaria are two examples. Nevertheless, the last half century has seen emerging illnesses quadruple, mainly because of a growing human interaction with the environment in hot areas worldwide, particularly in tropical regions. And with international air transport and a thriving animal trade industry, there is immense potential for a significant epidemic in major population centres. One research in the Amazon, for example, found that deforestation increased by 4%, increasing the incidence of malaria by nearly 50% because in recent deforestations mosquitoes that spread the disease thrive on the right combination of sunlight and water. Public health researchers have started to incorporate biodiversity into their models. For example, Australia has declared multimillion-dollar initiative to understand the ecology of Hendra viruses and bats. The West Nile virus came from Africa to the USA but has spread to the USA since the American robin is one of its favourite hosts and flourishes in the landscape of lawns and farm fields. 'The virus has had a huge effect on human health in the United States since it has taken advantage of animals that do well with humans', says Marm Kilpatrick, a biologist at Santa Cruz University of California. And the Eastern Coast epidemic, Lyme disease, is also the result of human changes in the environment: destruction and deterioration of the large adjacent forests. Technology has pursued predators – wolves, foxes, owls, and hawks – which resulted in a fivefold rise in white-footed mice, which are perfect 'reservoirs' for the Lyme bacteria, presumably due to weak immune systems. 'So, mice are developing huge numbers of tainted nymphs', says Lyme disease researcher Richard Ostfeld. Raina K. Plowright, a biotech scientist with a study into the ecosystem of disease at Pennsylvania State University, found that Hendra virus outbreaks were rare in rural flying foxes but much greater in urban and industrial animals. She believed that urbanized bats would be sedentary, that they lack regular access to the virus they used to enter the wild, and that infection would remain small. That means more bats – whether due to inadequate diet, habitat destruction, or other causes – get infected and bring some of the virus into the back-yards. Overall, the knowledge obtained from emerging diseases over the last couple of years helps us to sleep a little better, says EcoHealth veterinarian Dr. Epstein.

The growing frontier of ecology of diseases covers at least four key sectors: interactions between pathogens and strains (including mechanisms for in-host infections and interactions with the immune system), specific considerations of the spatial context, multi-host pathogens' mechanisms and drivers in spreading, and the evolution of hosts and parasites as a result of changes in the environment. Interactions among different pathogens have been investigated both within populations and indi-vidual hosts, establishing both immune-mediated and resource-based mechanisms by which different parasites communicate functionally (Graham 2008). A major challenge for the ecology of disease is understanding the spread of pathogens among hosts, particularly in a dry, acidifying, urbanizing area, and becoming more inter-relating in the aforementioned areas.

6.4   History of Development

We need to remember the lessons from previous outbreaks if we are to save the planet from the danger of infectious diseases. Each of these diseases was very dif-ferent, and our reactions to them were always very different as well. In part, this is because these pathogens are caused by very different species, but it is also because the social environment in which diseases occur is important. A brief history of nota-ble pandemic is illustrated in Fig. 6.2. Mosquito-borne diseases (MBDs) infect nearly 700 million people a year and are recognized in more than 100 countries, impacting all continents except Antarctica and causing millions of deaths per year (World Health Organization 2018a, b). In the zoonotic disease group, vector-borne diseases are evolving quicker with climate change and likely have a more complex epidemiology. The bulk of global emerging infectious diseases in the last 80 years have been zoonotic (Jones et al. 2008). More than 17% of infectious diseases are vector-borne, resulting in over 70,000 deaths each year (WHO 2014). Infectious disease studies have long been dominated by specialists on the taxation of infectious agents (e.g. bacteriologists, virologists), resistance mechanisms (e.g. immunolo-gists), human hosts (e.g. pathologists), host populations (epidemiologists), and results and diagnostics (e.g. physicians and veterinarians). Disease ecology evolved as scientists gradually realized that pathogen-host encounters might conceptually be related to other interspecific interactions, such as those between predator and prey, rivals, and mutualists. The emergence of the ecology of disease from the late twen-tieth century adds to the overall complexity, including knowledge of certain patho-gens infecting different host organizations, hosts possessing several pathogenic infections, and abiotic conditions (such as temperature and moisture) influencing the biotic transmission conditions and diseases at work. Consequently, a wider con-text than the simplistic host-pathogen paradigm is also required to understand the nature of the disease. Disease ecologists are also interested in the ecological roots

| 1334-1350 | Black Death (30 - 50 million deaths) |

- Bubonic plague
- Originated in hina, and spread to Europe along trade routes

| 1860 - 1903 | The modern plague (About 10 million deaths) |

- Bubonic plague
- Started in China and then spread to Hong Kong by 1894

| 1889-1890 | Russian flu (About 1 million deaths) |

- Influenza A
- The panademic was first recorded in Russia, and then spread through Europe, Asia and reached USA

| 1918-1919 | Spanish flu (About 50 million deaths) |

- Influenza A (H1N1)
- Estimated to have infected over 50 million people worldwide

| 1956-1958 | Asian flu (About 2 million deaths) |

- Influenza A (H3N2)
- Started in Hong Kong and then spread through Asia, Australia, Europe and the USA

| 1976 | Ebola Outbreak (280 deaths) |

- The first recorded outbreak of the Ebola Virus
- Contained within Zaire (now DRC)

| 1981 (ongoing) | HIV/AIDS (About 40 million deaths) |

- HIV
- Worldwide

| 2003 | SARS (775 deaths) |

- Viral respiratory illness (Coronavirus)
- The initial outbreak was reported in southern China and then quickly moved to Hong Kong, and other countries

| 2009 | Swine flu (Over 200000 deaths) |

- Influenza A (H1N1)
- WHO estimates around 18500 deaths, but other research suggests a death toll of over 2000000 deaths

| 2014 | Ebola (over 11,000 deaths) |

- Over 28000 cases of Ebola have been reported
- The worst hit countries are Liberia, Sierra Leone and Guinea in West Africa

| 2019-2020 | COVID-19 (Over 1,513,179 deaths till 05th December, 2020) |

- Severe acute respiratory syndrome coronavirus 2
- Started in Wuhan, Hubei, China
- Worldwide

Fig. 6.2 Pandemic history of infectious disease

of disease dynamics (e.g. how host population densities influence the rate of trans-mission) and the ecological impacts of disease (e.g. how host population dynamics change as epidemic develops). One seminal breakthrough was the production of a malaria statistical model shortly after the initial analysis of the malaria parasite's life cycle, *Plasmodium*, in Sir Ronald Ross' *Anopheles* mosquitoes (Ross 1915). One important breakthrough was the formation of a statistical malaria model shortly after the initial analysis of the malaria parasite's life cycle, *Plasmodium*, by Sir Ronald Ross (Ross 1915) at Anopheles mosquitoes. Ross' model separated sub-populations of mosquitoes and humans vulnerable from those infected and tracked vector and host latency to infection. The discovery by Thomas Park (Park 1948) that sporozoan parasite infection converted a higher rival into a lower one, reversing the outcome of competition between two species of flour beetle (*Tribolium* spp.), greatly affected population and group ecologists. A third main change was aware-ness on host cultures, economies, and habitats of the profound impact of infectious diseases. Examples include accounts of pathogens influencing human evolution by Diamond and Guns (1997) and Dobson and Carper (1996) and a study of rinderpest influencing African wildlife populations by Plowright (1982) and Daszak et al. (1999), which is an investigation of the worldwide relationship of chytrid fungus *Batrachochytrium dendrobatidis* with amphibian deteriorations.

6.5   Role of GeoComputation in Disease Ecology

Infectious disease epidemiology is continuously fluctuating in relation to changes in the ecosystem and evolving relationships between the organisms, habitats, vectors, and pathogens. Along with a rise in emerging zoonotic diseases, reservoir hosts, reservoirs, and pathogens they carry have been growing in numbers. Research also focuses on the relationships between at-risk individuals, with other species, and with aspects of their common ecosystems. Spatial influences thus also play a crucial role in disease ecology because of the effect of heterogeneous spatial distributions (e.g. host ranges, vector density, and communication landscape barriers) on the dis-semination of an infectious disease. GeoComputing technologies can aid in the analysis and response to epidemiological outbreaks and health hazards in popula-tions before and after epidemics. GeoComputing in public health is used for simple mapping which involves spatial study of disease outbreaks with environmental fac-tors. GeoComputation allows us in visual interpretation and exploratory research, as well as in estimating the relationship between disease occurrence measurements (Rytkonen 2004). GeoComputation's role in defining and mapping the danger of vector ecology and habitats is to provide appropriate proxy information in meteoro-logical and environmental variables linked to spatial variability. The root cause of the disease infection and source of infection can be determined by using GeoComputing techniques for modelling vector populations, presence of vectors, distribution and distance, risk assessment of vector-borne diseases, and spatial spread of disease. The areas vulnerable to spread of the epidemic are epidemiologically relevant for choosing the correct strategies for managing disease. The coastal habitat transition and the resulting environmental changes, including land-use changes, urban sprawl, and rapid expansion in humanization and industrial production, were motivated by the creation of an adequate ecosystem for vector-borne disease outbreaks.

6.5.1   Progress and Improvement in Remote Sensing Satellites and Sensors for Environmental Assessment

There are eight different remote sensing periods; some run parallel over time, while others are network-specific description of data management, scientific use, and system functions.

6.5.2   Airborne Remote Sensing Era

The period of airborne remote sensing was established during World War First and the Second (Avery and Berlin 1992). Remote sensing was primarily used during this period for observation, reconnaissance, mapping, and military surveillance purposes.

6.5.3   Rudimentary Space-Borne Satellite Remote Sensing Era

The space remote sensing era began with the flight of primitive satellites such as Russia's Sputnik 1 and the USA's Explorer 1 in the late 1950s (Devine and Divine 1993). The first meteorological satellite, Television and Infrared Observation Satellite-1 (TIROS-1), was soon followed by the USA in the late 1950s as well (House et al. 1986).

6.5.4   Spy Satellite Remote Sensing Era

At the end of the cold war, spy satellites like Corona (Dwayne et al. 1988) were widely used. The data were gathered for military reasons, almost entirely. Nevertheless, during the latter three periods, the spin-off of remote sensing evolved for military reasons, spilling over to mapping and eventually into applications of the climate and natural resources.

6.5.5   *Meteorological Satellite Sensor Remote Sensing Era*

The first satellite meteorological instruments included the Advanced Very High Resolution Radiometer (AVHRR), GOES (Geostationary Operational Environmental Satellites), and National Oceanic and Atmospheric Administration (NOAA) (Kramer 2002). This was also a period where global reporting was both rational and functional solutions for the environment.

6.5.6   *Landsat Era*

Landsat started in 1972 with the flight of Landsat 1 sensors with multispectral scanner (MSS), now known as 'Earth Resources Technology Satellite'. The trajectory was followed by satellites for Landsat 2 to 3 with MSS and the Thematic Mapper (TM) 4 and 5 in the pathway. Landsat 7 holds sensor Enhanced Thematic Mapper (ETM+). During flight, Landsat 6 crashed. Landsat 8 is scheduled to launch in 2011 with Operational Land Imager (OLI). Even in Landsat era, satellites such as France's Systeme Pour l'Observation de la Terre (SPOT) and India's Indian Remote Sensing (IRS) are equally important sun-synchronous satellites (Jensen 2000). Such satellites are high to medium resolution and have a regional capability (nominal 2.5–80 metres). Under this amendment, only Landsat collects currently regional wall-to-wall data.

6.5.7   *Earth Observing System Era*

The Earth Observing System (EOS) period (Melesse et al. 2007; Bailey et al. 2001) started in 1999 with the launch of Terra satellite and carried with it global reach, regular repeat reach, high level of processing (e.g. geo-rectified, at-satellite reflectance), and simple and mostly free data access. The Terra\Aqua satellites that hold sensors such as the Moderate Resolution Imaging Spectroradiometer (MODIS) and Measurement of Pollution in the Troposphere (MOPITT) have regular revisits and different processed results. Sensor data systems have been popular and expanding systems. In addition, the provision of validated data in terms of items such as leaf area index (LAI) and land use\land cover (LULC) has been standard. MODIS itself actually owns 40+ products. French Satellite Systeme Pour l'Observation de la Terre (SPOT); Indian Remote Sensing (IRS)-Linear Imaging Self-Scanning (LISS) I, LISS II, LISS III, and LISS IV; and CARTOSAT have been famous this time around. Also common during this time (and also during the Landsat era) is the launch of European Remote Sensing Satellite (ERS), Japan Earth Resources Satellite (JERS), RADARSAT, and Advanced Land Observation Satellite (ALOS). The Shuttle Radar Topographic Mission (SRTM) was used for optical elevation data collection.

6.5.8   New Millennium Era

The new millennium era applies to technologically sophisticated 'test-of-concept' spacecraft placed in space at the same time as the EOS era, although there are various principles and theories (Bailey et al. 2001). Those comprise Earth Observing-1 for the first hyperspectral data spatially carried. The concept of Advanced Land Imager (ALI) is still very enticing as an easier, technologically improved Landsat alternative.

6.5.9   Private Industry Era

The private business era started in the last thousand years and the beginning of the millennium. This age is comprised of many inventions: Firstly, very high-resolution data collection (<10 metres) is typified by the satellites IKONOS and QuickBird. Secondly, a new data-gathering method is characterized by a five-satellite constellation, RapidEye, which provides almost constant coverage in 5 spectral bands at every location on earth and a 6.5-meter resolution red edge band. Thirdly, the deployment of microsatellites, some under the Disaster Monitoring Constellation (DMC), was developed and launched by Surrey Satellite Technology Ltd. for Turkey, Nigeria, China, USGS, the UK, and others. Fourthly, Google Earth's breakthrough (http:/earth. google.com) is to allow VHRI for fast access to data for every part of the planet through streaming technologies, making it convenient for even non-specialists to zoom in and pan remote sensing data.

6.5.10   Analysis of Environmental Variables from Remote Sensing Data

Weather, topography, soil geology, and geographical details also constitute environmental data used to describe habitat requirement. Visual and digital interpretation and study of multi-spectral and multi-temporal satellite data ([Landsat TM (Thematic Mapper), French Satellite Systeme Pour l'Observation de la Terre (SPOT), Indian Remote Sensing (IRS) LISS I, LISS II, LISS III and panchromatic imagery IKONOS]) items originating from earth observation tool satellites and aerial images of red and infrared colours, and meteorological satellites, National Oceanic and Atmospheric Administration's Advanced Very High Resolution Radiometer (NOAA-AVHRR), or Moderate-resolution Imaging Spectrometer (MODIS) satellite sensors are used to delineate and chart vector ecosystems and vector ecology. The remote sensing-based land use/land cover satellite-based geographic mapping identification includes information on mosquitogenic conditions. Significant interaction was established in relation to changes in environmental variables such as water sources and vegetation in spatial changes to mosquito abundance. The National Oceanic and Atmospheric Administration (NOAA) and the National Aeronautics and Space Administration (NASA) use satellite-derived rainfall forecast estimate (RFEs) to support the water balance model. NASA provides daily rainfall data from the Tropical Rainfall Measuring Mission (TRMM) Project, which covers the globe at a resolution of 25 km between 60 ° north and south of the equator. The utility and durability of GDAS-PET-dependent PET for crop water balance studies have been demonstrated by recent validation of GDAS-PET using PET derived from station parameters in the USA.

For commercial field environmental research, the high spatial resolution, also known as fine spatial resolution, is typically less than 10 m and ranges from 0.5 to 10 m. The widely used devices for high-space optical sensors are IKONOS, QuickBird, OrbView-3, LISS IV, CARTOSAT 3, and SPOT-5 (Satellite Systeme Pour l'Observation de la Terre-5). The advantage of high-resolution spatial imagery is that it significantly improves the precision of detecting and characterizing tiny objects at spatial scales that were historically only possible from airborne platforms (Gillespie et al. 2008). Hyperspectral measurements usually with 100 or more adjacent spectral bands provide the potential to capture enough spectral information over a continuous spectrum. This is distinct from multispectral sensors where very few discrete bands are detected (Nagendra and Rocchini 2008). Hundreds of spectral bands with 10–20 nm spectral bandwidth give different ways of detecting small distinctions between objects of interest. The latest example is to classify on fine-scale, species-specific ground cover (Turner et al. 2003), such as divisions of vegetation or soil types (Aplin 2005), which make a remarkable contribution to the biodiversity trend analysis. Spectral signatures obtained from atmosphere-corrected hyperspectral data can be directly compared with the existing spectral database (e.g. Jet Propulsion Laboratory Spectral Database) to easily distinguish useful ground information for evaluating, characterizing, and defining land cover changes (Turner et al. 2003).

Shippert (2004) identified the present space-acquiring hyperspectral sensors, including the NASA EO-1 (National Aeronautics and Space Administration's Earth Observing-1) hyperion camera, the CHRIS (Compact High-Resolution Imaging Spectrometer) camera on the PROBA (PRoject for On-Board Autonomy) satellite of the European Space Agency, and the FTHSI (Fourier Transform Hyperspectral Imager) sensor on the US MightySat II satellite by Air Force Research Laboratory. The LEWIS Hyperspectral Imaging Instrument (HSI)-enhanced hyperion sensor measures visible light and other electromagnetic energy reflections at a resolution of 30 m in 220 spectral bands from 0.4 to 2.5μm (Campbell et al. 2007).

Thermal remote sensing measures the energy released from the surface of the Earth in the thermal infrared (3–15μm), where all objects above absolute zero are radiated. Theoretically, TIR sensors calculate the threshold surface temperature and thermal characteristics that are necessary to gain a deeper understanding of land surface energy balance interactions and more reliable models (Quattrochi and Luvall 2004). In fact, owing to the progress of ecological thermodynamics, TIR remote sensing is able to reveal the concepts of ecological patterns of form and function. Based on relative radiant temperatures (a thermogram), a thermal grey picture is produced, and light tones correspond to warmer temperatures and dark tones to cooler temperatures. The well-known TIR sensors include the Advanced Very High Resolution Radiometer (AVHRR) on board Polar-Orbiting Environmental Satellites (POES), the Landsat Thematic Mapper (TM) and ETM+, the Advanced Spaceborne Thermal Emission and Reflection Radiometer (ASTER) on the Terra Earth Observation Satellite Platform, etc.

Light Detection and Ranging (LiDAR), also known as Laser Altimetry, is an active remote sensing technology that uses a laser to illuminate a target object and a photodiode to monitor the radiation from the back scatter (Lim et al. 2003). The current LiDAR remote sensing can be subdivided into two general groups: non-scanning LiDAR and scanning LiDAR. The non-scanning LiDAR report measured the transit time between the signal transmitted and obtained from the surface of the target, and the scanning LiDAR monitor produces consistent wave spectrum, emitted in a sinusoidal signal and modulated in the intensity of the laser light (Wehr and Lohr 1999). For bathymetry, forestry, and other uses, airborne LiDAR remote sensing systems such as LVIS (Land, Vegetation, and Ice Sensor) were used (Hyyppa et al. 2009). The space-borne LiDAR has launched ICESat/GLAS (Ice, Clouds, and Land Elevation Satellite/Geoscience Laser Altimeter System), the first laser range for continuous global observational research in comparison with the large-scale, footprint-scale, and high-cost airborne LiDAR (Duncanson et al. 2010).

Remote sensing, the technology of collecting information by non-contact observation, has swept ecology, biodiversity, and conservation (EBC) fields open. In fact, the use of remote sensing deepens with the introduction of cutting-edge remote sensing devices and technologies. Remote sensing is believed to evolve in a course close to that of computer science which has infiltrated all facets of human life. As a propeller, EBC works to drive the birth of innovative remote sensing sensors and practices. For instance, the object-based image analysis (OBIA) is maturing in hopes of addressing the inquiry 'why are remote sensing and digital image processing still so focused on the statistical analysis of single pixels rather than on the spatial patterns that they make up' posed by Blaschke and Strobl (2001). In the EBC sector, these tools can be used to answer questions. Hence, having EBC practitioners and remote sensing specialists communicate effectively is still desperately needed.

6.6 Weather Variables and Disease Ecology: Case Study Through GeoComputation Technique

Climate is the normal state of the atmosphere, which is dictated by factors such as temperature, humidity, precipitation, and wind. In contrast, the environment refers to natural weather conditions for a certain period of time and can include information about the event, the magnitude, and other statistical characteristics in the atmosphere (usually at least a month). Spatial climatic variation involves well-known temperature gradients of the latitudinal and altitude. For example, the average surface air or temperature of the soil falls in normal conditions on mountain terrain by around 6.5 °C for every 1000 m elevation rise; the average surface temperature drops by approximately 5 °C over an equator to pole gradient of 1000 kilometres. Superimposed on these broad-scale gradients are more complicated regional trends of temperature, precipitation, storm frequency, and so on, and microtopographic shifts can produce large variations in surface temperature and soil moisture at very fine spatial scales.

Terminology used for literature selection is based on a study of the same paper over the last two to three decades, with three factors: infectious disease components, temperature variables, and selected infectious diseases. The climate and its changes in the season play a significant role in disease expansion. The environment for the vector or intermediate hosts determines whether infection in a region is likely to be a problem. Weather and climate may not be ideal for year-round transmission, and certain diseases occur seasonally when the weather is favourable for transmission. Vector-transmitted diseases, such as malaria and yellow fever, are related to rainy season (Bouma and Van der Kaay 1995).

After the beginning of the Industrial Revolution, greenhouse gas emissions such as CO_2, CH_4, and N_2O have risen due to the rising global population, increased consumption of energy per capita (primarily by burning fossil fuels), and uses of land including deforestation and agriculture. This greenhouse gases improve the normal atmospheric trapping of infrared (heat) radiation, a phenomenon known as radiative forcing, resulting in the increased greenhouse effect. Certain anthropogenic processes can affect the local and regional atmosphere, in addition to the effects of greenhouse gases. Enhanced land use and distribution of trees can also affect the environment on a number of spatial levels. Of course, trees can provide cover for the land below, but it can also influence the area cycles of precipitation, because the water vapour 'exhaled' by the woods is a major source of clouds and rainfall. Despite numerous debates on the causes of climate change, there is widespread awareness of a continuing global climate change and the non-minor role of human activity during this process (IPCC 2007). Throughout the twenty-first century, the IPCC projected an average global temperature rise of 1.5–5.8 °C, combined with rising extreme and anomalous weather events, including heatwaves, floods, and droughts (IPCC 2001).

Three components are important in most infectious diseases: an agent (or pathogen), a host (or vector), and the transmission ecosystem (Epstein 2001a, b). In most infectious diseases, three components are essential: an agent (or pathogen), a host (or vector), and an area of transmission (Epstein 2001a, b). Appropriate climatic and environmental conditions are required for pathogens, vectors, and hosts to thrive, replicate, spread, and transmit disease. Environmental changes can also impact infectious diseases by influencing pathogens, vectors, hosts, and living conditions (Epstein 2001a, b; Wu et al. 2014). Overall, climate conditions restrict the geographic and seasonal distribution of infectious diseases, and the time and the intensity of the outbreak of disease are affected by the climate (Kuhn et al. 2005; Wu et al. 2014). Growing and stagnant environments play an increasingly important role in driving economic changes, regeneration, and transition of infectious diseases (McMichael et al. 1996). Many of the more prevalent infectious diseases are highly vulnerable to temperature variability, especially those transmitted by insects (Tian et al. 2015). Many infectious diseases, such as salmonellosis (Chretien et al. 2015), cholera, and giardiasis, may have intensified outbreaks due to high temperatures and floods. The health implications of these factors persist in the spatial and seasonal patterns of infectious diseases and changes in their epidemic frequency and severity.

6.6.1   Climate and Pathogens

Pathogen refers to a broad variety of agents for the disease, including viruses, bacteria, parasites, germs, and fungi. The effect of climate change on pathogens may be overt, affecting pathogens' growth, development, and life cycle, or indirect, affecting the pathogens' habitat, ecosystem, or competitors (Fig. 6.3).

6.6.1.1   Temperature

Temperature can cause illness by affecting the pathogens' life cycle. First, for a pathogen to survive and establish a certain temperature range, for instance, the two thresholds, 22–23 °C maximum for mosquito production, 26–30 °C for sandfly growth for leishmaniasis transmission, and 25–26 °C minimum for Japanese encephalitis virus (JEV) transmission, play key roles in JEV ecology (Mellor and Leake 2000; Tian et al. 2015). In certain pathogens, extreme heat will increase mortality rates (Kuhn et al. 2005). Malaria parasite production (*Plasmodium falciparum* and *Plasmodium vivax*) stops when the temperature reaches over 33–39 °C (Patz et al. 1996). Second, increasing temperature can affect pathogen reproduction and extrinsic incubation duration (EIP) (Harvell et al. 2002). On the opposite, lower atmospheric temperatures are expected to increase EIP, which in turn may reduce disease transmission such as dengue as less mosquitoes will survive long enough. Third, extended hot weather intervals can increase the surface temperature of water sources and food system, which can provide an accommodating atmosphere for cycles of replication of microorganisms and algal blooms. For examples, *Vibrio* spp. bacteria, circulating in the Baltic and North Sea, demonstrated an improved growth rate during hot summers (Frank et al. 2013). *Salmonella* infection is a disease spread by food, and the bacteria are reproductive, because the temperature increases from 7 °C to 37 °C (IWGCCH 2010). Finally, rising temperatures could reduce a pathogen's propagation by favouring its competitors.

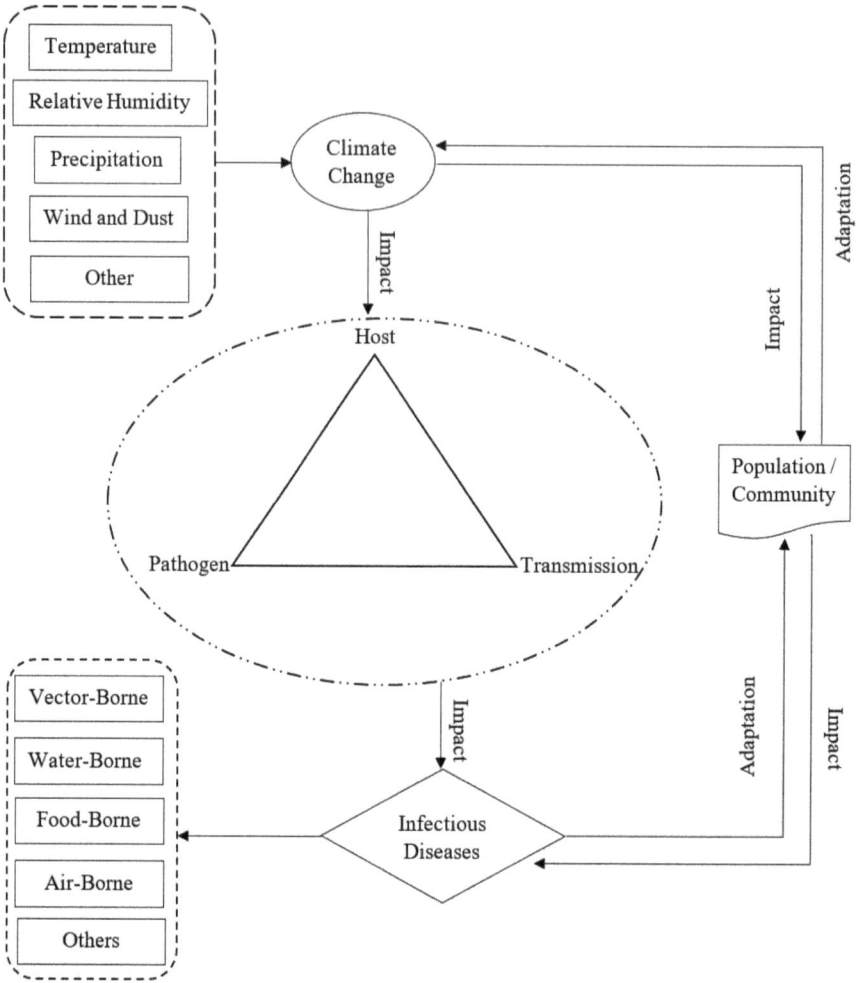

Fig. 6.3 Climate and its association with the disease (Modified after Wu et al. 2016). The first laid forth the elements of diseases: pathogen, host or vector, and propagation of diseases. The second collection explains environment and atmosphere, including climatic factors (such as temperature, precipitation, and humidity), unusual weather phenomena on broad scale (such as El Nino), and meteorological threats (such as drought, floods, and heatwaves). The third collection defines infectious diseases, comprising vector-borne diseases (e.g. malaria), waterborne diseases (e.g. cholera), respiratory diseases (e.g. influenza), or foodborne diseases (e.g. *Campylobacter*). More insights were received from the relevant experts to refine the search plan and find additional correlations

6.6.1.2   Precipitation

Changes in climate may cause changes in precipitation that affect the spread of pathogens in water. Rainfall plays a significant role in the development of pathogenic waterborne diseases. The rainy season correlates with the increase in faecal pathogens, because heavy rain can sediment in the water and therefore lead to the aggregation of faecal microorganisms (Jofre et al. 2009). Unusual precipitation after a long drought, however, can lead to increased pathogens, triggering an outbreak of disease (Wilby et al. 2005). Droughts/low runoff contribute to weak river flows resulting in the accumulation of waterborne contaminants in the effluent (Hofstra 2011).

6.6.1.3   Relative Humidity

Change in humidity also impacts infectious disease pathogens. Airborne infectious disease pathogens such as flu tend to be responsive to humidity. Absolute humidity and temperature, for example, have been shown to affect transmission and survival of influenza viruses (Xu et al. 2014). Lowen et al. (2007) suggested that cold weather and low relative humidity would be ideal for influenza virus transmission. Increase of humidity also has an effect on waterborne disease viruses. For example, because of the drying effect of surface water, the survival of waterborne viruses near the water surface is limited (Gerba 1999). Humidity in anopheline mosquito has been shown to affect the growth of malaria parasites (Patz et al. 2003). Lastly, vector-borne disease virus can be affected by an increase in humidity.

6.6.1.4   Sunshine

Sunshine is yet another significant climatic component which can affect infectious disease pathogens. For example, sunlight hours and temperatures work together during cholera cycles to determine a good situation with which to multiply *Vibrio cholera* in aquatic environments (Islam et al. 2009a, b).

6.6.1.5   Wind

The wind is a key factor which affects airborne disease pathogens. Desert dust has been reported to be associated with increased concentration of growing bacteria, cultivable fungi, and fungal spores in the atmosphere during Asian dust storms (ADSs) (Griffin 2007; Schlesinger et al. 2006). Chen et al. (2010) observed that the influenza A virus production during the ADS days was considerably higher than in usual days.

6.6.2 Climate and Vectors/Hosts

Host refers to live organisms or plants on or in which pathogenic diseases exist. Vectors are intermediate hosts and are transferred to live species hosts by pathogen. The geographical ranges and variations in population of insect vectors are directly related to climate conditions and variations. Climate change would also alter the distribution, timing, and intensity of infectious diseases by affecting disease vectors.

6.6.2.1 Temperature

Temperature affects the propagation of disease vectors spatially-temporally. As the temperature rises, insects in regions with low latitude in medium- or high-latitude and high-altitude areas may find new conditions that lead to disease expansion or change. Latest studies also shown that some human infectious diseases transmitted by mosquitoes, including measles, African trypanosomiasis, Lyme disease, tick-borne encephalitis, yellow fever, plague, and dengue, also spread to a broader range (Harvell et al. 2002). Most of these pathogens also spread to regions of higher latitude, with mosquitoes, ticks, and midge vectors expanding their habitat. For example, *Aedes aegypti* is the host of the yellow fever and dengue fever viruses (Epstein 2001a, b). When global warming progresses, the host organism like *A. aegypti* can disappear from some regions where temperature rises above its thresholds. Vectors/hosts of the disease can survive climate change by taking shields in small-scale environments where changes in atmospheric temperature do not prevail. For example, *A. aegypti* mosquito used underground pitchers or cement water tanks to escape from a high summer temperature of 40 °C in the town of Jalore in Rajasthan, India (Tyagi and Hiriyan 2004).

6.6.2.2 Rainfall

However, for vectors, rainfall isn't always pleasant. Improper precipitation can have devastating consequences on the mosquito population, as heavy rain can sweep away their breeding sites (Kuhn et al. 2005). On the contrary, drought in wet regions will decrease flow velocity in brooks and create more pools of stagnant water as breeding grounds for mosquitoes (Kovats et al. 2003). Drought induces the accumulation of decaying organic materials in these ponds, producing *Culex*'s favourite condition; heavy precipitation can wash the drains and flood the pools (Epstein 2001a, b), minimizing West Nile virus spread.

6.6.2.3   Relative Humidity

Most hosts of illnesses appear to react strongly to increases in humidity. Relative humidity impacts malaria transmission by affecting mosquito behaviour and survival. When the mean monthly relative humidity becomes less than 60%, the malaria vector mosquito's lifetime is too short for malaria transmission. High humidity will adversely affect A's survival in adults. Therefore, *aegypti* eliminates the spread of dengue fever (Christophers 1960). In general, low humidity creates unfavourable conditions for ticks and fleas (e.g. grasslands or forests), particularly when combined with high temperatures, limiting the spread of related infectious diseases (Gage and Kosoy 2005).

6.6.2.4   Wind

Wind has double effects on vectors/species of disease. Wind will adversely as well as positively affect the malaria process. Strong wind can decrease the mosquitoes' biting chances but may prolong their flight time. Wind can change the spatial distribution of mosquitoes during a monsoon season (Reid 2000).

6.6.2.5   Sunshine

The synergistic effect of sunlight can affect a host of disease. A time series study of cases of cholera in MATLAB, Bangladesh, has shown that elevated temperature and extended sunlight are directly linked to the incidence of monthly cholera (Islam et al. 2009a, b). Indeed, high temperatures and medium sunshine hours together form the best conditions for cholera outbreaks.

6.6.3   Climate and the Spread of Diseases

Transmission of the disease can be either overt or indirect, depending on the transmission pathway. Natural transmission refers to the transmission of a disease from human to human through droplet communication, direct physical communication, indirect physical contact, airborne transmission, or faecal-oral transmission. Indirect transmission refers to the spread from another cell, vector, or an indirect host of disease to humans. This effect can be immediate, as changes in temperature can alter the spread of disease by directly impacting the viability of pathogens. It can be indirect if human and vector/host response activities to climate change lead to a change in propagation routes.

6.6.3.1   Temperature

Temperature changes alone or in conjunction with other changes in the vector, such as rainfall, can alter disease transmission. Studies in African highlands have recorded a link between interannual temperature fluctuations and malaria transmission (Bouma 2003). Haemorrhagic fever with renal syndrome incidence is closely associated with meteorological factors, including temperature, precipitation, and rainfall (Xiao et al. 2014).

6.6.3.2   Wind and Dust Storms

Infectious diseases are spread by wind and dust storms. Wind may act as a way of transmitting pathogen and infectious disease virus. In winter months, human influenza virus may be transmitted from Asia to the Americas by predominant wind across the Pacific (Hamnett et al. 1999).

Climate change can influence infectious disease transmission by modifying the human-pathogen, human-vector, or human-host contact patterns. Evidences demonstrated that diseases spread by rodents often escalate due to altered habits of human-pathogen-rodent interaction during heavy rainfall and flooding events. For example, deer mice can invade human dwellings looking for food during hazardous periods and thus spread hantavirus to humans, leading to cases of hantavirus pulmonary syndrome (HPS) (Engelthaler et al. 1999). In Central and South America and South Asia, flood-associated outbreaks of leptospirosis (Weil's diseases) have been recorded (Ahern et al. 2005). Environment variability plays an important role in influencing habits in human and other host events and behaviours such as seasonal employment, migration, winter-summer diets, and physical activity (Viboud et al. 2004), which could in turn have a direct effect on habits in transmission of disease (Kuhn et al. 2005). Water shortage could become a larger and more serious problem with global warming, which may lead to more cases of diarrhoea worldwide (Lloyd et al. 2007).

6.6.4   Climate and Disease Transmission

Climate change is not just a 'normal' phenomenon in its contemporary nature but is primarily triggered by human activity. Climate change can weaken human immunity and disease resistance, thereby impacting the spread of disease. Climate change in effect impacts the transmission of the disease at three levels: Firstly, it impacts the physiology and development of bacteria, hosts, and vectors directly. Secondly, it affects the environments in an area, the population of hosts able to survive in it, and the life cycles or lifestyles of those hosts. Thirdly, it causes social and economic reactions, including adaptation and mitigation steps that modify land usage, travel habits, human population trends, and natural resource use and availability (Wilkinson et al. 2011). This would lead to the destruction of the environment, which may put pressure on agricultural production, causing problems such as crop loss, hunger, starvation, increased migration of populations, and disputes over land. Such stresses can help increase human vulnerability to infectious diseases. The early malaria forecasting models in the 1920s showed that the price of wheat, along with crop loss and starvation, caused harm to human immunity when food became low (Hay et al. 2002). Waterborne diseases in certain regions may prevail if there are still clean surface water crises caused by climate change (CDC 2010).

6.6.5   Extreme Weather Events and Disease Transmission

Extreme weather events refer to a climate or environment variable that reaches the top (or bottom) end of the usual value continuum (IPCC 2012), which include weather conditions on a global scale (e.g. El Nino, La Nina, and quasi-biennial oscillation (QBO)) and meteorological threats on a national or local basis (e.g. drought, heatwaves, flooding). Although these occurrences are uncommon and occur less than 5% of the time (Zhu and Toth 2001), their occurrence and scale have increased, reflecting a major element of global climate change (IPCC 2012; Lubchenco and Karl 2012). Several pieces of research investigated the effect of extreme weather hazards on human infectious diseases, as listed in Table 6.1. Nevertheless, all of these studies give a comprehensive insight into the mechanism – the ways in which the climate changes and how the cycles of diseases affect.

6.6.6   Role of GeoComputation for Climatic Data Acquisition and Analysis

To date, much of the conversation has concentrated on the attribution of historical increases in disease levels to climate change with the use of hypothetical situation-based models to predict potential improvements in disease risk. The International Research Institute for Climate and Society (IRI) is conducting collaboration and capacity building with academics, policy / policymakers, public health professionals and populations in tropical countries with lower middle-income disease to allow access and use of climate resources to first understand the principles that drive improvements in spread of disease. First, we seek to explain the association between disease and environment through the development of spatial and temporal inequalities of diseases and at-risk population (Thomson et al. 2019). When there is an interaction between diseases and the climate, the variability of diseases and the time of intervention are forecasted. Climate data and information – be it by satellite or station – can slowly be readily accessed online (del Corral et al. 2012). Station data (most usual measurements of rainfall and minimum and maximum temperatures)

Table 6.1 Key studies that assess the relationship between extreme weather events and infectious diseases

Extreme weather events	Disease type	Major findings	References
El Nino	Vector-borne disease	Rising new epidemic outbreaks is related to event El Nino	Epstein (1999)
		Outbreaks and malaria epidemics in many countries have been directly linked to El Nino events	Haines and Patz (2004)
		In the El Nino year, considerably less malaria was detected in the Usambara Mountains, Tanzania, than in the previous year	Lindsay et al. (2000)
		Evidence of hantavirus cardiopulmonary syndrome in the Colorado Plateau has been shown to be linked to El Nino cases	Hjelle and Glass (2000)
	Waterborne disease	The chance of diarrhoea-related effects is double the previous when introduced to coastal waters in southern California after a winter in El Nino	Dwight et al. (2004)
La Nina	Vector-borne disease	Chikungunya fever disease was related to the La Nina drought	Chretien et al. (2007)
		An outbreak of West Nile fever and Japanese encephalitis had occurred in La Nina year	Nicholls (1993)
	Waterborne disease	Over a winter in La Nina, concern increased over the symptom of diarrhoea	Bunyavanich et al. (2003)
Quasi-biennial oscillation (QBO)	Vector-borne disease	QBO was determined to be consistent with the outbreak of *Ross River virus* in south east Queensland	Dwight et al. (2004)
Heatwaves	Vector-borne disease	In 2000, heatwave was related with West Nile fever outbreak in Israel	Paz (2006)
	Airborne disease	The heatwave adds to the increased morbidity and mortality from respiratory infectious diseases	Kan (2011)
Hurricane	Vector-borne disease	Malaria and dengue fever in Honduras and Venezuela ensued following the storm	Epstein (2000)
Cyclone	Vector-borne disease	A cyclone threatens to cause more prevalent leptospirosis	Sanders et al. (1999)
	Waterborne/ food-borne disease	A cyclone threatens to heighten cholera occurrence	Shultz et al. (2005)

(continued)

Table 6.1   (continued)

Extreme weather events	Disease type	Major findings	References
Flood	Vector-borne disease	Malaria, typhoid, and cholera spread by floods in Mozambique	Epstein (1999)
		Heavy rain or flood can cause Ross River fever to erupt	Mackenzie et al. (2000)
		Droughts correlate with the chikungunya fever epidemic	Chretien et al. (2007)
		Such cases of diarrheal illness as cholera develop after a flood	Ahern et al. (2005)
		Increases in incidences of diarrhoea and malaria were observed following floods in Khartoum, Sudan, in 1988	Woodruff et al. (1990)
		Increases in lymphatic filariasis were recorded in several locations	Nielsen et al. (2002)
		Increases in arbovirus disease have also been recorded following flood	Cordova et al. (2000)
		Even HPS diseases can increase during flooding	CDC (2000)
		Leptospirosis diseases can also grow in different areas during floods	Leal-Castellanos et al. (2003)
	Waterborne disease	Flood encourages the spread of waterborne disease, such as *Cryptosporidium*	MacKenzie et al. (1994)
		In the town of Lewes in Southern England, a large rise in the incidence of gastroenteritis was associated with extent of rain	Reacher et al. (2004)
Drought	Vector-borne disease	Hantavirus pulmonary syndrome (HPS) has been shown to correlate dryness	Khasnis and Nettleman (2005)
		Growing threats for the West Nile virus are amid the drought	Wang et al. (2010)
		During droughts, the possibility of spreading St. Louis encephalitis virus will increase	Shaman et al. (2002)
		Droughts that correlate with the chikungunya fever epidemic	Chretien et al. (2007)
	Waterborne disease	Diarrheal diseases are common in refugee camps particularly during drought	Epstein (2001a, b)

(continued)

Table 6.1 (continued)

Extreme weather events	Disease type	Major findings	References
Climate change	Vector-borne disease	Leishmaniasis and climate change – case study: Argentina	Salomón et al. (2012)
		Change in global climate and prevalence of visceral leishmaniasis	Kumar and Kumar (2013)
		Impact of climate change on global malaria distribution	Caminade et al. (2014)
		Climate change and epidemiology of human parasitosis	Lotfy (2014)

Modified after Wu et al. (2016)

can generally be collected from a country's National Meteorological and Hydrological Service (NMHS). Satellite sensors for weather and environmental tracking collect the data which are stored constantly and cover widespread regions of the world. The raw data may in many cases be free, but the proper processing of the data necessitates skills and knowledge, and not all interfaces limit free access to their publicly available data.

6.6.6.1 Precipitation

There is still no satellite that can identify precipitation effectively and measure rainfall in all situations accurately. Satellite views the clouds above from below, but the cloud presence is not a good indicator of precipitation. Not all clouds produce precipitation, and the rainfall varies everywhere under those clouds that create precipitation. Scattered microwave radiation techniques have a more reliable estimation of the rainfall but have low spatial resolution and are revised twice a day. Techniques are being built to take benefit of improved microwave sensor precision and greater spatial and temporal performance of infrared sensors by integrating the two technologies in an optimized manner:

- The Global Precipitation Climatology Project (GPCP) incorporates data from satellites and stations (global monthly 2.5° and daily 1° rainfall estimates). Monthly data vary between 1979 and the daily output from 1996 to today. The software is available in IRI data library at a spatial resolution of 250 km (http://iridl.ldeo.columbia.edu/SOURCES/.NASA-/.GPCP/.V2p3/.CDR/-.precip/).
- The Climate Prediction Center (CPC) Merged Analysis of Precipitation (CMAP) integrates data from satellites and stations. This formula is somewhat similar to the GPCP but due to various methods used to predict precipitation, it has several variations. The software is available in the IRI Data Library at 250 km of spatial resolution (http://iridl.ldeo.columbia.edu/SOURCES/.NOAA/.NCEP/.CPC/.Merged_Analysis/.monthly/.latest/.ver2/.prcp_est/).
- The CPC MORPHing technique (CMORPH) offers projections of global precipitation at very high resolutions in space (25 km) and in time (3 h). This software is ideal for real-time rainfall reporting, because there is no need for a long

history because data would only be available from January 1998. The product can be found in the IRI Data Library on http://iridl.ldeo.columbia.edu/SOURCES/-.NOAA/.NCEP/.CPC/.CMORPH/.

- The Tropical Rainfall Measuring Mission (TRMM) offers precipitation forecasts for the tropics. Monthly totals make the results stronger. They will be available from January 1998 until May 31, 2015. If high-range (25 km) requirements are available, the product is of good quality, and real-time information is not important. The product can be found in the IRI Data Library on http://iridl.ldeo.columbia.edu/SOURCES/.NASA/.GES-DAAC/.TRMM_L3-/.TRMM_3B42/.v7/-.daily/.precipitation/.
- Global Precipitation Measurement (GPM) offers regional precipitation figures. They are accessible to contemporary beginning in March 2014. The GPM is a rain-sensing package extension of the TRMM. The product is provided on https://gpm1.gesdisc.eosdis.nasa.gov/data/-GPM_L3/GPM_3IMERGDF.05/.

6.6.6.2 Air Temperature

The air temperature in weather stations estimated at 2 m elevation is usually derived from synoptic measurements. Estimating the temperature of the near-surface air temperature (Ta) is important for a wide variety of safety applications. The derivation of Ta from the satellite-derived land surface temperature (LST) is far from candid. Though night-time satellite products have fair evaluations of minimum temperatures, it is troublesome to estimate maximum temperatures (Vancutsem et al. 2010). MODIS land surface temperature (LST) offers forecasts of the soil-surface temperature. Estimates of the average and smallest air temperature can be obtained from the land surface temperatures.

6.7 Case Study: Association with Kala-Azar Incidence and Climate

Climate is a key component of the ecosystem and plays a significant role in the periodic cycle or temporal spread of diseases acquired by vectors, since they also propagate different climatic conditions. According to current evidence, kala-azar is determined by climate components like temperature, precipitation, and relative humidity (Bhunia et al. 2010a, b; Bhunia 2014). Temperature especially affects the multiplication rate in the insect. This in effect increases the risk of contamination of the salivary secretions and thus the probability of successful transmission to another host. Increases in temperature, rainfall, and relative humidity due to annual weather variability are likely to directly influence VL occurrence by altering the behaviour and geographic distribution of *P. argentipes* and by changing the duration of the *Leishmania donovani*'s life cycle (Lodge et al. 2006). The commonly considered climatic variables were collected from the nearest Indian Meteorological Department (IMD) stations of the study area. Temperature variables such as average mean temperature (°C), minimum temperature (°C), high temperature (°C), relative humidity (per cent), precipitation (cm), sea level (mb), and mean wind speed (kmph) were acquired between 2005 and 2011. For this time, monthly reports of notified VL cases were collected from district malaria offices (DMOs) and public health centres (PHCs) and also registered. In this study, an association between weather parameters and VL incidence has been highlighted. Descriptive statistics were calculated for each climatic parameter for both the study districts separately. Spearman's univariate rank correlation analysis was initially undertaken to establish the relationship between climatic variables and kala-azar incidences.

6.7.1   Temperature and Kala-Azar Incidence

High variation in monthly temperature distribution is typical of the region. The average annual temperature ranges from 13.5 °C to 34.57 °C across the area. The coldest months of the year are December and January, while April and May are the hottest. However, the highest known temperature reported during the time of analysis in this region was 39.95 °C (June 2005). In this region, the lowest known minimum temperature recorded was 7.4 °C (January 2011). Coefficient test of the Spearman rank correlation was used to analyse the association between temperature and frequency of events. There was a strong low association between the VL incidences and the median annual average ($rho = 0.35$; $P < 0.00$), the minimum annual average ($rho = 0.20$; $P < 0.07$), and the mean annual average temperatures ($rho = 0.29$; $P < 0.007$) during the 2005–2011 time frame. By comparison, a consistently important and clear association was formed for 2005, 2006, 2010, and 2011 when monthly estimates were correlated with the VL incidences with the observed mean and maximum temperature variables separately (Fig. 6.4). This indicates that

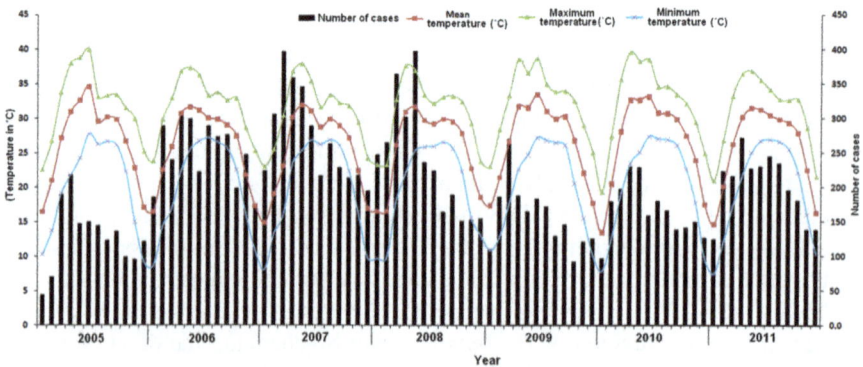

Fig. 6.4 Association between kala-azar incidences and temperature variables

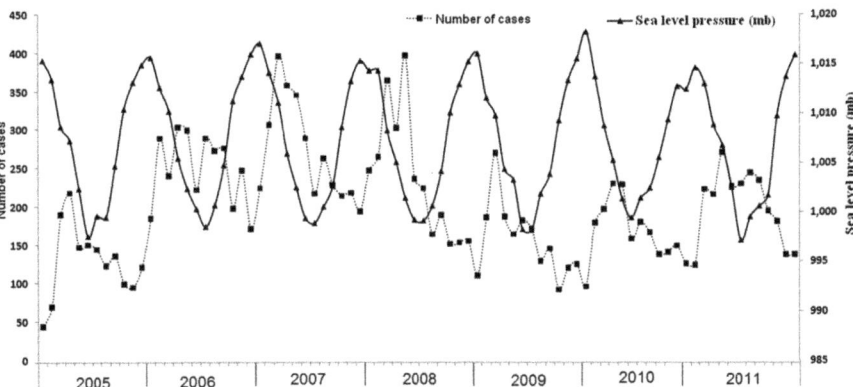

Fig. 6.5 Association between kala-azar incidences and sea level pressure variables

temperature can work by effecting favourable conditions on growth. In addition, contradictory relationship with the minimum temperature variables in the field of analysis was found. During 2007–2009, poor and negligible association between temperature variables and occurrence of disease was found.

6.7.2   Relative Humidity and Kala-Azar Incidence

Seasonal differences in humidity are distinguished by high monsoon values and low summer values. The area's average annual relative humidity was 66.39%. The highest RH was recorded in November 2005 (87.25%), while the lowest RH was reported in April 2009 (33.90%) during the 2005–2011 era. Figure 6.5 indicates how the kala-azar cases may be linked to relative humidity in the Vaishali District. Aside from 2006, 2007, and 2011, there is a clear and negative association between kala-azar incidences and relative humidity. The cumulative relationship estimation between annual average RH and case occurrence, however, also reflects negative correlation ($rho = -0.27$; $P < 0.01$) during 2011.

6.7.3   Association Between Rainfall and Kala-Azar Incidence

According to the effect of monsoon onset and withdrawal, the rainfall is unequally spread over the district. The district's mean annual rainfall for the period 2005–2011 is 657 mm. The district gets about 85% of total monsoon rainfall. The monthly rainfall data and kala-azar outbreaks indicate an insignificant relationship in Vaishali District ($rho = 0.18$; $P < 0.10$). When monthly number of reported cases was compared for each year separately, there was no consistent strong relationship, except in

2011 (*rho* = 0.67; *P*<0.01). Moreover, weak and negative relationship was found between case incidence and rainfall in the study area during 2007–2011 (Fig. 6.6).

The association between kala-azar incidences and relative humidity remains high and negative aside from 2006, 2007, and 2011. Nevertheless, the cumulative relationship estimation between the annual average RH and case occurrence also reflects negative correlation (*rho* = −0.27; *P* < *0.01*) over the 2005–2011 time frame. This may be due to the variations in kala-azar development in the incubation phase.

The findings of this analysis suggest that climatic factors may have played a significant part in the kala-azar transmission process. Scientific data shows that in highly endemic areas, a demand for kala-azar changes seasonally (Bern et al. 2005). The relation between kala-azar and weather parameters in the Indian subcontinent has long been the subject of research (Napier 1926). The additional factors may explain the April–July and May–November peaks in VL coverage and *P. argentipes* (Picado et al. 2010), respectively; these are not symmetric. This may be attributed to the unrecognized incubation time of *L. donovani*, but it hypothetically varies from 2 to 6 months (Pearson et al. 1999). Minute increases in temperature significantly influence the probability of transmission as maximum values for kala-azar transmission capacity are located in the ranges 25–29 °C, provided a certain vector density. When RH is less than 60%, the sand fly life is reduced, which in effect reduces the spread of disease. RH is considered optimum for successful kala-azar transmission between 60% and 80% (Fig. 6.7). Rainfall also influences kala-azar transmission as it raises relative humidity and reduces temperature and also affects kala-azar vector breeding and distribution (Sharma and Singh 2008). In addition, heavy rainfall will cause vector species to flood and decrease by destroying larval habitats and generating undesirable conditions for vertebrate hosts and reservoirs (Bhunia et al. 2011).

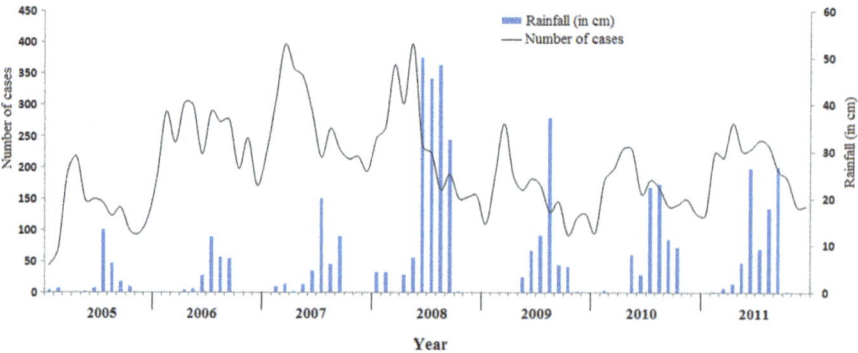

Fig. 6.6 Association between kala-azar incidences and rainfall variables

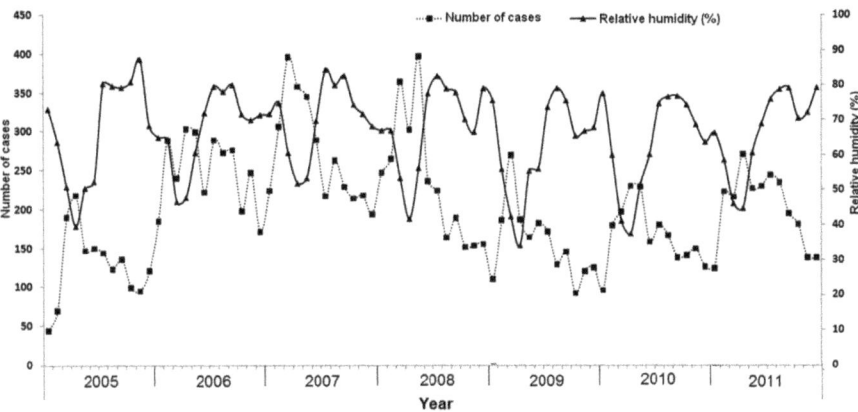

Fig. 6.7 Association between kala-azar incidences and relative humidity variables

6.8   Topography and Disease Ecology: Case Study Through GeoComputation Technique

Areas of high altitude, here described as >2500 m, do not per se harbour particular infectious agents. The adjacent lowlands also reveal pathogens found in the mountains. The high mountains lie in different ecosystems, occupy nearly one-fifth of the Earth's atmosphere, and sustain 1300 million inhabitants and tens of millions of tourists traveling each year in lower altitudes (Zafren and Honigman 1997), typically, travellers to high-altitude hiking or traveling through several environments, from tropical forest or desert to higher alpine and tundra regions at the base of the climb. Environments of high altitude create stressors in the form of elevated UV radiation, hypobaria, hypoxemia, adverse weather conditions, failure to sustain proper personal hygiene, crowded living environments, and isolation from adequate medical treatment. Such environmental stressors can in theory affect insect vectors and microbial pathogens, but these possible connections are little apparent except that vector mosquitoes are insufficiently present at higher altitudes.

   The highland topography consists of slopes, valleys, and plateaus. Typical features are rivers and streams which flow in the ecosystem of the valley and the swamps along the valley bottoms. Topographic characteristics that affect local distributions are likely to play an important role in disease transmission (Zevenbergen and Thorne 1987). Primary geomorphological attributes, such as height, slope, and angle, or related attributes, such as topographical curvature and radiation indices, were extracted from simulated terrain models of high resolution. It has been shown, for example, that the distribution of large herbivores is influenced by abiotic factors such as water supply, environment, and topography in addition to the quantity and nature of the forage (Lachica et al. 1997). In this context, main topographical features such as angle, slope, solar radiation, or visibility range have been identified as features of habitat selection (Hopewell et al. 2005). In order to explain the inoculum pressure (foci density and distance to the closest target) for *Phytophthora cinnamomi*, topography (distances from lakes, solid, rye forage crops, and roads), geomorphology (slope, aspect, and terrain curvature plan), and soil-, water-, and temperature-related proxy variables (topographic wetness index and spring potential radiation) were considered (Cardillo et al. 2018).

In high altitudes, the function of T-lymphocytes is slightly diminished, and protection against bacterial infection may be impaired, although tolerance to viral infection is not disrupted and immunization response is preserved (Meehan 1987). Granulocyte function experiments at high altitude during physical activity identify the rapid reversal of granulocytosis after initial extravasation and the reduction of superoxide anion development (Demetz et al. 2001). Additionally, hypoxia per se can cause systemic changes in inflammatory factors and affect high-altitude illness (Hartmann et al. 2000).

6.8.1 Vector-Borne Disease Transmission vs. Topography

Contrary to lowland plains with inadequate drainage and large-scale breed habitats, most breed habitats in the hilly highlands are restricted to the valley floors because they have sufficient drainage through the gradients of hillside (Minakawa et al. 2005). Differences in local terrain form can also play an important role in the determination of areas suitable for smaller spatial scale mosquito breeding (Balls et al. 2004). The pattern of malaria transmission in highland environments may be less distinct as the topography and more complex hydrology derived from multiple strands. Given the flat topography and the more complex hydrology arising from multiple streams, the pattern of malaria transmission in the highland plateau habitats may be less distinct. A retrospective cohort research was undertaken with the primary emphasis on a comparative spatial-temporal evaluation of infection susceptibility using immunological markers in different environments (Wanjala and Kweka 2016). The highland topography was divided in three ecosystems in this analysis: the flat lowland valleys (U-shaped), the narrow valley on the bottom (shaped in V), and the plateau. Previous studies have shown that highland residents usually lack immunity to *Plasmodium falciparum* and are especially susceptible to malaria infection (Frosch and John 2012; John et al. 2002). In highlands, the proportion of asymptomatic individuals is typically smaller than in high-transmission areas where the incidence of *P. falciparum* and parasite densities is limited in the inter-season; therefore, a slight rise in the abundance of vectors will contribute to a major outbreak of malaria in the highlands (Smith et al. 1992). In high altitudes in mountains and hills where malaria is low, people are poorly immune to malaria, making infections uncommon. Individuals are vulnerable to severe infectious diseases and complications from *Plasmodium* infection (Snow et al. 2003). While *Plasmodium* species are not commonly transmitted at altitudes of 12,000 m (Murphy and Oldfield 1996), febrile disease from acute or reactivated malaria infection at lower altitudes occurs at higher altitudes (Bishop and Litch 2000). Therefore, people who live in areas with low to moderate transmission intensities are at high risk for serious malaria infection (Snow et al. 1997). Bhunia et al. (2010a, b) recorded the highest occurrence of visceral leishmaniasis (VL) to be below 150 m of altitude, with very few cases over 300 m. *Phlebotomus* is the vector of bartonellosis from *Bartonella bacilliformis* in the Colombian, Ecuadorian, and Peruan Andes at a height of 600–2500 m (1500–100 ft) (Maguina and Gotuzzo 2000). Hemorrhagic dengue fever occurs in natives at lower altitudes and is rarely observed, if ever, in mountain travellers. Typhus is potentially an undiagnosed cause of fever in highland travellers although the prevalence is sporadic, like most other diseases at higher altitudes.

6.8.2   Waterborne Disease vs. Topography

The spatial structure of the Harare epidemic has been explained by previous historical, geographical, social, and environmental risk factors. Cholera is meticulously correlated with low environmental conditions in developed countries and a shortage of adequate services. In this respect, high population densities and inadequate access and additional environmental factors contribute to an outbreak of cholera (Luque Fernandez et al. 2012). Downstream rainfall follows hilly areas throughout the rainy season when no waste drainage or a working sewer network exists, entering the lower sections of the cities where shallow wells are easily polluted by excreta. Spatial knowledge on topographical elevation may help drive preventive measures in cholera-endemic areas (Osei and Duker 2008).

6.8.3   Neurological Infections vs. Topography

Rabies infection is a big concern for people travelling to high altitudes who are at elevated risk because wild animals cannot escape healthy and cannot be treated right after exposure. In rural Southern Asian regions and parts of the Indian subcontinent, Japanese encephalitis (JE) is endemic. JE must be differentiated from tuberculosis meningitis, bacterial meningitis, and typhoid encephalopathy, all seen in the local population, although no high-altitude tourists were reported, probably due to the rareness of the *Culex* mosquito vector at high altitudes. Tick-borne encephalitis is widespread in the deciduous forest below 1200 m and can be rare in inhabitants of the Alps and the Ural Mountains when the disease has been assimilated at a lower altitude. Although prevalent in many developed world environments, such as the Indian subcontinent, bacterial meningitis is rare in high-altitude travellers.

6.8.4   Respiratory Infections vs. Topography

At high altitudes, respiratory problems are common (Murdoch 1995; Basnyat et al. 2000). Over 4000 m of cough and sore throat are usual. Unlike the omnipresent Khumbu cough in Nepal's Himalayas, bronchitis in climbers in Aconcagua was observed in 13 of 19 climbers at 4300 m (Rabold 1987). Typical signs of cough, tachycardia, tachypnoea, and shortness of breath are caused by respiratory infections and pulmonary oedema at high altitudes. Non-specific protection of respiratory symptoms is necessary at higher altitudes; such protection involves keeping the head dry, adequate hydration, the use of nasal decongestants, and breathing through a silk scarf to keep the air moist (Hackett 2001).

6.8.5   Other Infections vs. Topography

Mountain expeditions also demand that the base camp move through water, thereby providing future visibility (Basnyat et al. 2001). Moreover, pain may occur at the site of dental caries, likely due to reduced air pressure at high altitude with gas expansion in the cavity. Sexually communicated infections, infections of the yeast, and diseases of the urinary tract are widespread at sea level and high altitude. Gonorrhoea, chlamydia, trichomonas, candidiasis, genital herpes, and an acute HIV infection can result among other infections. Essentially, high-altitude pathogens and respiratory diseases also mirror those found in neighbouring lowland ecosystems. Further, research on high-altitude pathogens is obviously needed to better understand their pathogenic processes and epidemiology and to improve diagnosis and prevention.

6.8.6   Identification of Topographic Variables
and GeoComputational Technology

Topography is a significant land surface feature that influences most aspects of a catchment's water balance including surface and subsurface runoff production, flow paths pursued by water as it passes down and over hillslopes, and water movement velocity. Digital elevation models (DEM) are a grid graphical terrain representation; each pixel value corresponds to an elevation above a location. DEMs may be generated from field surveys, digitization of current topographic hardcopy maps, or remote sensing techniques. Remote sensing strategies involve photogrammetry

(Uysal et al. 2015; Coveney and Roberts 2017), airborne and space-borne interfero-metric synthetic aperture radar (InSAR), and light detection and ranging (LiDAR). In field analysis, topographical attributes obtained from digital elevation models (DEMs) and automatic terrain analysis are widely used. Traditional surface charac-teristics which can be measured from a DEM contain slope gradient, slope dimen-sion, slope curvature, upslope volume, different catchment area, compound topographic index (CTI), etc. DEM is useful for describing the Earth's constantly changing topographical landscape and is a common source of data for terrain inter-pretation and other spatial uses. In a given landscape, DEM of various resolutions could be used to extract DEM-based properties which could be used to explore and analyse properties such as soil, water, vegetation, etc. Digital elevation models (DEMs) are key to the effective study of satellite imagery in mountain areas. DEMs offer topographical knowledge in computer-compatible formats which can be used to more accurately explain the propagation of terrain constituents that contribute to the spectral response. DEMs may also be used to direct field research and homog-enize difficulties with image processing and are used to normalize and adjust satellite-observed radiance or reflection in planimetric shape to true physical values. With their present existence as results of stereo-processing satellite data for nearly every place on Earth, elevation models may be chosen to be used in some applica-tions at the same time as the original spectral response; their use would undoubtedly become more prevalent.

After Miller and Laflamme's groundbreaking research (1958), DEMs have developed into an important part of a number of systematic applications. Advances in satellite technologies, image processing and data storage capacities, make it increasingly possible to build modern, more precise DEMs (Table 6.2). Despite its acquisition in 2000, SRTM remains the world's most common DEM due to its sim-plicity, feature resolution, vertical precision, and subordinate objects and noise equated to substitute global DEMs (Hu et al. 2017). Two worldwide DEMs have recently been published, mainly optically generated ALOS AW3D30 (Tadono et al. 2014) (almost 30 m resolution) and TanDEM-X 90 (Rizzoli et al. 2017) derived from SAR (almost 90 m resolution). ASTER and SRTM, for example, were mashed and composed to produce the global EarthEnv DEM (Robinson et al. 2014). High resolution (<10 m) of LiDAR open-access data is progressively accessible through projects including OpenTopography (Hawker et al. 2018). In addition, Ghuffar (2018) has proven that a 5 m DEM can be produced using semi-global matching from Planet Labs CubeSat imagery obtained from PlanetScope. Despite this prom-ising move, we conservatively predict that open-access LiDAR data encompasses only 0.005% of the Earth's land area based on OpenTopography data and worldwide LiDAR coverage is somewhat off, with the partial amount of LiDAR data presently obtainable almost entirely in advanced nations.

Table 6.2 Existing global DEM

Dataset	Coverage	Sensor/satellite	Resolution (m)	Vertical accuracy	References
Free sources of Global Digital Elevation Model (DEM)					
LOS AW3D30	82 °S–82 °N	Optical	30	4.4 m (RMSE)	Tadono et al. (2016)
ASTER GDEM	83 °S–83 °N	Optical	30	17 m (95% Conf.)	Tachikawa et al. (2011)
Bare Earth DEM	56 °S–60 °N	SRTM	90	5.9 m (RMSE)	O'Loughlin et al. (2016)
EarthEnv	60 °S–83 °N	ASTER and SRTM	90	4.15 m (RMSE)	Robinson et al. (2014)
GMTED2010	Entire Earth	SRTM and ten other sources	250, 500, 1000	26 m (RMSE)	Danielson and Gesch (2011)
MERIT	Entire Earth	AW3D3O, SRTM, and Viewfinder Panorama	90	5 m (LE90)	Yamazaki et al. (2017)
Commercial source of Global Digital Elevation Model data					
ALOS AW3D	82 °S–82 °N	Optical	5	2.7 m (RMSE)	Takaku and Tadono (2017)
PlaneDEM 30 Plus	Entire Earth	SRTM	30	Not reported	Planet Observer (2017)
NEXTMap World 10	Entire Earth	Not reported	10	10 m (LE95)	InterMap (2018)
WorldDEM	Entire Earth	Optical	12	<1.0 m (RMSE)	Rozzoli et al. (2017)

6.9   Case Study: Spatial Association Between Kala-Azar Incidence and Absolute Relief

Topography has long been reported as one of the variables linked to visceral or kala-azar leishmaniasis (Bhunia et al. 2010a, b). Higher altitudes are usually distinguished by high temperatures and lower relative humidity, and these measures appear to have larger variability that adversely influences the vector distribution. Knowing the epidemiology of kala-azar transmission and relief inequalities that exist in close contact would help strengthen an area-specific regional prevention and transmission management action programme. The DEM method was selected because it was assumed to provide a valuable basis on which to model the gathered risk of kala-azar/visceral leishmaniasis (VL) results. The ASTER Global Digital Elevation Model (GDEM) is rendered with posts of 30 meters and is structured as GeoTIFF files in 1 × 1-degree tiles. A 5-year average annual occurrence rate for the period 2007–2011 was estimated for the purpose of this analysis. The 5-year annual average occurrence numerator was the total of all kala-azar cases reported in 2007–2011, and the denominator was the estimated number of at-risk persons per 10,000 populations in the 2007–2011 period. The recorded cases were overlaid on the map, but each point was added to the polygon attributes it fell under. To examine f region, the numbers of cases were counted manually.

The ASTER created a raster-based DEM including 7.5-min DEM. The 7.5-min DEMs have elevation data on a grid measured in UTM coordinates at a range of 30 metres. Level 2 DEMs that had the vertical precision of ±7 meters are used in this study. However, in this analysis, due to its higher spatial resolution and improved precision, ASTER-inspired DEM is used to demarcate the micro-relief characteristics of the study site. The findings are described here, based on automated morphometric analysis using ASTER DEM.

Absolute relaxation is an important physical factor affecting the frequency of the delivery of kala-azar. The regions of low total relief support transmission and dissemination of VL. Approximately 54.75% of the sample region is limited to regions with total relief values under 30 m. As such, in areas with low absolute relief, 56.66% of VL distributions are found. It is also noticed that VL's propagation frequency decreases with an increase in absolute relief values. A quantitative analysis of the VL distribution in relation to the category of absolute relief reveals a very strong converse connection ($r = -0.79$; $p < 0.000$) in the study area. In comparison, from the absolute relief range of 30–45 m (27.40% of the study area), the maximum VL density per 2 km^2 area was observed. In elevation values of >45 m, the least VL density is observed (Fig. 6.8).

6.10 Surface Water Bodies and Disease Ecology: Case Study Through GeoComputation Technique

Waterborne infections (e.g. mainly transmitted by polluted water) are associated globally with a major risk of sickness. Since these diseases transmit either directly or by flies or filth, water is the primary source for these diseases to transmit. Waterborne diarrhoeal diseases, for example, are responsible for two million deaths worldwide, with the majority happening in children under 5 (https://www.who.int/sustainable-development/housing/health-risks/waterborne-disease/en/). Around one-third of the world's population resides in countries of moderate to high water tension, and water shortage issues are rising, owing in part to pollution and degradation of the environment. Globally, 80% of wastewater spills directly into the environment without treatment or reuse, leading to a situation in which approximately 1.8 billion people use faecal-contaminated drinking water, putting them at risk for cholera, dysentery, typhoid, and polio contracts (WHO 2002). The relationships between pathogens and their hosts include dynamic processes at genomic, biochemical, phenotypic, environmental, and community levels, whereas the proliferation and distribution of naturally occurring microorganisms and their microbial pathways are predisposed by both biotic and abiotic factors functioning on various levels. One of the most severe environmental issues is contamination of rivers, streams, ponds, and other bodies of water (Table 6.3). This originates from untreated or

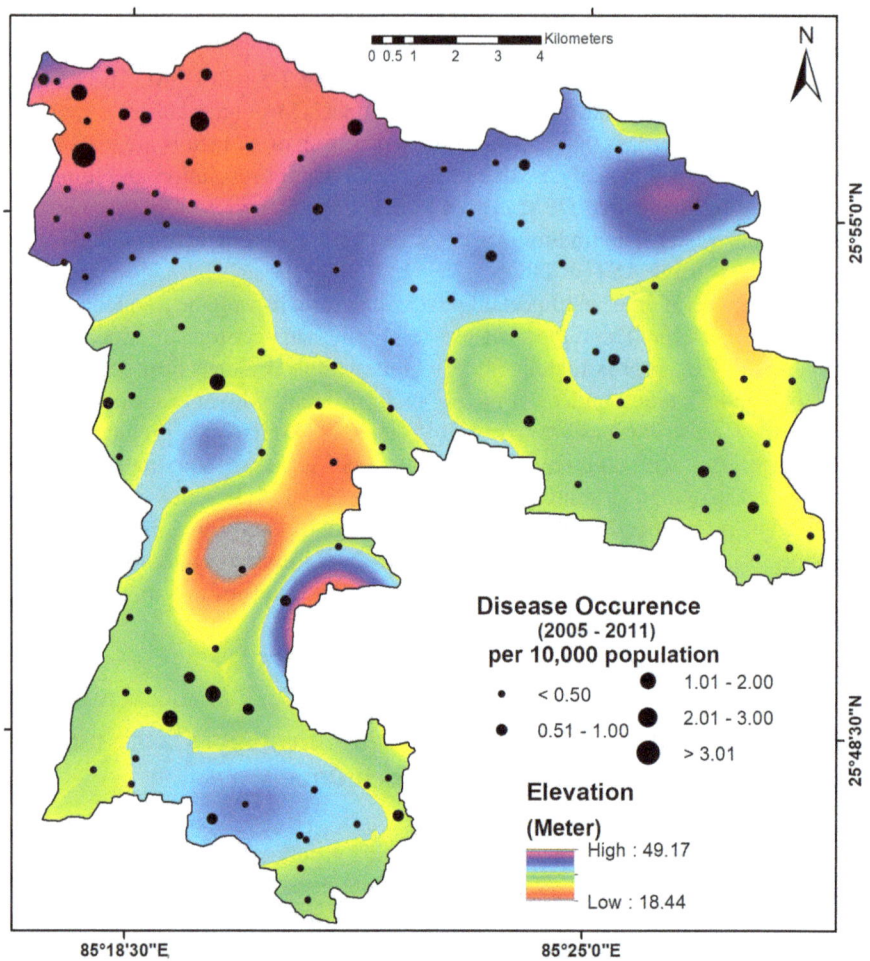

Fig. 6.8 Spatial association between kala-azar incidence and absolute relief characteristics

Table 6.3 Causes and types of waterborne disease

Causes	Types
Bacterial infections	Typhoid
	Cholera
	Paratyphoid fever
	Bacillary dysentery
Viral infection	Infectious hepatitis
	Poliomyelitis
Protozoal infections	Amoebic dysentery

improper handling of residential and/or industrial waste discharges. Such superficial waters typically contain a wide array of organic and inorganic contaminants, such as suspended solids, solvents, fats, grease, chemicals, plasticisers, phenols, heavy metals, and pesticides. Trace amounts of metals found in the ecosystem are the function of specific food chains and growth in plants and animals to levels above allowable for both humans and other living species.

Large-scale or regional environmental or climate changes are expected to have major impacts on waterborne or vector-borne diseases (Patz and Reisen 2001). Regarding waterborne pathogens, the evolutionary principle of 'source/sink' (Pulliam and Danielson 1991) must be better understood. For example, the development and spread of waterborne diseases may be influenced by some marine habitat conservation efforts that require regulations to create wetlands (Table 6.4). Andradottir and Nepf (2000) proposed that coastal wetlands would potentially increase the temperature of inflow during summer, causing surface encroachments rather than dipping inflows. In other words, variations in density between surface and underlying water will allow warm water to overwhelm the colder layers. Consequently, pollutants, toxins, and bacteria typically found in the natural water penetrate to the surface layer, further raising the risk of human contamination in outdoor water environments. Lebaron et al. (1999) showed that changing nutrient concentrations in seawater directly and indirectly influence bacterial populations by inducing either bacteria or various protozoans selectively feeding on the bacterial assembly.

Table 6.4 Water-related infection and its sources

Disease	Most common sources
Legionellosis	Drinking water, water heater, cooling tower
Gastroenteritis – viral	Drinking water, swimming area, spa
Cryptosporidiosis	Drinking water, swimming pool
Hepatitis A	Drinking water, sauna
Campylobacteriosis	Drinking water
Leptospirosis	Drinking water, outdoor recreational area
Rotavirus	Drinking water
Shigellosis	Drinking water, fountain
Typhoid and other enteric fever	Drinking water
Tularaemia	Drinking water
E. coli diarrhoea	Drinking water, swimming pool
Giardiasis	Drinking water
Cercarial dermatitis	Outdoor swimming and bathing areas
Adenovirus	Drinking water, swimming pool
Mycobacteriosis – nontuberculous	Swimming pool
Yersiniosis	Drinking water
Aeromonas and marine *Vibrio* infection	Swimming area
Blastocystis hominis infection	Drinking water

Many human viruses may invade the entry or respiratory tract, or both, and are a risk of polluted water and other environmental media exposures. While inert in the atmosphere, viruses may be resilient, remain in environmental media for long periods of time, and withstand numerous physical and chemical agents, including disinfectants. Moreover, viruses are so small that they are distributed readily in water and waste and can migrate through soils and other porous media. The presence and movement of human enteric viruses in water and other environmental media is a public health issue as the viruses will sustain their infection and induce human infection if the environment media ingest or otherwise come into contact with them. Inorganic mercury was used in the industrial development of acetaldehyde in Minamata, Japan. Fish and other marine species quickly became infected, and gradually inhabitants of this region who ingested the fish died from MeHg (methylmercury) poisoning, later referred to as Minamata disease.

In tropical and subtropical regions, the occurrence of excess water (due to the lack of sufficient drainage) and the spread of water-related vector-borne diseases are closely associated. Malaria, schistosomiasis (bilharziasis), and lymphatic filariasis are important water-related vector-borne diseases. Water-related vector-borne diseases cause bacterial, viral, and parasitic (protozoa and helminths) spread by water-related transmission organisms, also called as vectors or intermediate hosts. But in the case of schistosomiasis, such an organism does not deliberately spread a pathogen, like freshwater snails. Vectors and intermediate hosts represent essential elements of parasitic water-associated diseases in different disease transmission processes. Misuse and lack of maintenance are the two main reasons why drainage systems are often associated with environmental quality issues (road drainage ditches, culverts, dam site runoff or runoff channels in irrigation schemes, as well as surface water collection and disposal infrastructure). Waterborne epidemics and threats to health in the marine ecosystem are primarily caused by inadequate water supply control. Proper water resource management has been the need for the hour, as this will eventually lead to a safer and healthier world.

6.11   Identification of Surface Water Bodies Using GeoComputation

Surface water leads to water on Earth's surface, such as a river, stream, wetland, ocean, etc. In the spatio-temporal domain, water bodies which play a vital role in the global carbon cycle and climate fluctuations are analysed to evaluate the degree and rate of their deterioration and disappearance. Surface water bodies are complex in nature as they expand, extend, or adjust their position or flow rate over time due to multiple natural and human factors (Karpatne et al. 2016). Change of the amount of surface water typically results to extreme consequences. In severe situations, the sudden surface water rise will lead to flooding. It is also important that the presence of surface water is effectively observed, its magnitude measured, its volume quantified, and its dynamics tracked.

Usually, there are two types of sensors that can do the measurement of surface water – the optical sensor and the microwave sensor. Because of the high data quality and the correct spatial and temporal resolutions, the optical sensors which are the focus of this study were widely used in this field (Huang et al. 2016). Microwave sensors, due to their use of long wavelength radiation, have the ability to penetrate cloud space and some vegetation cover. Schumann and Moller (2015) undertook a thorough analysis of microwave remote flood sensing and considered synthetic aperture radar (SAR) the most effective form of microwave flood monitoring sensor. Visual satellite data analysis offers the best demarcation of water sources of varying sizes but is time-consuming, particularly when dealing with data of high resolution (Kupidura 2013). The use of the normalized differential water index (NDWI) approach maximizes water reflectance characteristics by reducing low near-infrared reflectance (NIR) and optimizing green wavelength reflectance (Xu 2006). The threshold procedure is among the algorithms most commonly used to separate water bodies from the satellite imagery. Figure 6.9 showed a perception model for automatic extraction of water bodies from the remote sensing data. The approach is based on the assumption that in short-wave infrared (SWIR) band, the reflected radiance of water is smaller than that of other entities such as trees, houses, bare earth, and roads. Each pixel passing the threshold test is marked as a large body of water along with certain other items that are not really a water entity, resulting in false positives.

The theory of obtaining surface water from optical images is the relatively lower appearance of water in infrared channels, relative to those of other forms of ground cover. A basic method for deriving a water map is to density slice a single infrared band (Frazier and Page 2000). Decision trees have also been assembled using multispectral bands to demarcate other water distributions (Acharya et al. 2016). Many indexes were established to identify areas of surface water or level of flood inundation (Table 6.5). For example, Crist (1985) proposed the tasseled cap wetness

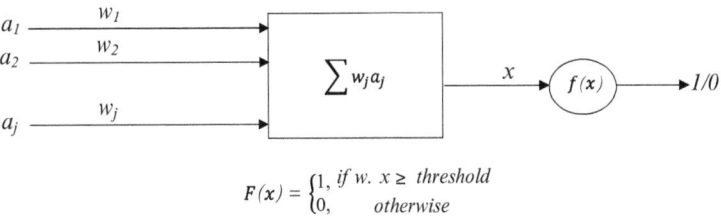

$$F(x) = \begin{cases} 1, & \text{if } w.\ x \geq \text{ threshold} \\ 0, & \text{otherwise} \end{cases}$$

Where, w is a real valued weight vector, $w.x$ is the dot product

Fig. 6.9 Perceptron model of automatic extraction of surface water bodies. (Modified after Mishra and Prasad 2015)

Table 6.5 Remotely sensed indices to extract surface water bodies

Water index	Equation	References
NDWI	$\dfrac{(GREEN - NIR)}{(GREEN + NIR)}$	McFeeters (1996)
mNDWI	$\dfrac{(GREEN - SWIR)}{GREEN + SWIR)}$	Xu (2006)
AWEI	$AWEI_{nsh} : 4 * (GREEN - SWIR_1) - (0.25 * NIR + 2.75 * SWIR_2)$ $AWEI_{sh} : BLUE + 2.5 * GREEN - 1.5 * (NIR + SWIR_1) - 0.25 * SWIR_2)$	Feyisa et al. (2014)
WI_{2015}	$1.7204 + 171 * Green + 3 * Red - 70 * NIR - 45 * SWIR_1 - SWIR_2$	Fisher et al. (2016)

(TCW) index based on six bands of surface reflectance data and set a zero threshold to distinguish water artifacts from non-water objects. For this reason, more useful indices are those that can help illuminate water bodies, such as the normalized difference water index (NDWI (McFeeters 1996)) and the upvdated NDWI (mNDWI (Xu 2006)). Xu (2006) later discovered that the short-wave infrared (SWIR) band could reflect certain different water characteristics, replace the near-infrared (NIR) band in NDWI with the SWIR band, and propose the mNDWI band. This is now commonly accepted that mNDWI is more robust and effective than NDWI as the SWIR band is less sensitive than the NIR band to salt concentrations and other optical active constituents within the atmosphere. Depending on the temperature, NDWI, and mNDWI reflectance properties, the temperature values are normally greater than zero. Middle-infrared, near-infrared, and red reflections from Terra MODIS are usable at 250 m resolution for 16-day intervals from April 2000 onwards. Landsat mid-infrared, near-infrared, and red reflections are available at a spatial resolution of 30 m every 16 days. You can view the items via Google Earth Engine. Inundation fraction products with a spatial resolution of 25 km are available for daily, 6-day and 10-day periods for the entire globe (Schroeder et al. 2010).

Jarihani et al. (2014) attempted to combine Landsat and MODIS photos in order to produce multispectral indices and performed a thorough analysis of the strategies 'index-then-blend' (IB) and 'blend-then-index' (BI). In general, most indicators use surface reflectance as data, given that the surface reflectance is the difference between soil coverings, in particular digitally numbered (DN) and high air reflectance, more accurate than other remotely measured measurements. Fisher et al. (2016) performed a detailed analysis on the efficiency of numerous common water index methods for water classification in Landsat TM/ETM+/OLI imagery with 30 m resolution. They observed that the precision of each index was extremely reliant on the validity pixel structure, with no index doing better over all forms of water and non-water pixels. Therefore, new technologies for surface water production are expected to begin to evolve in the immediate future, with greater efforts likely to be placed into creating highly automated and internationally pertinent processes.

6.11.1   Case Study: Automatic Detection of Open Water Bodies to Map Malaria Incidence

It was estimated that there were 219 million cases worldwide in 2017, with about 435,000 deaths (WHO 2018a, b). The provision of timely and precise surface water maps is useful for modelling malaria risk and planning strategies to combat diseases. Studies have shown that cases of malaria rise with declining distances to the shores of large bodies of water (Lautze et al. 2007). Dam building will change the seasonal transmission to permanent transmission. Therefore, it is rational to hypothesize that despite the year-round supply of water, the climate associated with the surface water body supports a high density of malaria vectors (Minakawa et al. 2012). In addition, satellite imagery has been widely used to delineate and classify wetland zones, providing geographical information for wetland habitat monitoring and management (Hardy et al. 2019). Landsat 8 OLI image is used to map open water bodies. The modified normalized difference water index (mNDWI) was used by the transition method that incorporates the short-wave infrared band I (1.55–1.75μm) and the green band (0.525–0.605μm) to acquire water bodies. The accuracy evaluation showed good cooperation with optical validation results, with a 95% average overall accuracy (kappa = 0.90). Those methods, however, are concerned only with mapping open water. From the histogram of the NDPI image and the corresponding map relative to real positions in the study area, the threshold values (which vary in different areas, different temporal regimes, and even different images) were chosen. The village-wise malaria incidence and population data of Khargram Block of Murshidabad District, West Bengal (India), for the year 2017 were collected from the Public Health Centre of Murshidabad District (West Bengal). The epidemiological parameter like malaria annual parasite index (API) was calculated and considered to identify the malaria-prone zone in the study site. A point layer was created which includes all villages affected by malaria in the study areas. The areas close to the surface water bodies are also related positively to the *Plasmodium falciparum* API (Fig. 6.10). Obviously, these areas have a high propensity for malaria transmission, but our research also shows that regions marked by a comparatively small body of water do not influence the vector in comparison to the disease ratio. Identifying the location and geographic range of the water sources vis-à-vis the API in this case will have a greater effect on the prevalence of the malaria. Thus, the approaches studied here may provide a valuable resource for public health in malaria-ecological settings found in the district of Murshidabad and other tropical regions.

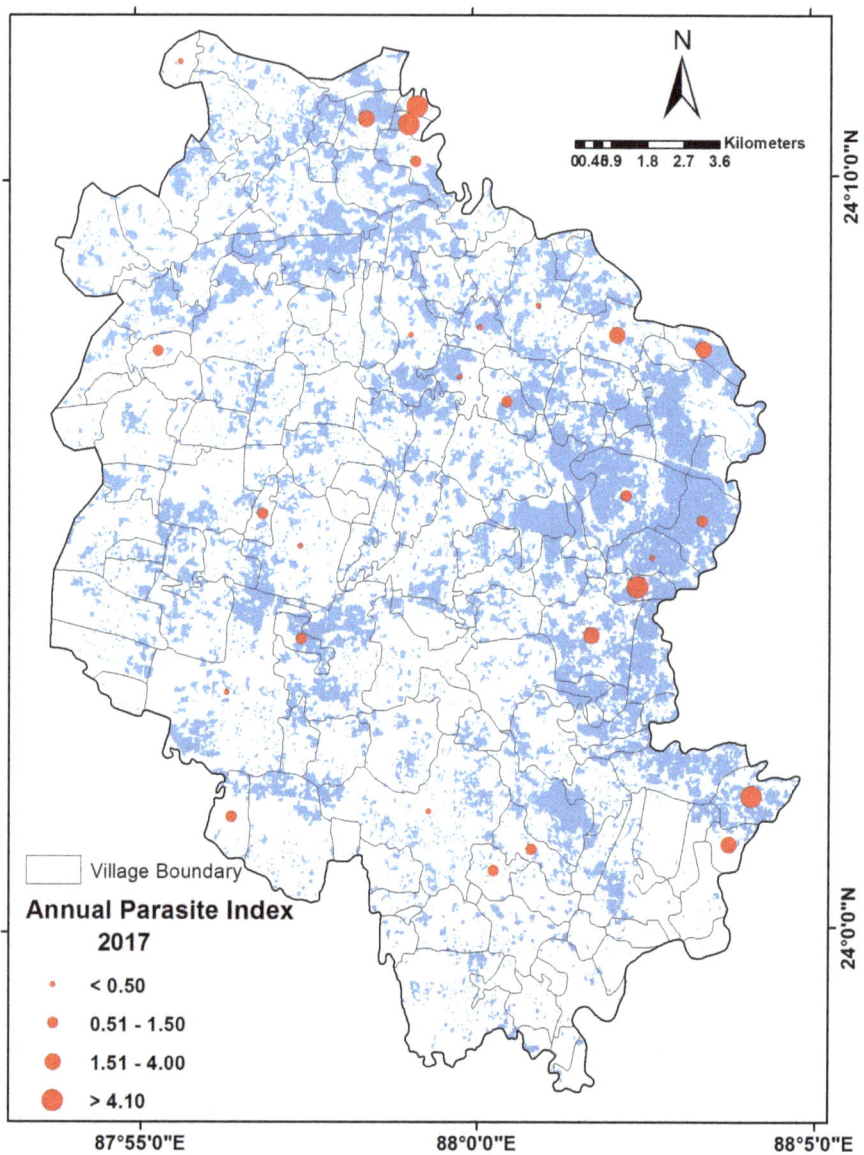

Fig. 6.10 Spatial correlation between surface water bodies and annual parasite index of Khargram Block of Murshidabad District, West Bengal (India)

6.12   Green Biomass and Disease Ecology: Case Study Through GeoComputation Technique

Knowing the correlation between the spread of infectious diseases and host ecology is a long-standing biological science objective and is central to the control of emerging infectious diseases and the prevention of new diseases. While infectious diseases include encounters between at least two species, habitat biodiversity has been considered to play a pivotal role in disease risk for a number of years. Two main hypotheses with differing estimates link biodiversity to risk of disease (Keesing et al. 2006). The hypothesis of 'amplification effect' suggests that variability will be positively associated with disease incidence, since this would result in an enhanced proliferation of sources of inoculum for a focal host. The hypothesis of the 'dilution effect' suggests a negative association between biodiversity and disease risk, because a decline in diversity could lead to an expanded abundance of focal host species promoting spread of disease. Other risk factors may arise on a landscape scale, such as forest degradation, habitat heterogeneity, or geophysical features.

6.12.1   Malaria Transmission and Green Biomass Ecology

It includes essential plant and mosquito interactions, for terrestrial and aquatic tropical species as well as influencing the malaria spread capacity of a vector. Invasive plants tend to have longer flowering times, have more intensive growth, and can increase the biomass, especially in areas where vegetation was previously limited. Some indigenous alien plants protect or rest adult mosquitoes and are attractive hosts for the production of nectar, increasing their vector ability (Stone et al. 2018).

It has been established for over a decade that adult mosquitoes are mainly phytophages that prey on nectar and other vegetable juices from sugar sources (Nyasembe et al. 2012). Many females require blood to mature their eggs (Nayar and Sauerman 1971), while in most species some protein is derived from the blood itself (Van Handel 1984). *Aedes albopictus* females, for example, tend to favour oviposition at sites adjacent to flowering plants (*Buddleja davidii* Franch, Scrophulariaceae) (Davis et al. 2016). In measures of preference, *An. gambiae* have a better risk of ovipositing on bare soil than on water around grassy vegetation (Huang et al. 2006), whereas *An. minimus* (s.l.) favours ovipositing on water around leavened plants (Overgaard 2007). Although women of several species of mosquitoes are sometimes blood hunters (Harrington et al. 2001), they also feed sugar at a certain time and reproductive age or when a host or oviposition site has been missing (Stone et al. 2012). Many specific species of plants are fed on, but not all plants have the same appeal or impact on survival and flight (Müller et al. 2010). However, laboratory studies on *An. gambiae* have shown that every plant is safer than no plant (Manda et al. 2007), including *Lantana camara L.* (Verbenaceae), a genus with mosquito-repellent properties. Plant preference may also display seasonal, diel, and even sexual variations (Müller et al. 2010).

The effect of exposure to different species of plants and their nectar on the survival of mosquitoes was analysed in depth. For example, although exposure to nectar improves longevity reliably, there are wide variations between different plant species in terms of their impact on longevity (Stone et al. 2012). In addition to the impact of nectar content, availability, and accessibility, certain plants also provide habitat for mosquitoes: several plants can establish more appropriate microclimates that promote the activity of resting mosquitoes and their diurnal preservation. The degree to which this contributes to the survival of mosquitoes that rely on the species or the ecosystem in question (e.g. need or desire to rest outdoors). In certain studies, the biting rate decreases when mosquitoes have contact with sugar sources (Stone et al. 2012). In other situations, with exposure to nectar-bearing plants which is more enticing and/or yields more copious quantities of nectar, the biting rate is unchanged or higher. For example, mosquitoes are more likely to feed on nectar and consume greater meals while blood hosts are present only for a brief duration at night or at times that do not match the highest piercing behaviour of the species (Stone et al. 2012).

They conclude that while many questions also need to be asked, these plants can increase the risk of malaria transmission in some conditions to decide how much this hypothesis holds true (Stone et al. 2018). A population's growth rate can also be affected by plant effects on male mosquitoes. However, current data in the case of aquatic tropical plants indicates that management of these plants may lead to malaria prevention. Feeding on nectar is the primary source of nutrients for male mosquitoes. Without nectar, their life and breeding chances dwindle. Laboratory cage and mesocosm studies suggest that insufficient females can become inseminated in the absence of nectar to support a population (Gary et al. 2009). Finally, very little is understood about the influence of biomass or the availability of various plants on the choices taken by mosquitoes to forage. Presumably (because there are no specific preferences for various plant species), plants that are fed on in a particular environment would be a result of both their beauty or acceptability to mosquitoes and their occurrence or how many plant species are seen. Therefore, changes in plant abundance or productivity affect not only which plants are being used but also the rate at which humans are bitten.

6.12.2 Filariasis and Green Biomass Ecology

Much focus has been put over the last few decades on the production by many researchers of anti-filarial agents from plant or natural products and on the development of new plant-based medicinal preparations for complementary or alternative medicines (CAM), backed by clinical evidence of the effectiveness and health and quality assurance of the preparations (Sahare et al. 2008; Comley 1990). Reliable plant supply that can cause filarial parasite apoptosis may likely offer a new path for the production of a new class of anti-filarials (Table 6.6). Mukherjee et al. (2013) measured the anti-filarial potency of an engineered ethanol-rich polyphenol extract of *Azadirachta indica* leaves. While there are few detailed studies on the anti-filarial effects of plant extracts or drugs, such signs are not unique in a broad collection of medicinal plants (Burkill 1985). Several plants are documented to be effective against tissue-dwelling nematodes and specific filarial species and are used in conventional medicinal structures (Kiuchi et al. 1987).

Table 6.6 Plant products with activity in filariasis or against the parasites

Name of the plant	Family	Part used/product	Activity against
Adenia gummifera	Passifloraceae	Root	Filariasis, hydrocoele
Aegle marmelos Corr.	Euphorbiaceae	Leaf	Mf of *B. malayi* (in vitro)
Alstonia boonei	Apocynaceae	Bark, fresh latex, fresh stem bark	Loiasis, filarial swellings
Agrimonia eupatoria	Rosaceae	Agrimophol	*Schistosoma* sp.
			Taenia sp.
Alstonia congensis	Apocynaceae	Latex	Loiasis, filarial swellings (bandaged along with crushed bark of *Erythrophleum guineense*)
Alstonia scholaris	Apocynaceae	Latex, bark	Filariasis, elephantiasis
Aloe barteri	Liliaceae	Leaf	Guinea worm, disease causing white skin patches (onchocerciasis?)
Ammannia multiflora	Lythraceae	Leaf	Sight problems, including those caused by filariasis
Andrographis paniculata	Acanthaceae	Dried leaf	Filariasis, mf of *D. reconditum* in dogs (in vivo and in vitro) and adults of *B. malayi* in rodents
Argyreia speciosa	Convolvulaceae	Whole plant	Filariasis, parasitic skin diseases, active in vitro against *S. cervi*
Azadirachta indica	Meliaceae	Leaf, flower	*S. digitata* (in vitro)
Boerhavia repens	Nyctaginaceae	Immature shoots	Elephantiasis
Botryocladia leptopoda	Rhodymeniaceae	Red algae	*L. sigmodontis, A. viteae, B. malayi* (in vivo)
Butea monosperma	Leguminosae-Papilionaceae	Root and leaf	Mf of *B. malayi* (in vitro)
Caesalpinia bonducella	Caesalpiniaceae	Seed kernel	*S. digitata* (in vitro), *L. sigmodontis, B. malayi* (in vivo)
Calotropis gigantea	Asclepiadaceae	Leaf, latex	Filariasis, elephantiasis, skin changes, *S. digitata* (in vitro)
Calotropis procera	Asclepiadaceae	Whole plant/ milky juice, dried aerial parts, root, bark, latex	Guinea worm, filariasis, elephantiasis
Carapa procera	Meliaceae	Dried fruit, seed	Filariasis, *O. volvulus* (in vitro), parasitic skin disease
Cardiospermum halicacabum	Sapindaceae	Plant	*B. pahangi* adult and Mf (in vitro)
Cassia alata	Leguminosae	Fresh leaf juice	Parasitic skin disease

(continued)

Table 6.6 (continued)

Name of the plant	Family	Part used/product	Activity against
Cassia aubrevellei	Leguminosae	Root, bark	Onchocerciasis, skin microfilaricidal, active in vitro against *O. volvulus* mf
Cassia occidentalis	Leguminosae	Leaf, seed	Guinea worm, parasitic skin diseases acute lymphedema, skin changes
Cassia tora	Leguminosae	Dried leaf	Parasitic skin diseases
Cayaponia martiana	Cucurbitaceae	Root	Elephantiasis
Cedrus deodara	Pinaceae	Plant extract	*S. digitata* adults (in vitro)
Centratherum anthelminticum	Asteraceae	Plant extract	*S. cervi, S. digitata* adults (in vitro)
Cinnamomum culilawan	Lauraceae	Bark	Rubefacient for filarial lymphangitis
Cleistopholis glauca	Annonaceae	Dried bark	Filariasis, inactive in vitro against *O. volvulus*
Clerodendrum capitatum	Verbenaceae	Root	Elephantiasis
Crossopteryx febrifuga	Rubiaceae	Fresh fruit juice	Eye filariasis
Cyrtomium fortunei	Polypodiaceae	Dried rhizome	Filariasis
Daniella thurifera	Leguminosae	Gum	Parasitic skin diseases
Delonix elata	Leguminosae	Whole plant	Filariasis, elephantiasis
Dichrostachys cinerea D. glomerata	Leguminosae	Dried stem bark, inner bark	Elephantiasis
Dombeya amaniensis	Sterculiaceae	Root	Filariasis/lymphatic disorders
Eclipta alba	Compositae	Dried whole plant	Elephantiasis
Elaeophorbia drupifera	Euphorbiaceae	Leaf	Guinea worm, filariasis
		Leaf	Guinea worm, used with *Hilleria latifolia*
Elephantopus scaber	Compositae	Dried root	Filariasis
Enicostema littorale	Gentianaceae	Whole plant	Filariasis, microfilaricidal in vitro against *Conispiculum guindiensis*
Erythrophleum guineense	Leguminosae	Crushed bark	Loiasis (filarial swellings), used with *Alstonia congensis*
Erythrophleum ivorense	Leguminosae	Dried stem bark	Loaiasis (filarial swellings) used in *O. volvulus*
Eucalyptus robusta	Myrtaceae	Leaf	Microfilariasis
Guiera senegalensis	Combretaceae	Leaf	Parasitic skin diseases, guinea worm inflammatory swellings

(continued)

Table 6.6 (continued)

Name of the plant	Family	Part used/product	Activity against
Hilleria latifolia	Phytolaccaceae	Whole plant, leaf	Guinea worm, used with *Elaeophorbia drupifera*, filariasis, *O. volvulus* (in vitro)
Jatropha curcas	Euphorbiaceae	Seed oil, leaf, whole plant	Guinea worm, rubefacient for parasitic skin diseases
Kigelia africana	Bignoniaceae	Whole plant	Elephantiasis of scrotum
Lantana camara	Verbenaceae	Stem	*A. viteae, B. malayi*
Limeum pterocarpum	Molluginaceae	Aerial parts	Filariasis
Lawsonia inermis	Lythraceae	Leaf	*S. digitata* (in vitro)
Lycopodium rubrum	Lycopodiaceae	Whole plant	Elephantiasis
Mallotus philippensis	Euphorbiaceae	Leaf	*S. cervi* (in vitro)
Melia azadirachta	Meliaceae	Bark	Filariasis, component (15%) of Filarin
Microglossa afzelii	Compositae	Dried leaf	Filariasis, *O. volvulus* (inactive in vitro)
Mussaenda elegans	Rubiaceae	Leaf	Elephantiasis
Myrianthus arboreus	Moraceae	Dried stem bark	Filariasis, *O. volvulus* (inactive in vitro)
Newbouldia laevis	Bignoniaceae	Root and leaf	Elephantiasis, scrotal elephantiasis, orchitis
Neurolaena lobata	Asteraceae	Leaf	*B. pahangi* adults (in vitro)
Ocimum sanctum	Lamiaceae	Leaf	*S. digitata* (in vitro)
Ochrocarpus africanus	Guttiferae	Root/resinous sap	Parasitic skin diseases
Odyendea gabunensis	Simaroubaceae	Dried stem bark	Filariasis, *O. volvulus* (inactive in vitro)
Pachyelasma tessmanii	Leguminosae	Dried fruit	Filariasis, *O. volvulus* (in vitro)
Pachylobus edulis	Buseraceae	Bark	Parasitic skin diseases
Pachypodanthium staudtii	Annonaceae	Dried stem bark	Filariasis, *O. volvulus* (in vitro)
Physedra longipes	Cucurbitaceae	Whole plant	Elephantiasis of the scrotum
Phychotria tanganyikensis	Rubiaceae	Leaf	Elephantiasis
Raphia farinifera	Palmae	Dried fruit	Filariasis, *O. volvulus* (in vitro)
		Dried leaf	Filariasis, *O. volvulus* (in vitro)
Ricinus communis	Euphorbiaceae	Plant extract, leaf	*S. digitata* adults (in vitro), mf of *B. malayi* (in vitro)
Richiea caparoides	Capparidaceae	Leaf, root	Filariasis, guinea worm
Rhynchosia hirta	Leguminosae	Whole plant	Filariasis, elephantiasis
Sargentodoxa cuneata	Sargentodoxaceae	Dried stem	Filariasis

(continued)

Table 6.6 (continued)

Name of the plant	Family	Part used/product	Activity against
Senecio nudicaulis	Asteraceae	Leaf	*S. cervi* mf (in vitro)
Sphaeranthus indicus	Asteraceae	Plant extract	*S. digitata* adults (in vitro)
Streblus asper	Rutaceae	Stem bark	Filarial lymphedema, microfilaricidal and macrofilaricidal, *S. cervi, L. carinii, B. malayi, A. viteae, B. malayi*
Trachyspermum ammi	Apiaceae	Fruit	*S. digitata* adults (in vitro), *B. malayi* (in vivo)
Terminalia chebula	Combretaceae	Not known	Filariasis
Tinospora cordifolia	Menispermaceae	Not known	Filariasis (acute lymphedema, skin changes)
Xerodermis stuhlmannii	Leguminosae	Root	Elephantiasis
Vitex negundo L.	Euphorbiaceae	Root, leaf	Mf of *B. malayi* (in vitro)
Zingiber officinale	Zingiberaceae	Fresh rhizome	Filariasis, *D. immitis* (microfilaricidal) (acute lymphedema)

Source: Murthy and Joseph (2011)

6.13   Satellite-Derived Vegetation Indices and Disease Pattern

The plant is usually determined by vegetation indices that are involved in the reaction to chlorophyll activity. Vegetation index levels are typically higher for trees and fewer for grasslands. The vital importance of remote sensing in providing extensive environmental analyses and understanding of disease patterns has long been recognized (Bhunia et al. 2012). Time sequence of plant indices help to track seasonal fluid fluctuations, spring/autumn detection, and the development cycle. The vegetation indices can be used to determine habitat suitability for different mosquito species, in particular the Normalized Difference Vegetation Index (NDVI) (Lourenĩço et al. 2011; Brown et al. 2008). These studies find a correlation between NDVI changes in time-space and the occurrence of vector-borne diseases. The Normalized Difference Water Index (NDWI), based on the reflectance bands used in the database, is used as a supplement to plant water quality (Gao 1996) and the open water supply concept (Xu 2006). The vegetation-related water quality version can be paired with NDVI and other environmental parameters for modelling vector distribution (Eisen and Eisen 2011). Spatio-temporal variations in water indices allow for the identification of droughts that can lead to disease vector agglomerations and corresponding disease outbreaks (Epstein 2001a, b). The Disease Water Stress Index (DWSI) was another water index that was found to contribute to disease vector spread (Brown et al. 2008). The components of the MODIS surface reflectance will have these water indices. Like the MODIS NDVI tests, cloud cover variations can be resolved by means of harmonious time series analyses (Roerink et al. 2000).

Using the traditional Penman-Monteith equation (Allen et al. 1998), USGS/EROS measures average PET values for the globe at 1.0-degree (almost 100 km) spatial resolution from 6-h computational meteorological model output fields of GDAS (Global Data Assimilation System). The analysis of the NDVI data available from 1981 to 2004 has been successful in identifying plant patterns in certain areas (but not all) and should be used with precaution (Ceccato 2005). For 16-day cycles starting in April 2000, Terra MODIS NDVI and the Enhanced Vegetation Index (EVI) have 250 m resolution. The EVI, which can complement the NDVI, is also used in forecasting flora (Huete et al. 2002).

6.13.1   Case Study: Acute Encephalitis Syndromes (AES) and Green Biomass

Japanese encephalitis virus transmission can be asymptomatic or may cause febrile fever, meningitis, myelitis, or encephalitis. Clinically, the condition is characterized as an abrupt onset of fever, tachycardia, decerebrate stiffness, headache, sometimes vomiting, nausea, and change in mental state such as agitation, disorientation, inability to talk, and lastly convulsion that goes to coma. Several acute encephalitis syndrome (AES) outbreaks were registered in various parts of India including West Bengal, Uttar Pradesh, Tamil Nadu, Assam, Manipur, Andhra Pradesh, Pondicherry, Karnataka, Maharashtra, and recently Bihar. AES in Muzaffarpur District of Bihar, India, is considered a significant public health issue. The outbreak of this high case fatality AES disease (63.3%) was recorded in 2011 in children from Bihar District of Muzaffarpur (Sahni 2012). Here, we discussed the effect of green biomass and the distribution of AES cases in space.

The epidemiological data was recorded from the S.K. Medical College and Hospital, Muzaffarpur, (Bihar) and K.D.K.M Hospital, Juran Chhapra, Muzaffarpur (Bihar). The Muzaffarpur District Administrative Boundary Chart at the village level was obtained from the Census Office (Census of India, Patna, Bihar). A point layer of each patient's position inside the village was identified based on the patient's reporting record. Based on the flight range of the vectors, appropriate buffers with a distance of 1 km from the location sites of the patient were built taking into account all environmental factors. Landsat 8 OLI data which has 30 m spatial resolution is used to extract the green biomass using Normalized Difference Vegetation Index (NDVI). It is the ratio between the red (0.63–$0.69\mu m$) and the near-infrared (1.55–$1.75\mu m$) band, measured from channel$_6$ and channel$_4$ of Landsat 8 OLI. The ratio is a measure of the discrepancies between the total absorption of chlorophyll in a plant range and the infrared unit and therefore a direct indicator of the amount of photosynthetically active green biomass (Tucker and Sellers 1986). Negative NDVI values (values approximating to -1) refer to vapour, though. Values below zero (-0.1 to 0.1) typically equate to rocky, dry fallow regions. Last but not least, low positive values reflect shrub and grassland (about 0.2 to 0.4), while high values suggest thick vegetation and woodland. Nonetheless, for each village impacted by the AES, a 1 km buffer zone was established, and the buffer zones for each village were subset separately. For each village, the minimum, average, mean, and standard deviation of NDVI were determined to understand the vegetation vigour character-istics of the villages affected by AES.

Figure 6.11 demonstrates the plant vigour of the sample region deriving from Landsat5 TM. However, in the central part of the district, the vegetation density is high, and in the study section, the rural north and southeast portion also shows higher vegetation density. Overlaying the NDVI map displaying the spatial distribu-tion of AES incidence shows that regions of 0.40 and above higher NDVI values typically correlated with AES incidence zones. In addition, NDVI's minimum, aver-age, mean, and standard deviation are deteFFrmined for each village affected by AES (Table 6.7). The average value FRof the AES-affected villages for high NDVI was 0.448 (S.D. ± 0.035), while the avRFFerage value of low NDVI was 0.015 (S.D. ± 0.037). Our test findings have demonstrated the correlation of higher NDVI with the spatial distribution of the AES. No forest was found in Muzaffarpur District; however, the district is filled with mango and litchi plantation land that reflects the higher NDVI importance. Nonetheless, the plantation lands may act as suitable rest-ing places in their natural environment, and vice versa, where vegetation density influences the temperature and relative humidity that may also influence the durabil-ity and vector ability of the vectors.

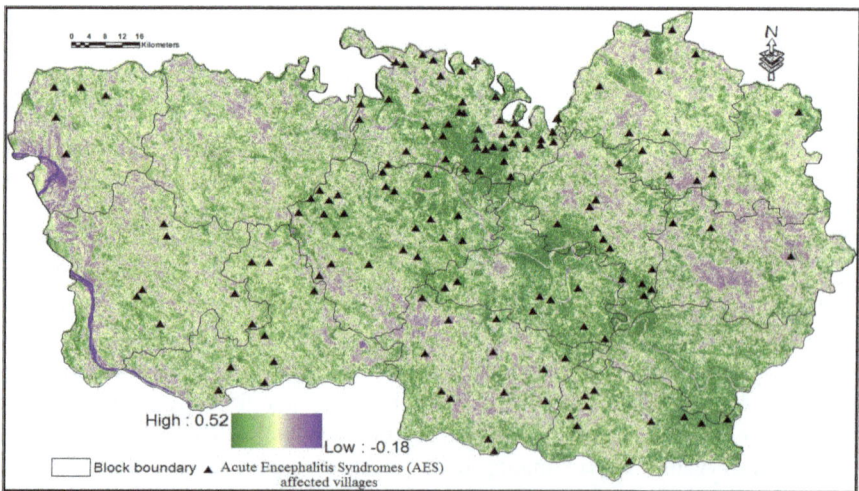

Fig. 6.11 Distribution of vegetation vigour in relation to acute encephalitis syndromes (AES) in the affected villages of Muzaffarpur District, Bihar

Table 6.7 Characteristics of vegetation vigour (AES) of the villages affected by acute encephalitis syndromes in the study sites

NDVI variables	Average	Standard deviation	Skewness	Kurtosis	95% CI	
Minimum	0.015	0.037	−0.551	0.158	−0.078	0.085
Maximum	0.448	0.035	0.362	0.003	0.385	0.551
Mean	0.229	0.030	0.056	−0.818	0.167	0.289
Standard deviation	0.116	0.009	0.263	−0.292	0.099	0.142

6.14 Land Use/Land Cover and Disease Ecology: Case Study Through GeoComputation Technique

The landscape consists of the surrounding larger physical structures. Generally, such constructs are normal, but they may be man-made. Land cover is characterized by land surface and specific subsurface properties of the earth comprising biota, vegetation, topography, surface water and groundwater, and human structures. They provide mountains, deserts, rivers, jungle, deforested areas, and artificial water reservoirs. Land use is characterized by the activities for which people have to take use of the land cover. Although focusing on land use enables to consider the existence of vectors and hosts, an emphasis on land use determines which places visitors visit for particular events, at which time of day and year, and at what level. The attractiveness of different sites for a particular activity relies on their characteristics, such as usability and utility, for that activity, e.g. the existence of recreational facilities and readily accessible woods with facilities like good trails. An interactive landscape scale analysis provides a deeper understanding of the relationships between variations in vegetation and climate, land use and human activity, and the ecology of infectious agent vectors and animal hosts. Various land uses alter the interaction of vulnerable humans with infectious vectors (Vanwambeke et al. 2007) and alter the spread of human cases.

Habitat destruction due to changes in ground use is potentially the main environmental source of altered exposure for infectious diseases. Change in habitat can impact disease vector breeding sites or the vector or reservoir host biodiversity. Key drivers of land use reform involve proposals for agricultural or water production, urbanization and sprawl, and deforestation. Such developments in effect cause a series of factors that intensify the proliferation of infectious diseases, such as degradation of the environment, invasion of pathogen, contamination, hunger, and human migration.

6.15   Landscape Ecology and Vector-Borne Disease Transmission

Development of water supplies, deforestation, wetland agriculture, crop distribution and land use improvements for agricultural purposes in the highlands, and urban farming extend all areas for mosquito-carrying malaria, resulting in increased malaria transmission in different locations (Patz et al. 2004; Klinkenberg et al. 2005). The consequences of deforestation on malaria transmission can differ geographically and rely heavily on the distributions of the vectors (Guerra et al. 2006). For instance, the re-emergence of malaria in western Kenya's highlands was heavily accused for cutting the forests for the production of tea estates. In different parts of the world, indeed, poor environmental control and peri-urban agriculture, particularly fish ponds, offer ideal conditions for mosquitoes, thus increasing malaria propagation due to the increased adult *Anopheles* densities and more outbreaks of malaria than in non-agricultural areas (Klinkenberg et al. 2005). Eliminating the papyrus and reclaiming the swamps also resulted in higher temperatures, promoting the reproduction and proliferation of the mosquitoes and thus increasing malaria propagation in the infected areas (Githeko et al. 2003). The propensity of female vectors to migrate through hosts from their breeding places enhances interactions between host and vector. These activities are primarily regulated by the landscape features (Raffy and Tran 2005). The study of RVF in the semi-arid area of Ferlo in Senegal found that mammal herds living around seasonal water sources are at higher risk of RVF if they are situated near vegetated ponds (Tran et al. 2007). Bhunia et al. (2012) undertook at four levels (national, state, district, and village) examining the significance of land use/land cover (LULC) in the transmission of leishmaniasis culminating in a system emphasizing the linkages between LULC and disease endemic areas. The significance of LULC for the dissemination of sandfly is emphasized by authors who use vegetation and other ecological parameters, particularly meteorological and altitude records, to determine geographical limits for different species of vectors (Kolaczinski et al. 2008; Bhunia et al. 2010a, b).

6.15.1   Landscape Ecology and Waterborne Disease Transmission

The relationship between humans and surface water bodies can have complicated links to health through interrelated processes in the watercourse (Vrebos et al. 2017). All point-source and non-point-source water contamination from cropland and soil-borne disease arise from animals (Ribolzi et al. 2011). Evidence of diarrheal diseases like cholera and dysentery with respect to land use and water quality is gone up and down year after year. Throughout most cases, water acts as a medium for the propagation of infectious diseases. Land usage, in all ways and sizes, along with sanitation and hygiene activities is important risk factor that underlies the harm. Highly pathogenic avian influenza viruses (HPAIV) and their dissemination are closely related to the spatial spread and movement as habitats of domestic ducks and wild water fowls use water-dependent ground cover types (e.g. open surface water body, natural wetland, and rice paddy). The analyses established proved mechanisms like storm water runoff, floods, and overt drainage of residential and industrial wastewater into surface water supplies as the main sources of water contamination. Such processes are outlined in regard to the classes of land use and the water quality parameters provided. Owing to the uncertainties inherent in propagation mechanisms of pathogens, it is often difficult to comprehend the full nature of the disease and the fundamental causes of infection. Changes in water temperature, turbidity, and pH have been documented to affect the longevity and proliferation of the *Anopheles* mosquito which transmits malaria (Lindblade et al. 2000). Related to this, diarrheal diseases (cholera and dysentery) were associated with higher proportions of *E. coli* and faecal matter in drinking water (Saravanan 2013). Similarly, schistosomiasis was linked with contaminated surface water structures in Kenya and Senegal due to the existence of faecal matter in the water, which facilitates the transitional snail host's survival (Anyona et al. 2014). However, water acts a noteworthy role in the development and propagation of infectious diseases by means of direct and indirect pathways. The shortage of sanitation and hygiene services caused households in rural areas to conduct open evacuation in and near water sources, which give rise to high concentrations of faecal coliforms in the water and raised the risk of infectious disease mortality.

6.15.2   Anthropogenic Land Use Change and Emerging Infectious Disease

Change in land use leads to variations or full alteration of land into another form, because of differences in human actions. Anthropogenic land use shifts may have a detrimental effect on ecological stability and biodiversity by upsetting the structure and operation of the food chain, altering biogeochemical terrestrial and aquatic processes, changing habitat resources, and adding non-native organisms, including pathogens. LULC is affected at various levels by multiple drivers of urban and rural transformation processes, such as population growth, culture, climate change, migration, and democracy, which influence the quality (and quantity) of water resources (Henderson et al. 2014). The global concern about EIDs has been centred on the time since 1973. Changes in land use (LU) are documented to be a significant driver of evolving infectious diseases (Keesing et al. 2010). In addition, LUC including agriculture, forestry, mining, and other types of urban and industrial use make up approximately half of all global zoonotic EIDs collectively. The advent of zoonotic diseases may be affected by changes in the number or distribution of hosts, vectors, and pathogens, changes in human population geometry, variations in life cycles and transmission pathways, and competitive pressure for improved pathogen potency. Enhanced regional LUC and natural degradation of the environment are a common function of the second quarter of the last century (Ellis 2011). These variations in the host and pathogen are caused by habitat change, most commonly human involvement (i.e. LU), and may occur at various spatial and temporal scales, in which cycles vary in response to a particular abiotic and biotic factor underlying it (McCallum and Dobson 2002). Nonetheless, data on LUCC at the emergency site may not be of sufficient magnitude to understand the appearance. The rate at which ecological change influences appearance is not known for other diseases (McFarlane et al. 2011). Association between ecological alteration and cycles of infectious diseases is dynamic and an epidemiological concern. The diverse linkages between LUC and EIDs were mainly investigated in limited disease research. The hosts and vectors were benefitting from the proliferation of irregular open surfaces with water and manufactured and relatively shallow dams linked to agrarian and pastoral expansion. Host and vector have been enhanced by the development of uneven open surface areas with water and handmade and relatively low dams that can contribute to agricultural and pastoral production. Some infections have arisen from ecotones, i.e. edge effects, including clearings of forests where historically isolated organisms come into contact and can share pathogens. Nevertheless, most EIDs are the consequence of anthropogenic landscape development that promotes improvements in the success of a natural host or organism, not always at the same place (e.g. increased mating conditions for vectors, progress in the urban forest for relocated wildlife hosts from natural habitats, dissemination of ecological pathogens).

The microclimate, the availability of water, and the forms of plants are features of the environment that can most affect disease transmission. Man-made ecosystem improvements also raise the possibility of spreading disease by establishing a vector-favourable habitat or intermediate hosts. Landscape and human environments are deeply connected, and simply separating the two is complicated. The distinction is that of scale; although individuals usually cannot change or enhance the ecosystem, individuals may alter the human environment. Though the geography would usually be equivalent for all individuals living in a city, people living in the same country, village, or even household have a very various human environment. Most diseases are related to particular conditions, and persons with different professions, social class, ethnicity, or religion may be even more at risk than others.

6.15.3 GeoComputational Technology in Land Use/Land Cover Information Extraction

Numerous spatial data appear, differing in nature and resolution, but remote sensing data provides broad coverage and a rich content of information. They are being used effectively in epidemiology and ecology of vector-borne diseases (Herbreteau et al. 2007), especially for mapping the dissemination of vectors or disease cases. The collection of data generated from remote sensing varies from high (such as CARTOSAT, IKONOS, and QuickBird) to regional datasets produced at regular intervals (such as LISS III, TM/ETM+, SPOT) to lower spatial resolution (>250 m) images now produced daily throughout the Earth (such as AVHRR, MODIS). Commonly used in remote sensing images, classification methods include maximum likelihood (Rahman 2016); minimum distance (Grobler et al. 2012); object-oriented (Xie et al. 2016), spectral angle mapper (SAM); support vector machine (SVM); and neural networking classifier. In addition to creating more complex classifiers, multi-classifier variations are one means of increasing classification precision, since various classifiers may offer extra knowledge about the items to be categorized. Since the changes in land use are more or less unidirectional, without any oscillation, it is possible to extrapolate adjustments in spatial scale and even to quantify transition pace. GeoComputing is a versatile resource for collection, organizing, and retrieval of spatial information in a user-friendly environment.

6.16   Case Study: Malaria Incidences and Microhabitat Characteristics

Malaria spread is dictated by temperature and other regional factors affecting the growth of mosquitoes and plasmodium at a given time but is also affected by changes in the atmosphere over time. Changes in the ecosystem resulting from natural events or human activity on a local or global scale may alter the environmental equilibrium and context in which vectors and their parasites produce and spread the disease. Several observations and studies indicate that these changes in LULC affected the quality, production, density, and distribution of the larval malaria vector environment in the world (Krefis et al. 2011; Stefani et al. 2013). The use of larvicides and biological control proved successful for global malaria control. The prevalence of adult vectors is positively correlated with the availability of aquatic environments required for egg deposition, and areas with the highest risk of malaria are often located within only a few hundred meters of these larval environments (Shililu et al. 2003). Extensive maize cultivation has been suggested to affect mosquito larval growth, pupation success, and adult size in the vicinity (Ye-Ebiyo et al. 2000). The spread and prevalence of vector-borne diseases such as malaria is heavily affected by spatial and temporal changes in the environment as defined by geographic information systems (GIS) and remote sensing (RS) data over the last 20 years.

   The association between LULC and the occurrence of malaria has been analysed at village level in Murshidabad District of West Bengal (India). At village level, the risk factor recognition was based on the 30 m spatial resolution Landsat 8 OLI sensor data, collected for the year 2017. The buffer zones used for the study at the village levels consisted of areas protected by rings with a circumference of 500 m. For each heavily infected area, these buffer zones were based (more than five cases of malaria) to estimate the ground cover groups suitable for vector ecosystems within the region. The LULC map was created using a supervised classification approach based on the maximum likelihood classification of Gaussians. Relevant LULC groups suitable for malaria transmission were identified by surveying literature and gathering information during field visits to the affected area. The LULC groups were classified according to the suitability of their *Anopheles* sandfly, i.e. suggesting their possible risk of malaria transmission. The scores were determined by the following equation, literature, expert opinion, and knowledge interest (I_j), the latter measured by:

$$I_j = \log_{10} \frac{Class\,density}{Map\,density}$$

where the class density is the number of LULC classes within a buffer zone separated by the number of these classes within the study area and the region density is the proportion of the overall LULC region affected by endemic malaria. The *Ij* values were rated in this study with an index varying from 4 to 0, corresponding to five specific categories of suitability, i.e. extremely small (>1.51), small (1.01 to 1.50), moderate (0.51 to 1.00), low (0.0 to 0.50), and very low or negligible (<0.0).

Figure 6.12 displays the land use/land use characteristics of the study area obtained from Landsat 8 OLI satellite data. In order to identify ideal habitat areas, LULC were divided into elevated types as seen in Fig. 6.12. The study area is characterized by built-up area, marshy land, moist fallow, mixed settlement, scrub land, surface water body, river, dense forest, crop land, and agricultural fallow. 75% of the total area comprised agricultural fallow fields and crop land. The findings for the study of the *Anopheles* mosquito habitat show the affected malaria areas as strongly impacted by built-up environment, mixed settlement, surface water body, and dense forest (Table 6.8). Interestingly, the study highlights the high suitability of certain forms of LULCs occupying marshy land and cropland-based areas. Surface water bodies (such as deep water, shallow and shade water, ponds, and fish pools) were a primary risk factor for malaria transmission, since they could influence vector breeding sites.

It is understood that with crop lands and built-up environments typically having greater access to sunshine than those in natural moist areas, the average air temperature was slightly higher in the cultivated land and built-up areas. However, the connection between the characteristics of LULC and the malaria data (such as incidence and prevalence) is not direct. Satellite imagery, however, can have benefits for environmental health research because it allows for a spatially accurate and almost continuous representation of the Earth's surface at extremely high spatial and temporal resolution. Consequently, the approach to landscape epidemiology will greatly benefit from botanical experience in offering improved characterization of landscape areas.

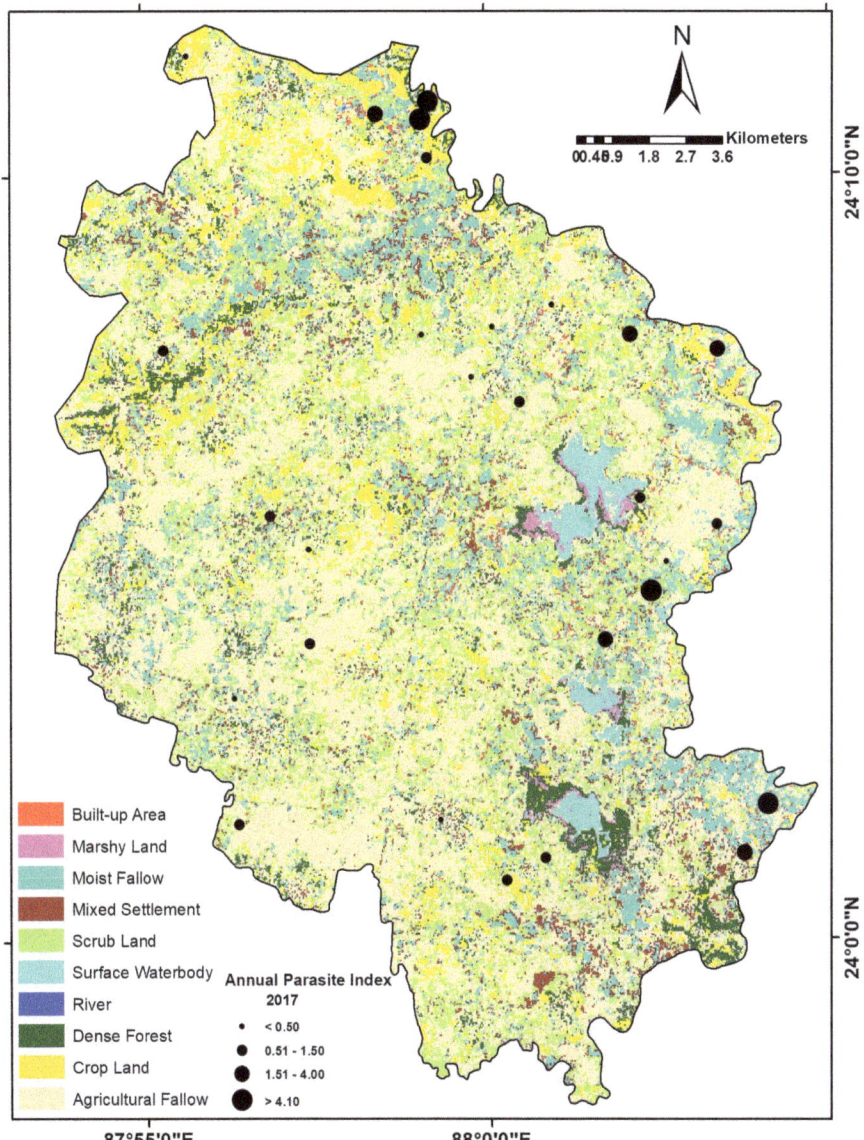

Fig. 6.12   Land use/land cover and its association with malaria incidence

6.17   Conclusion

Change in climate and land use may affect numerous human infectious diseases, operating either spontaneously or synergistically. While disease resurgence has been linked in isolated cases to current patterns in warming, a few of the long-term

Table 6.8 Calculation of information value in relation to malaria incidence

LULC class	Borai	Dhanigram	Digha	Eroali	Jadavpur	Kalgram	Kalidaspur	Khesar	Rahigram	Sujapur
Built-up area	1.59	1.20	1.79	2.29	2.05	2.56	1.55	1.14	1.64	1.35
Marshy land	2.84	0.00	0.00	0.00	0.00	0.00	0.00	0.00	0.00	0.00
Moist fallow	0.52	0.70	0.77	1.01	1.17	1.38	0.34	0.54	0.83	0.37
Mixed settlement	0.80	2.21	1.43	1.41	1.91	1.60	1.04	1.42	2.15	2.07
Scrub land	0.72	0.43	0.50	0.51	0.94	0.41	0.12	0.35	0.71	0.78
Surface water body	1.57	1.95	2.27	1.90	1.62	1.55	0.00	1.75	1.73	1.24
River	0.00	0.00	0.00	3.44	2.30	0.00	0.00	0.00	3.11	0.00
Dense forest	1.55	1.24	1.47	1.50	0.54	1.66	0.00	1.78	1.79	2.59
Crop land	1.56	0.48	0.29	1.66	0.39	0.61	2.36	1.45	0.54	1.34
Agricultural fallow	0.21	0.48	0.55	0.24	0.54	0.21	0.31	0.34	0.28	0.20

and nuanced problems presented by climate change may not be easily recognizable from other causative factors. Health risks are just one of the markets predicted to be impacted by climate and ecological change, and they reflect the complex framework in which policies will be adopted by policy-makers. To maximize preventive capacities, upstream solutions to the ecosystem must be part of any action, rather than attacking single disease agents. In 'open-source' data and the advent of modern satellite sensors, the gap from high spatial resolution to high time resolution would be well solved. Remote sensing is commonly used in epidemiology as distant sensors are controlled by host-pathogen networks and are suspect of environmental factors that can contribute to the spread of multiple disease vectors and consequently pathogens. In addition, as a host's infectivity ranges between several days and his existence, spatio-temporal trends are important for transmission of disease. Biotic data (e.g. tick and host profusion) and abiotic data (environmental restrictions) are widely used to conduct the probability assessment of exposure to vector-borne diseases. Using geospatial technology and satellite data may be analysed to collect information from time series on environmental factors, niches, events, and patterns.

Standard epidemiologic surveillance schemes report clinical infections but never take spatial trends into account. GeoComputation can lead to monitoring systems by incorporating a spatial dimension that allows for a finer grain evaluation of risk. Early warning services may be applied with the detection of critical criteria for environmental parameters to forecast potential outbreaks in space and time. While quantitative time series data based on remote sensing have been accessible for more than a decade, their complete capacity for epidemiological tracking and modelling has yet to be utilized. The prosperity of usable satellite product is a valuable tool for research and public administration, particularly in terms of the ability to detect trends from time series.

References

Acharya BK, Cao C, Lakes T, Chen W, Naeem S (2016) Spatiotemporal analysis of dengue fever in Nepal from 2010 to 2014. BMC Public Health 16:849. https://doi.org/10.1186/s12889-016-3432-z

Ahern M, Kovats RS, Wilkinson P, Few R, Matthies F (2005) Global health impacts of floods: epidemiologic evidence. Epidem Rev 27(1):36–46

Albuquerque MDFMD (2001) What do public health researchers expect of geocomputation? Cad Saúde Pública 17:1077–1078

Allen RG, Pereira LS, Raes D, Smith M (1998) Crop evapotranspiration-Guidelines for computing crop water requirements-FAO Irrigation and drainage paper 56. Fao Rome 300(9):D05109

Andradóttir HÓ, Nepf HM (2000) Thermal mediation by littoral wetlands and impact on lake intrusion depth. Water Resour Res 36(3):725–735

Anyona DN, Matano A, Abuom PO, Adoka SO, Ouma C, Kanangire CK, Owuor PO, Ofulla AVO (2014) Distribution and abundance of schistosomiasis and fascioliasis host snails along the Mara River in Kenya and Tanzania. Inf Ecol Epidemiol 1:1–7

Aplin P (2005) Remote sensing: ecology. Prog Phys Geog 29(1):104–113

Avery E, Berlin GL (1992) Fundamentals of remote sensing and air photo interpretation, vol 472, 5th edn. Macmillan Publishing Company, New York

Bailey TC, Gatrell AC (1995) Interactive spatial data analysis, vol 413, No. 8. Longman Scientific & Technical, Essex

Bailey GB, Lauer DT, Carneggie DM (2001) International collaboration: the cornerstone of satellite land remote sensing in the 21st century. Space Policy 17(3):161–169

Balls MJ, Bødker R, Thomas CJ, Kisinza W, Msangeni HA, Lindsay SW (2004) Effect of topography on the risk of malaria infection in the Usambara Mountains, Tanzania. Trans R Soc Trop Med Hyg 98(7):400–408

Basnyat B, Subedi D, Sleggs J, Lemaster J, Bhasyal G, Aryal B, Subedi N (2000) Disoriented and ataxic pilgrims: an epidemiological study of acute mountain sickness and high-altitude cerebral edema at a sacred lake at 4300 m in the Nepal Himalayas. Wilderness Environ Med 11(2):89–93

Bhunia GS (2014) An appraisal of microclimatic determinants of proliferation of the disease visceral leishmaniasis using remote sensing & GIS techniques: case study of Vaishali and Muzaffarpur district, Bihar. PhD Thesis. Department of Geography, University of Calcutta, Kolkata, India

Basnyat B, Cumbo TA, Edelman R (2001) Infections at high altitude. Clin Infect Dis:1887–1891

Bern C, Hightower AW, Chowdhury R, Ali M, Amann J, Wagatsuma Y et al (2005) Risk factors for kala-azar in Bangladesh. Emerg Infect Dis 11(5):655

Bhunia GS, Kesari S, Jeyaram A, Kumar V, Das P (2010a) Influence of topography on the endemicity of Kala-azar: a study based on remote sensing and geographical information system. Geospat Health 4(2):155–165

Bhunia GS, Kumar V, Kumar AJ, Das P, Kesari S (2010b) The use of remote sensing in the identification of the eco–environmental factors associated with the risk of human visceral leishmaniasis (kala-azar) on the Gangetic plain, in north–eastern India. Ann Trop Med Parasitol 104(1):35–53

Bhunia GS, Kesari S, Chatterjee N, Pal DK, Kumar V, Ranjan A, Das P (2011) Incidence of visceral leishmaniasis in the Vaishali district of Bihar, India: spatial patterns and role of inland water bodies. Geospat Health:205–215

Bhunia GS, Kesari S, Chatterjee N, Kumar V, Das P (2012) Localization of kala-azar in the endemic region of Bihar, India based on land use/land cover assessment at different scales. Geospat Health:177–193

Bishop RA, Litch JA (2000) Malaria at high altitude. J Travel Med 7(3):157

Blaschke T, Strobl J (2001) What's wrong with pixels? Some recent developments interfacing remote sensing and GIS. Zeitschrift für Geoinformationssysteme:12–17

Bouma MJ (2003 Mar-Apr) Methodological problems and amendments to demonstrate effects of temperature on the epidemiology of malaria. A new perspective on the highland epidemics in Madagascar, 197289. Trans R Soc Trop Med Hyg 97(2):133–139. https://doi.org/10.1016/s0035-9203(03)90099-x. PMID: 14584363

Bouma MJ, Van Der Kaay HJ (1995) Epidemic malaria in India's Thar desert. The Lancet 346(8984):1232–1233

Brown H, Diuk-Wasser M, Andreadis T, Fish D (2008) Remotely-sensed vegetation indices identify mosquito clusters of West Nile virus vectors in an urban landscape in the northeastern United States. Vector Borne Zoonotic Dis 8(2):197–206

Bunyavanich S, Landrigan CP, McMichael AJ, Epstein PR (2003) The impact of climate change on child health. Ambul Pediatr 3(1):44–52

Burkill HM (1985) The useful plants of West Tropical Africa. Vol. 1. Families AD (No. Ed. 2). Royal Botanic Gardens

Caminade C, Kovats S, Rocklov J, Tompkins AM, Morse AP, Colón-González FJ et al (2014) Impact of climate change on global malaria distribution. Proc Nat Acad Sci 111(9):3286–3291

Campbell PKE, Middleton EM, McMurtrey JE, Corp LA, Chappelle EW (2007) Assessment of vegetation stress using reflectance or fluorescence measurements. J Environ Qual 36:832–845. https://doi.org/10.2134/jeq2005.0396

Cardillo E, Acedo A, Abad E (2018) Topographic effects on dispersal patterns of Phytophthora cinnamomi at a stand scale in a Spanish heathland. PloS One 13(3):e0195060

CDC (2010) Waterborne diseases. Climate and Health Program: Centers for Disease Control and Prevention

Ceccato PN (2005) Operational early warning system using SPOT-vegetation and Terra-MODIS to predict Desert Locust outbreaks

Centers for Disease Control and Prevention (CDC) (2000) Hantavirus pulmonary syndrome--Panama, 1999–2000. MMWR. Morb Mortal Wkly Rep 49(10):205

Chen PS, Tsai FT, Lin CK, Yang CY, Chan CC, Young CY, Lee CH (2010) Ambient influenza and avian influenza virus during dust storm days and background days. Environ Health Perspect 118(9):1211–1216

Chretien JP, Anyamba A, Bedno SA, Breiman RF, Sang R, Sergon K, Linthicum KJ (2007) Drought-associated chikungunya emergence along coastal East Africa. Am J Trop Med Hyg 76(3):405–407

Chretien JP, Anyamba A, Small J, Britch S, Sanchez JL, Halbach AC, Tucker C, Linthicum KJ (2015) Global climate anomalies and potential infectious disease risks: 2014–2015. PLoS Curr 7

Christophers SR (1960) The yellow fever mosquito. Its life history, bionomics and structure

Comley JCW (1990) New macrofilaricidal leads from plants? Trop Med Parasitol 41(1):1–9

Cordova S, Smith D, Broom A, Lindsay M, Dowse G, Beers M (2000) Murray Valley encephalitis in Western Australia in 2000, with evidence of southerly spread. Commun Dis Intell 24(12):368–372

Coveney S, Roberts K (2017) Lightweight UAV digital elevation models and orthoimagery for environmental applications: data accuracy evaluation and potential for river flood risk modelling. Int J Remote Sens 38(8–10):3159–3180

Crist EP (1985) A TM tasseled cap equivalent transformation for reflectance factor data. Remote Sens Environ 17(3):301–306

Daszak P, Berger L, Cunningham AA, Hyatt AD, Green DE, Speare R (1999) Emerging infectious diseases and amphibian population declines. Emerg Infect Dis 5(6):735

Davis TJ, Kline DL, Kaufman PE (2016) Aedes albopictus (Diptera: Culicidae) oviposition preference as influenced by container size and Buddleja davidii plants. J Med Entomol 53(2):273–278

Del Corral J, Blumenthal MB, Mantilla G, Ceccato P, Connor SJ, Thomson MC (2012) Climate information for public health: the role of the IRI climate data library in an integrated knowledge system. Geospat Health 6(3):S15–S24

Demetz F, Chouker A, Bauer A et al (2001) Changes in microcirculatory and immunological responses during physical exercise at high altitude. Wilderness Environ Med 12:139–142

Diamond J, Guns G (1997) Steel: the fates of human societies

Divine RA, Divine RA (1993) The sputnik challenge. Oxford University Press

Dobson AP, Carper ER (1996) Infectious diseases and human population history. Bioscience 46(2):115–126

Duncanson LI, Niemann KO, Wulder MA (2010) Estimating forest canopy height and terrain relief from GLAS waveform metrics. Remote Sens Environ 114(1):138–154

Dwayne AD, Logsdon JM, Latell B (1988) Eye in the sky: the story of the corona spy satellites. Smithsonian Books, Washington, DC: 29–85, 143–156. ISBN: 1560988304 (ISBN 1-56098-773-1 (paperback) or ISBN 1-56098-830-4

Dwight RH, Baker DB, Semenza JC, Olson BH (2004) Health effects associated with recreational coastal water use: urban versus rural California. Am J Public Health 94(4):565–567

Eisen L, Eisen RJ (2011) Using geographic information systems and decision support systems for the prediction, prevention, and control of vector-borne diseases. Ann Rev Entomol 56:41–61

Ellis EC (2011) Anthropogenic transformation of the terrestrial biosphere. Philos Trans R Soc A Mathematical Phys Eng Sci 369(1938):1010–1035

Engelthaler DM, Mosley DG, Cheek JE, Levy CE, Komatsu KK, Ettestad P, Davis T, Tanda DT, Miller L, Frampton JW, Porter R (1999) Climatic and environmental patterns associated with hantavirus pulmonary syndrome, Four Corners region, United States. Emerg Infect Dis 5(1):87–94

Epstein PR (1999) Climate and health. Science 285(5426):347–348

Epstein PR (2000) Is global warming harmful to health? Sci Am 283(2):50–57

Epstein PR (2001a) West Nile virus and the climate. J Urban Health 78(2):367–371

Epstein PR (2001b) Climate change and emerging infectious diseases. Microbes Infect 3(9):747–754

Feyisa GL, Meilby H, Fensholt R, Proud SR (2014) Automated Water Extraction Index: A new technique for surface water mapping using Landsat imagery. Remote Sens Environ 140:23–35

Fisher A, Flood N, Danaher T (2016) Comparing Landsat water index methods for automated water classification in eastern Australia. Remote Sens Environ 175:167–182

Frank B, Fulton R, Weimar C, Lees K, Sanders R (2013) Use of paracetamol in ischaemic stroke patients: evidence from VISTA. Acta Neurol Scand 128:172–177. https://doi.org/10.1111/ane.12094

Frazier PS, Page KJ (2000) Water body detection and delineation with Landsat TM data. Photogramm Eng Remote Sens 66(12):1461–1468

Frosch AE, John CC (2012) Immunomodulation in Plasmodium falciparum malaria: experiments in nature and their conflicting implications for potential therapeutic agents. Expert Rev Anti-Infect Ther 10(11):1343–1356

Gage KL, Kosoy MY (2005) Natural history of plague: perspectives from more than a century of research. Annu Rev Entomol 50:505–528

Gao B (1996) NDWI - a normalized difference water index for remote sensing of vegetation liquid water from space. Remote Sens Environ 58(3):257–266. https://doi.org/10.1016/s0034-4257(96)00067-3

Gary RE, Cannon JW, Foster WA (2009) Effect of sugar on male Anopheles gambiae mating performance, as modified by temperature, space, and body size. Parasit Vectors 2(1):19

Gerba CP (1999) Virus survival and transplant in groundwater, (Volume 24). J Ind Microbiol Biotechnol 22

Gillespie TW, Foody GM, Rocchini D, Giorgi AP, Saatchi S (2008) Measuring and modelling biodiversity from space. Prog Phys Geogr 32(2):203–221

Githeko AK, Munga SO, Afrane YA, Guiyun Y (2003) Land use changes and micro climate change: Lessons for highland malaria and climate change. Kenya Medical Research Institute (KEMRI), Nairobi

Graham AL (2008) Ecological rules governing helminth–microparasite coinfection. Pro Nat Acad Sci 105(2):566–570

Griffin DW (2007) Atmospheric movement of microorganisms in clouds of desert dust and implications for human health. Clin Microbiol Rev 20(3):459–477

Grobler TL, Ackermann ER, Olivier JC, van Zyl AJ, Kleynhans W (2012) Land-cover separability analysis of modis time-series data using a combined simple harmonic oscillator and a mean reverting stochastic process. IEEE J Sel Top Appl Earth Observ Remote Sens 5(3):857–866

Guerra CA, Snow RW, Hay SI (2006) A global assessment of closed forests, deforestation and malaria risk. Annals Trop Med Parasitol 100(3):189

Ghuffar S (2018) DEM generation from multi satellite planetscope imagery. Remote Sens 10:1462. https://doi.org/10.3390/rs10091462

Hackett PH (2001) High-altitude medicine. Wilderness Med:2–43

Haines A, Patz JA (2004) Health effects of climate change. Jama 291(1):99–103

Hamnett MP, Anderson CL, Guard CP (1999) The Pacific ENSO Applications Center and the 1997–98 ENSO warm event in the US-affiliated Micronesian Islands: minimizing impacts through rainfall forecasts and hazard mitigation. Pacific ENSO Applications Center, Honolulu

Hardy A, Ettritch G, Cross DE, Bunting P, Liywalii F, Sakala J, Silumesii A, Singini D, Smith M, Willis T, Thomas CJ (2019) Automatic detection of open and vegetated water bodies using Sentinel 1 to map African malaria vector mosquito breeding habitats. Remote Sens 11(5):593

Harrington LC, Edman JD, Scott TW (2001) Why do female Aedes aegypti (Diptera: Culicidae) feed preferentially and frequently on human blood? J Med Entomol 38(3):411–422

Hartmann G, Tschöp M, Fischer R, Bidlingmaier C, Riepl R, Tschöp K, Hautmann H, Endres S, Toepfer M (2000) High altitude increases circulating interleukin-6, interleukin-1 receptor antagonist and C-reactive protein. Cytokine 12(3):246–252

Harvell CD, Mitchell CE, Ward JR, Altizer S, Dobson AP, Ostfeld RS, Samuel MD (2002) Climate warming and disease risks for terrestrial and marine biota. Science 296(5576):2158–2162

Hawker L, Bates P, Neal J, Rougier J (2018) Perspectives on digital elevation model (DEM) simulation for flood modeling in the absence of a high-accuracy open access global DEM. Front Earth Sci 6:233

Hay SI, Cox J, Rogers DJ, Randolph SE, Stern DI, Shanks GD et al (2002) Climate change and the resurgence of malaria in the East African highlands. Nature 415(6874):905–909

Henderson L, Mahoney C, McClelland C, Myers A (2014) The effect of land use and land cover on water quality in urban environments. Natural Resources and Environmental Sciences (NRES), Kansas State University

Herbreteau V, Salem G, Souris M, Hugot JP, Gonzalez JP (2007) Thirty years of use and improvement of remote sensing, applied to epidemiology: from early promises to lasting frustration. Health Place 13(2):400–403

Hjelle B, Glass GE (2000) Outbreak of hantavirus infection in the Four Corners region of the United States in the wake of the 1997–1998 El Nino—Southern Oscillation. J Infect Dis 181(5):1569–1573

Hofstra N (2011) Quantifying the impact of climate change on enteric waterborne pathogen concentrations in surface water. Curr Opin Environ Sustain 3(6):471–479

Hopewell L, Rossiter R, Blower E, Leaver L, Goto K (2005) Grazing and vigilance by Soay sheep on Lundy island: influence of group size, terrain and the distribution of vegetation. Behav Pro 70(2):186–193

House FB, Gruber A, Hunt GE, Mecherikunnel AT (1986) History of satellite missions and measurements of the Earth radiation budget (1957–1984). Rev Geophys 24(2):357–377

Hu Z, Peng J, Hou Y, Shan J (2017) Evaluation of recently released open global digital elevation models of Hubei, China. Remote Sens 9(3):262

Huang J, Walker ED, Otienoburu PE, Amimo F, Vulule J, Miller JR (2006) Laboratory tests of oviposition by the African malaria mosquito, Anopheles gambiae, on dark soil as influenced by presence or absence of vegetation. Malar J 5(1):1–5

Huang C, Chen Y, Zhang S, Li L, Shi K, Liu R (2016) Surface water mapping from Suomi NPP-VIIRS imagery at 30 m resolution via blending with Landsat data. Remote Sens 8(8):631

Hudson PJ, Rizzoli AP, Grenfell BT, Heesterbeek JAP, Dobson AP (2002) Ecology of wildlife diseases.

Huete A, Didan K, Miura T, Rodriguez EP, Gao X, Ferreira LG (2002) Overview of the radiometric and biophysical performance of the MODIS vegetation indices. Remote Sens Environ 83(1–2):195–213

Hyyppä J, Wagner W, Hollaus M, Hyyppä H (2009) Airborne laser scanning. SAGE, London, pp 199–211

InterMap (2018) NextMap world 10. Available at: https://www.intermap.com/data/nextmap

IPCC (2001) Climate change (2001). Synthesis report RT Watson, CW Team (eds) A Contribution of Working Groups I, II, and III to the Third Assessment Report of the Intergovernmental Panel on Climate Change, Cambridge University Press, Cambridge, UK, and New York

IPCC (Intergovernmental Panel on Climate Change) (2007) Impacts, adaptation and vulnerability. In: Contribution of Working Group II to the Fourth Assessment Report of the Intergovernmental Panel on Climate Change

IPCC (2012) Managing the risks of extreme events and disasters to advance climate change adaptation. In: Field CB, Barros V, Stocker TF, Qin D, Dokken DJ, Ebi KL, Mastrandrea MD, Mach KJ, Plattner G-K, Allen SK, Tignor M, Midgley PM (eds) A special report of working groups I and II of the intergovernmental panel on climate change. Cambridge University Press, Cambridge, UK/New York, NY, USA, 582 pp

Islam MS, Sharker MAY, Rheman S, Hossain S, Mahmud ZH, Islam MS, Uddin AMK, Yunus M, Osman MS, Ernst R, Rector I (2009a) Effects of local climate variability on transmission dynamics of cholera in Matlab, Bangladesh. Trans R Soc Trop Med Hyg 103(11):1165–1170

Islam MS, Sharker MAY, Rheman S, Hossain S, Mahmud ZH, Islam MS, Uddin AMK, Yunus M, Osman MS, Ernst R, Rector I (2009b) Effects of local climate variability on transmission dynamics of cholera in Matlab, Bangladesh. Trans R Soc Trop Med Hyg 103(11):1165–1170

IWGCCH (2010) A human health perspective on climate change: a report outlining the research needs on the human health effects of climate change.. Environmental Health Perspectives and the National Institute of Environmental Health Sciences

Jarihani AA, McVicar TR, Van Niel TG, Emelyanova IV, Callow JN, Johansen K (2014) Blending Landsat and MODIS data to generate multispectral indices: A comparison of "Index-then-Blend" and "Blend-then-Index" approaches. Remote Sens 6(10):9213–9238

Jensen JR (2000) Remote sensing of the environment an earth resource perspective. Prentice Hall. Upper Saddle River (NJ), USA

Jofre J, Blanch AR, Lucena F (2009) Water-borne infectious disease outbreaks associated with water scarcity and rainfall events. In: Water scarcity in the Mediterranean. Springer, Berlin/Heidelberg, pp 147–159

John CC, Ouma JH, Sumba PO, Hollingdale MR, Kazura JW, King CL (2002) Lymphocyte proliferation and antibody responses to Plasmodium falciparum liver-stage antigen-1 in a highland area of Kenya with seasonal variation in malaria transmission. Am J Trop Med Hyg 66(4):372–378

Jones KE, Patel NG, Levy MA, Storeygard A, Balk D, Gittleman JL, Daszak P (2008) Global trends in emerging infectious diseases. Nature 451:990–993

Kan H (2011) Climate change and human health in China. 119:A60–A61

Karpatne A, Khandelwal A, Chen X, Mithal V, Faghmous J, Kumar V (2016) Global monitoring of inland water dynamics: State-of-the-art, challenges, and opportunities. Computational sustainability. Springer, Cham, pp 121–147

Keesing F, Holt RD, Ostfeld RS (2006) Effects of species diversity on disease risk. Ecol Lett 9(4):485–498

Keesing F, Belden LK, Daszak P, Dobson A, Harvell CD, Holt RD, Hudson P, Jolles A, Jones KE, Mitchell CE, Myers SS (2010) Impacts of biodiversity on the emergence and transmission of infectious diseases. Nature 468(7324):647–652

Khasnis AA, Nettleman MD (2005) Global warming and infectious disease. Arch Med Res 36(6):689–696

Kiuchi F, Miyashita N, Tsuda Y, Kondo K, Yoshimura H (1987) Studies on crude drugs effective on visceral larva migrans. I. Identification of larvicidal principles in betel nuts. Chem Pharm Bull 35(7):2880–2886

Klinkenberg E, McCall PJ, Hastings IM, Wilson MD, Amerasinghe FP, Donnelly MJ (2005) Malaria and irrigated crops, Accra, Ghana. Emerg Infect Dis 11(8):1290

Kolaczinski JH, Reithinger R, Worku DT, Ocheng A, Kasimiro J, Kabatereine N, Brooker S (2008) Risk factors of visceral leishmaniasis in East Africa: a case-control study in Pokot territory of Kenya and Uganda. Int J Epidemiol 37(2):344–352

Kovats RS, Bouma MJ, Hajat S, Worrell E, Haines A (2003) El Niño and health. Lancet 361: 1481–1489. Find this article online.

Kramer HJ (2002) Observation of the Earth and its Environment: Survey of Missions and Sensors. Springer Science & Business Media

Krefis AC, Schwarz NG, Nkrumah B, Acquah S, Loag W, Oldeland J, Sarpong N, Adu-Sarkodie Y, Ranft U, May J (2011) Spatial analysis of land cover determinants of malaria incidence in the Ashanti Region, Ghana. PloS one 6(3):17905

Kuhn K, Campbell-Lendrum D, Haines A, Cox J (2005) Using climate to predict infectious disease outbreaks: A review. World Health Organization

Kumar R, Kumar S (2013) Change in global climate and prevalence of visceral leishmaniasis. Int J Sci Res Publ 3(1):2250–3153

KuPidura P (2013) Distinction of lakes and rivers on satellite images using mathematical morphology. BiuletynWojskowejAkademiiTechnicznej 62(3):57–69

Lachica M, Prieto C, Aguilera JF (1997) The energy costs of walking on the level and on negative and positive slopes in the Granadina goat (Capra hircus). Br J Nutr 77(1):73–81

Lautze J, McCartney M, Kirshen P, Olana D, Jayasinghe G, Spielman A (2007) Effect of a large dam on malaria risk: the Koka reservoir in Ethiopia. Trop Med Int Health 12(8):982–989

Leal-Castellanos CB, Garcia-Suarez R, Gonzalez-Figueroa E, Fuentes-Allen JL, Escobedo-De La Pena J (2003) Risk factors and the prevalence of leptospirosis infection in a rural community of Chiapas, Mexico. Epidemiol Infect 131(3):1149–1156

Lebaron P, Servais P, Troussellier M, Courties C, Vives-Rego J, Muyzer G, Bernard L, Guindulain T, Schäfer H, Stackebrandt E (1999) Changes in bacterial community structure in seawater mesocosms differing in their nutrient status. Aquat Microb Ecol 19(3):255–267

Lim K, Treitz P, Wulder M, St-Onge B, Flood M (2003) LiDAR remote sensing of forest structure. Prog Phys Geogr 27(1):88–106

Lindblade KA, Walker ED, Onapa AW, Katungu J, Wilson ML (2000) Land use change alters malaria transmission parameters by modifying temperature in a highland area of Uganda. Trop Med Int Health 5(4):263–274

Lindsay SW, Bødker R, Malima R, Msangeni HA, Kisinza W (2000) Effect of 1997–98 El Niño on highland malaria in Tanzania. The Lancet 355(9208):989–990

Lloyd SJ, Kovats RS, Armstrong BG (2007) Global diarrhoea morbidity, weather and climate. Climate Res 34(2):119–127

Lodge R, Diallo TO, Descoteaux A (2006) Leishmania donovani lipophosphoglycan blocks NADPH oxidase assembly at the phagosome membrane. Cell Microbiol 8(12):1922–1931

Lotfy WM (2014) Climate change and epidemiology of human parasitosis in Egypt: a review. J Adv Res 5(6):607–613

Lourenço PM, Sousa CA, Seixas J, Lopes P, Novo MT, Almeida APG (2011) Anopheles atroparvus density modeling using MODIS NDVI in a former malarious area in Portugal. J Vector Ecol 36(2):279–291

Lowen AC, Mubareka S, Steel J, Palese P (2007) Influenza virus transmission is dependent on relative humidity and temperature. PLoS Pathog 3(10):e151. https://doi.org/10.1371/journal.ppat.0030151

Lubchenco J, Karl TR (2012) extreme weather events. Phys Today 65(3):31

Luque Fernandez MA, Schomaker M, Mason PR, Fesselet JF, Baudot Y, Boulle A, Maes P (2012) Elevation and cholera: an epidemiological spatial analysis of the cholera epidemic in Harare, Zimbabwe, 2008–2009. BMC Public Health 12(1):1–8

May RM, Anderson RM (1979) Population biology of infectious diseases: Part II. Nature 280(5722):455–461

Mackenzie J, Lindsay M, Daniels P (2000) The effect of climate on the incidence of vector-borne viral diseases in Australia: the potential value of seasonal forecasting. In: Applications of seasonal climate forecasting in agricultural and natural ecosystems. Springer, Dordrecht, pp 429–452

Mac Kenzie WR, Hoxie NJ, Proctor ME, Gradus MS, Blair KA, Peterson DE et al (1994) A massive outbreak in Milwaukee of Cryptosporidium infection transmitted through the public water supply. N Engl J Med 331(3):161–167

Maguiña C, Gotuzzo E (2000) Bartonellosis: new and old. Infect Dis Clin N Am 14(1):1–22

Manda H, Gouagna LC, Foster WA, Jackson RR, Beier JC, Githure JI, Hassanali A (2007) Effect of discriminative plant-sugar feeding on the survival and fecundity of Anopheles gambiae. Malar J 6(1):113

McCallum H, Dobson A (2002) Disease, habitat fragmentation and conservation. Proc R Soc Lond B Biol Sci 269(1504):2041–2049

McFarlane R, Becker N, Field H (2011) Investigation of the climatic and environmental context of Hendra virus spillover events 1994–2010. PloS one 6(12):28374

McFeeters SK (1996) The use of the Normalized Difference Water Index (NDWI) in the delineation of open water features. Int J Remote Sens 17(7):1425–1432

McMichael AJ, Haines JA, Slooff R, Sari Kovats R, World Health Organization (1996) Climate change and human health: an assessment (No. WHO/EHG/96.7). World Health Organization

Meehan RT (1987) Immune suppression at high altitude. Ann Emerg Med 16(9):974–979

Melesse AM, Weng Q, Thenkabail PS, Senay GB (2007) Remote sensing sensors and applications in environmental resources mapping and modelling. Sensors 7(12):3209–3241

Mellor PS, Leake CJ (2000) Climatic and geographic influences on arboviral infections and vectors. Revue Scientifique et Technique-Office International des Epizooties 19(1):41–48

Minakawa N, Dida GO, Sonye GO, Futami K, Njenga SM (2012) Malaria vectors in Lake Victoria and adjacent habitats in western Kenya. PloS one 7(3):e32725

Minakawa N, Munga S, Atieli F, Mushinzimana E, Zhou G, Githeko AK, Yan G (2005) Spatial distribution of anopheline larval habitats in Western Kenyan highlands: effects of land cover types and topography. Am J Trop Med Hyg 73(1):157–165

Mishra K, Prasad P (2015) Automatic extraction of water bodies from Landsat imagery using perceptron model. J Comput Environ Sci 2015

Mukherjee N, Mukherjee S, Saini P, Roy P, Babu SPS (2013) Antifilarial effects of polyphenol rich ethanolic extract from the leaves of Azadirachta indica through molecular and biochemical approaches describing reactive oxygen species (ROS) mediated apoptosis of Setaria cervi. Exp Parasitol 136:41–58

Müller GC, Beier JC, Traore SF, Toure MB, Traore MM, Bah S, Doumbia S, Schlein Y (2010) Field experiments of Anopheles gambiae attraction to local fruits/seedpods and flowering plants in Mali to optimize strategies for malaria vector control in Africa using attractive toxic sugar bait methods. Malar J 9(1):262

Murdoch DR (1995) Symptoms of infection and altitude illness among hikers in the Mount Everest region of Nepal. Aviat Space Environ Med 66(2):148

Murphy GS, Oldfield EC III (1996) Falciparum malaria. Infect Dis Clin N Am 10(4):747–775

Murthy PK, Joseph SK (2011) Plant products in the treatment and control of filariasis and other helminth infections and assay systems for antifilarial/anthelmintic activity. Planta Med 77(06):647–661

Nagendra H, Rocchini D (2008) High resolution satellite imagery for tropical biodiversity studies: the devil is in the detail. Biodivers Conserv 17(14):3431

Napier LE (1926) An Epidemiological Consideration of the Transmission of Kala-azar in India. An Epidemiological Consideration of the Transmission of Kala-azar in India. 4

Nayar JK, Sauerman DM Jr (1971) Physiological effects of carbohydrates on survival, metabolism, and flight potential of female Aedes taeniorhynchus. J Insect Physiol 17(11):2221–2233

Nicholls N (1993) El Niño-Southern Oscillation and vector-borne disease. Lancet (British edition) 342(8882):1284–1285

Nielsen NO, Makaula P, Nyakuipa D, Bloch P, Nyasulu Y, Simonsen PE (2002) Lymphatic filariasis in Lower Shire, southern Malawi. Trans R Soc Trop Med Hyg 96(2):133–138

Nyasembe VO, Teal PE, Mukabana WR, Tumlinson JH, Torto B (2012) Behavioural response of the malaria vector Anopheles gambiae to host plant volatiles and synthetic blends. Parasit Vectors 5(1):1–11

O'Loughlin FE, Paiva RC, Durand M, Alsdorf D, Bates P (2016) A multi-sensor approach towards a global vegetation corrected SRTM DEM product. Remote Sens Environ 182:49–59. https://doi.org/10.1016/j.rse.2016.04.018

Osei FB, Duker AA (2008) Spatial dependency of V. cholera prevalence on open space refuse dumps in Kumasi, Ghana: a spatial statistical modelling. Int J Health Geogr 7(1):62

Overgaard HJ (2007) Effect of plant structure on oviposition behavior of Anopheles minimus sl. J Vector Ecol 32(2):193–197

Park T (1948) Interspecies competition in populations of *Tribolium confusum* Duval and *Tribolium castaneum* Herbst. Ecol Monogr 18:265–307

Patz JA, Githeko AK, McCarty JP, Hussain S, Confalonieri U, de Wet N (2003) Climate change and infectious diseases: Word Health Organization

Patz JA, Daszak P, Tabor GM, Aguirre AA, Pearl M, Epstein J et al (2004) Unhealthy landscapes: policy recommendations on land use change and infectious disease emergence. Environ Health Perspect 112(10):1092–1098

Patz JA, Epstein PR, Burke TA, Balbus JM (1996) Global climate change and emerging infectious diseases. Jama 275(3):217–223

Patz JA, Reisen WK (2001) Immunology, climate change and vector-borne diseases. Trends Immunol 22(4):171–172

Paz S (2006) The West Nile Virus outbreak in Israel (2000) from a new perspective: the regional impact of climate change. Int J Environ Health Res 16(1):1–13

Pearson RD, Jeronimo SMB, de Queiroz Sousa A (1999) Leishmaniasis. In: Guerrant RL, Walker DH, Weller PF (eds) Tropical infectious diseases: principles, pathogens, and practice. Churchill Livingstone, Philadelphia, pp 797–813

Picado A, Das ML, Kumar V, Dinesh DS, Rijal S, Singh SP, Das P, Coosemans M, Boelaert M, Davies C (2010) Phlebotomus argentipes Seasonal Patterns in India and Nepal. J Med Entomol 47(2):283–286

Planet Observer (2017) PlanetDEM 30 Plus. Available at: https://www.planetobserver.com/products/planetdem/planetdem-30/

Plowright W (1982) The effects of rinderpest and rinderpest control on wildlife in Africa. In: Symposia of the Zoological Society of London, vol 50, pp 1–28

Pulliam HR, Danielson BJ (1991) Sources, sinks, and habitat selection: a landscape perspective on population dynamics. Am Nat 137:S50–S66

Quattrochi D, Luvall J (2004) Thermal remote sensing in land surface processing. CRC Press, Boca Raton, FL, USA. isbn:978-0-415-30224-1

Rabold MB (1987) High altitude bronchitis on Cerro Aconcagua (Aspen, Colorado, 1987). Program and abstracts of the Wilderness Medical Society (WMS). WMS, Colorado Springs

Raffy M, Tran A (2005) On the dynamics of flying insects populations controlled by large scale information. Theor Popul Biol 68(2):91–104

Rahman MT (2016) Detection of land use/land cover changes and urban sprawl in Al-Khobar, Saudi Arabia: An analysis of multi-temporal remote sensing data. ISPRS Int J Geo-Inf 5(2):15

Reacher M, McKenzie K, Lane C, Nichols T, Kedge I, Iversen A, Hepple P, Walter T, Laxton C, Simpson J (2004) Health impacts of flooding in Lewes: a comparison of reported gastrointestinal and other illness and mental health in flooded and non-flooded households. Commun Dis Public Health 7:39–46

Reid C (2000) Implications of climate change on malaria in Karnataka, India (Doctoral dissertation, Brown University)

Ribolzi O, Cuny J, Sengsoulichanh P, Mousquès C, Soulileuth B, Pierret A, Huon S, Sengtaheuanghoung O (2011) Land use and water quality along a Mekong tributary in Northern Lao PDR. Environ Manag 47(2):291–302

Rizzoli P, Martone M, Gonzalez C, Wecklich C, Tridon DB, Bräutigam B, Bachmann M, Schulze D, Fritz T, Huber M, Wessel B (2017) Generation and performance assessment of the global TanDEM-X digital elevation model. ISPRS J Photogramm Remote Sens 132:119–139

Robinson N, Regetz J, Guralnick RP (2014) EarthEnv-DEM90: a nearly-global, void-free, multi-scale smoothed, 90m digital elevation model from fused ASTER and SRTM data. ISPRS. Journal of Photogrammetry and Remote Sensing 87:57–67. Available at http://www.sciencedirect.com/science/article/pii/S0924271613002360

Roerink GJ, Menenti M, Verhoef W (2000) Reconstructing cloudfree NDVI composites using Fourier analysis of time series. Int J Remote Sens 21(9):1911–1917

Ross R (1915) Some a priori pathometric equations. Br Med J 1(2830):546

Rytkönen MJ (2004) Not all maps are equal: GIS and spatial analysis in epidemiology. Int J Circumpolar Health 63(1):9–24

Sahare KN, Anandharaman V, Meshram VG, Meshram SU, Gajalakshmi D, Goswami K, Reddy MVR (2008) In vitro effect of four herbal plants on the motility of Brugia malayi microfilariae. Indian J Med Res 127(5):467–472

Sahni GS (2012) Recurring epidemics of acute encephalopathy in children in Muzaffarpur, Bihar. Indian Pediatr 49(2012):502–503

Salomón OD, Quintana MG, Mastrángelo AV, Fernández MS (2012) Leishmaniasis and climate change—case study: Argentina. J Trop Med 2012

Sanders EJ, Rigau-Pérez JG, Smits HL, Deseda CC, Vorndam VA, Aye T et al (1999) Increase of leptospirosis in dengue-negative patients after a hurricane in Puerto Rico in 1996 [correction of 1966]. Am J Trop Med Hyg 61(3):399–404

Saravanan VS (2013) Urbanizing diseases: Contested institutional terrain of water-and vector-borne diseases in Ahmedabad, India. Water Int 38(7):875–887

Schlesinger P, Mamane Y, Grishkan I (2006) Transport of microorganisms to Israel during Saharan dust events. Aerobiologia 22(4):259–273

Schroeder R, Rawlins MA, McDonald KC, Podest E, Zimmermann R, Kueppers M (2010) Satellite microwave remote sensing of North Eurasian inundation dynamics: development of coarse-resolution products and comparison with high-resolution synthetic aperture radar data. Environ Res Lett 5(1):015003

Schumann GJP, Moller DK (2015) Microwave remote sensing of flood inundation. Phys Chem Earth Parts A/B/C 83:84–95

Shaman J, Day JF, Stieglitz M (2002) Drought-induced amplification of Saint Louis encephalitis virus, Florida. Emerg Infect Dis 8(6):575

Sharma U, Singh S (2008) Insect vectors of Leishmania: distribution, physiology and their control. J Vector Borne Dis 45(4):255–272

Shililu J, Ghebremeskel T, Seulu F, Mengistu S, Fekadu H, Zerom M, Ghebregziabiher A, Sintasath D, Bretas G, Mbogo C, Githure J (2003) Larval habitat diversity and ecology of anopheline larvae in Eritrea. J Med Entomol 40(6):921–929

Shippert P (2004) Why use hyperspectral imagery? Photogramm Eng Remote Sens 70(4):377–396

Shultz JM, Russell J, Espinel Z (2005) Epidemiology of tropical cyclones: the dynamics of disaster, disease, and development. Epidemiol Rev 27(1):21–35

Smith H, Crandall I, Prudhomme J, Sherman I (1992) Optimization and inhibition of the adherent ability of Plasmodium falciparum-infected erythrocytes. Memórias do Instituto Oswaldo Cruz 87:303–312

Snow RW, Craig MH, Newton CRJC, Steketee RW (2003) The public health burden of Plasmodium falciparum malaria in Africa. Working paper 11. Disease Control Priorities Project, Bethesda, Maryland, USA: Fogarty International Center, National Institutes of Health

Snow RW, Omumbo JA, Lowe B, Molyneux CS, Obiero JO, Palmer A, Weber MW, Pinder M, Nahlen B, Obonyo C, Newbold C (1997) Relation between severe malaria morbidity in children and level of Plasmodium falciparum transmission in Africa. The Lancet 349(9066):1650–1654

Stefani A, Dusfour I, Corrêa APS, Cruz MC, Dessay N, Galardo AK, Galardo CD, Girod R, Gomes MS, Gurgel H, Lima ACF (2013) Land cover, land use and malaria in the Amazon: a systematic literature review of studies using remotely sensed data. Malar J 12(1):1–8

Stone CM, Jackson BT, Foster WA (2012) Effects of bed net use, female size, and plant abundance on the first meal choice (blood vs sugar) of the malaria mosquito Anopheles gambiae. Malar J 11(1):3

Stone CM, Witt AB, Walsh GC, Foster WA, Murphy ST (2018) Would the control of invasive alien plants reduce malaria transmission? A review. Parasites & vectors 11(1):1–18

Tachikawa T, Kaku M, Iwasaki A (2011) ASTER GDEM version 2 validation report, 2nd edn, Japan, pp A1–A24

Tadono T, Ishida H, Oda F, Naito S, Minakawa K, Iwamoto H (2014) Precise global DEM generation by ALOS PRISM. ISPRS Ann Photogramm Remote Sens Spatial Inf Sci 2(4):71

Tadono T, Nagai H, Ishida H, Oda F, Naito S, Minakawa K, Iwamoto H (2016) Generation of the 30 m-mesh global digital surface model by ALOS PRISM. In: The international archives of the photogrammetry, remote sensing and spatial information sciences, volume XLI-B4. XXIII ISPRS Congress; July 12–19. Czech Republic, Prague

Takaku J, Tadono T (2017) Quality updates of 'AW3D' global DSM generated from ALOS PRISM. In: Proceedings of the IEEE international geoscience and remote sensing symposium (IGARSS). IEEE, Fort Worth, TX. https://doi.org/10.1109/IGARSS.2017.8128293

Thomson M, Lyon B, Ceccato P (2019) Climate matters in health decision-making. In: The Palgrave handbook of global health data methods for policy and practice. Palgrave Macmillan, London, pp 263–281

Tian H, Zhou S, Dong L, Van Boeckel TP, Cui Y, Newman SH, Takekawa JY, Prosser DJ, Xiao X, Wu Y, Cazelles B (2015) Avian influenza H5N1 viral and bird migration networks in Asia. Pro Nat Acad Sci 112(1):172–177

Tran A, Gaidet N, L'Ambert G, Balenghien T, Balança G, Chevalier V, Soti V, Ivanes C, Etter E, Schaffner F, Baldet T (2007) The use of remote sensing for the ecological description of multi-host disease systems: a case study on West Nile virus in southern France. Veterinaria Italiana 43(3):687–697

Tucker CJ, Sellers PJ (1986) Satellite remote sensing of primary production. Int J Remote Sens 7(11):1395–1416

Turner W, Spector S, Gardiner N, Fladeland M, Sterling E, Steininger M (2003) Remote sensing for biodiversity science and conservation. Trends Ecol Evol 18(6):306–314

Tyagi BK, Hiriyan J (2004) Breeding of dengue vector Aedes aegypti (Linnaeus) in rural Thar Desert, north-western Rajasthan, India

Uysal M, Toprak AS, Polat N (2015) DEM generation with UAV Photogrammetry and accuracy analysis in Sahitler hill. Measurement 73:539–543

Vancutsem C, Ceccato P, Dinku T, Connor SJ (2010) Evaluation of MODIS land surface temperature data to estimate air temperature in different ecosystems over Africa. Remote Sens Environ 114(2):449–465

Van Handel E (1984) Metabolism of nutrients in the adult mosquito. Mosq News 44:573–579

Vanwambeke SO, Lambin EF, Eichhorn MP, Flasse SP, Harbach RE, Oskam L, Somboon P, Van Beers S, Van Benthem BH, Walton C, Butlin RK (2007) Impact of land-use change on dengue and malaria in northern Thailand. Eco Health 4(1):37–51

Viboud C, Pakdaman K, Boelle PY, Wilson ML, Myers MF, Valleron AJ, Flahault A (2004) Association of influenza epidemics with global climate variability. Eur J Epidemiol 19(11):1055–1059

Vrebos D, Beauchard O, Meire P (2017) The impact of land use and spatial mediated processes on the water quality in a river system. Sci Total Environ 601:365–373

Wang G, Minnis RB, Belant JL, Wax CL (2010) Dry weather induces outbreaks of human West Nile virus infections. BMC Infect Dis 10(1):38

Wanjala CL, Kweka EJ (2016) Impact of highland Topography changes on exposure to Malaria Vectors and immunity in Western Kenya. Front Public Health 4:227

Wehr A, Lohr U (1999) Airborne laser scanning—an introduction and overview. ISPRS J Photogramm Remote Sens 54(2–3):68–82

World Health Organization (2002) The world health report 2002: reducing risks, promoting healthy life. World Health Organization

Wilby RL, Hedger M, Orr H (2005) Climate change impacts and adaptation: a science agenda for the Environment Agency of England and Wales. Weather 60(7):206–211

Wilkinson, K., Grant, W.P., Green, L.E., Hunter, S., Jeger, M.J., Lowe, P., Medley, G.F., Mills, P., Phillipson, J., Poppy, G.M. and Waage, J., 2011. Infectious diseases of animals and plants: an interdisciplinary approach.

Woodruff BA, Toole MJ, Rodrigue DC, Brink EW, Mahgoub ES, AHMED MM, Babikar A (1990) Disease surveillance and control after a flood: Khartoum, Sudan, 1988. Disasters 14(2):151–163

World Health Organization (2014) A global brief on vector-borne diseases (No. WHO/DCO/WHD/2014.1). In: World Health Organization

World Health Organization (2018a) World Malaria Report 2018 [cited 2018 Dec 15]. World Health Organization, Geneva

World Health Organization. (2018b).Mosquito-borne diseases. Available online: https://www.who.int/neglected_diseases/vector_ecology/mosquito-borne-diseases/en/. Accessed on 20 Nov 2018

Wu X, Lu Y, Zhou S, Chen L, Xu B (2016) Impact of climate change on human infectious diseases: Empirical evidence and human adaptation. Environ Int 86:14–23

Wu X, Tian H, Zhou S, Chen L, Xu B (2014) Impact of global change on transmission of human infectious diseases. Sci China Earth Sci 57(2):189–203

Xiao H, Tian HY, Gao LD, Liu HN, Duan LS, Basta N, Cazelles B, Li XJ, Lin XL, Wu HW, Chen BY (2014) Animal reservoir, natural and socioeconomic variations and the transmission of hemorrhagic fever with renal syndrome in Chenzhou, China, 2006–2010. PLoS Negl Trop Dis 8(1):2615

Xie Z, Shi R, Zhu L, Chen X (2016) Comparison of pixel-based and object-oriented land cover change detection methods. International Archives of the Photogrammetry, Remote Sensing & Spatial. Inf Sci 41

Xu H (2006) Modification of normalised difference water index (NDWI) to enhance open water features in remotely sensed imagery. Int J Remote Sens 27(14):3025–3033

Xu B, Jin ZY, Jiang ZB, Guo JP, Timberlake M, Ma XL (2014) Climatological and geographical impacts on global pandemic of influenza a (H1N1). In: Weng Q (ed) Global urban monitoring and assessment through earth observation. Taylor & Francis/CRC Press, Florida, U.S.A

Yamazaki D, Ikeshima D, Tawatari R, Yamaguchi T, O'Loughlin F, Neal JC et al (2017) A high-accuracy map of global terrain elevations. Geophys Res Lett 4:5844–5853. https://doi.org/10.1002/2017gl072874

Ye-Ebiyo Y, Pollack RJ, Spielman A (2000) Enhanced development in nature of larval Anopheles arabiensis mosquitoes feeding on maize pollen. Am J Trop Med Hyg 63(1):90–93

Zafren K, Honigman B (1997) High altitude medicine. Emerg Med Clin N Am 15(1):191–222

Zevenbergen LW, Thorne CR (1987) Quantitative analysis of land surface topography. Earth Surf Pro Landforms 12(1):47–56

Zhanwei Du, Wu Y, Wang L, Cowling BJ, Meyers LA (2020) COVID-19 serial interval estimates based on confirmed cases in public reports from 86 Chinese cities. medRxiv

Zhu Y, Toth Z (2001) Extreme weather events and their probabilistic prediction by the NCEP ensemble forecast system. In: The 81st American Meteorological Society Annual Meeting, Albuquerque, NM. Available from www.emc.ncep.noaa.gov. Accessed on Jun (Vol. 19, p. 2010)

Chapter 7
GeoComputation and Spatial Modelling for Decision-Making

Abstract The extent to which computational modelling has influenced and been a part of decision-making in geography. Over the past decade, GeoComputing as the primary visualization tool has matured into a tool that a wider range of users can increasingly use. With the ever-increasing availability of spatial data, there has been a growing demand for translation into usable information which supports many aspects of spatial decision-making. The ability of GeoComputation-based decision-making to manage vast data sets and integrate satellite images improves the feasibility of researching disease transmission determinants and promptly provides data for monitoring and policy formulation. This chapter offers an overview of some of the methods used, as well as two examples of online GeoComputational mapping tools that support spatial mapping and modelling in the control strategy for epidemiology.

Keywords Epidemiology · Multi-criteria decision approach · Weighted linear combination · Analytical hierarchy process · Ordered weighted averaging · Hotspot analysis · Space scan statistics

7.1 Introduction

At the beginning of the twentieth century, 'epidemiology' was almost exclusively concerned with the investigation of epidemics of diseases such as smallpox, malaria, and typhoid fever which were transmitted by bacterial and virus infection. By 2050, the current 7.6 billion people worldwide are projected to hit 9.8 billion, with more and more people living in densely populated areas. The amalgamation of high population density and augmented human agility can lead to increased exposure to dangerous zoonotic diseases. About 60 per cent of the overall human health pathogens that have adverse effects come from animals, and a new evolving illness emerge about every 8 months (Jones et al. 2008). With each new introduction, 60–80% of known growing diseases in animals were found to be animal diseases (Morens and Fauci 2013). Nevertheless, technical advancements provide the ability to better understand disease transmission trends, underlying factors, and vulnerable

© The Author(s), under exclusive license to Springer Nature
Switzerland AG 2021
G. S. Bhunia, P. K. Shit, *GeoComputation and Public Health*,
Springer Geography, https://doi.org/10.1007/978-3-030-71198-6_7

population distribution. Tools to support spatial decisions will provide health officers with data, research, knowledge, capacity modelling, and visual tools in order to effectively make decisions and recommendations on policy for changes in public health outcomes (Beard et al. 2018). The literature review shows that opportunities, mistakes, and gaps in the development of such techniques have been missed.

The World Health Organization (WHO) describes public health surveillance as the continuous and systematic collection, review, and evaluation of health-related data needed for public health practice planning, implementation, and assessment (WHO 2005). A variety of monitoring objectives can be driven, including outbreak detection, situation recognition, and identification of long-term patterns and device target-specific statistical and analytical methods (Robertson et al. 2014). Many spatial-epidemiological studies are concerned with the determination of factors leading to the distribution of spatial risk. This group includes most models of regression using either the generalized linear modelling paradigm or Bayesian hierarchical modelling. A wide range of modelling techniques, such as Maxent (Phillips et al. 2004), random forests, and similar approaches (Breiman et al. 1984), and general additive models using a number of non-linear maps are now used in relation to landscape hazards (Hastie and Tibshirani 1990).

In the USA, the federal government and separate states track communicable diseases, but as a result of available data, resources, and training, they are constrained in their research by the increasing advancement and implementation of digital technologies. This advancement involves the implementation of digital disease monitoring systems like HealthMap (Brownstein et al. 2008), whereby current disease outbreaks are presented using online data collection, such as ProMED (Madoff and Woodall 2005), RSS feeds, and Twitter updates. Recently, public health authorities have started addressing disease surveillance through the integration into epidemic modelling and use of GeoComputations of spatial and temporal elements of reportable disease data (Joyce 2009; Bhunia et al. 2013). These include statistical tools and GIS tools for generating disease maps with a variety of data for clinical, physical, or human mobility data to classify disease outbreaks or clusters. Although GeoComputation and geostatistics have advanced conception and decision-making services for disease sensing, similar technological advancements in other IT-related fields such as spatial decision support system (SDSS) have been implemented and fused into integrated systems. SDSS originated from a wider application of public health, governance, and environmental management systems, focused on general decision support systems (DSSs). For instance, DSS may be used to assist medical staff in a variety of tasks such as treatment plans or anti-medicinal alerts Electronic health records (Fraccaro et al. 2015).

SDSS are computer-based schemes that permit policymaking to use data available in a more interactive, integrated interfacing mechanism that enables data processing, analysis, and visualization to solve spatially related issues (Eisen and Eisen 2011). Like Hopkins and Armstrong, Jerome Dobson coined the term 'spatial decision support network' in a sequence of conference proceedings from 1983 to 1985 (Dobson 1983). Attributes that discriminate SDSS from other similar systems, together with methods for supporting clinical decisions and GIS applications, are extracted from a dynamic decision-making process involving all users in many arenas. Enactment of these systems has the probability to allow decision-makers in the field of public health to identify a high-risk site for influenza epidemics or to allocate medical amenities, vaccinations, and personnel through the population distribution (Owens 2002). As technology and analytical tools have become more sophisticated, it has become increasingly possible for decision-makers to explore more fully accessible data to build successful, evidence-based planning.

Because of the scenarios of this complex type, 'semi-structured' space problems were referred to as they were not clearly specified and the methods used in the decision-making process were not consistent between investigations. Technology also focuses on the creation of methods that are too restrictive and mostly for a predetermined decision-making process. Many instances of the SDSS concept projects prior to the term exist, such as the GADS (Geodata Analysis and Display System) of IBM that allows users to analyse and view geographical data (Mantey and Carlson 1980). Over several years, however, SDSS has been in a transitional period where decision support systems and GIS technologies have become more coherent. During the 1990s, the number of architectural designs frequently produced using Esri's ArcGIS software increased. The first deployment of the modern SDSS took place in the early 2000s and continues today, taking advice from the global advances in web service, data storage, and advanced analysis (Sugumaran and Degroote 2010). Huang et al. (2012) developed a global disease outbreak risk assessment SDSS for airline travel. The purpose of this tool was to assist decision-making and easily quantify and compare the levels of risks to assist end users with priority routes for prevention of propagation or mitigation. In comparison, Bouden et al. (2008) built a framework that allows users to control outbreak model parameters such as the environment in order to help policymakers imagine the process of spread of infectious disease. The online group has started by making electronic data repositories of various states of the organization easier to collect diverse types of data for analytical purposes. The incorporation of these sources is important if all the factors leading to zoonotic diseases are to be explored. Recent SDSS has also made it possible to automate data collection integration from various related sources, including hospitals, laboratories, and physics studies.

7.2   GeoComputation and Decision-Making for Epidemiological Analysis

Decision-making is a major challenge in a complex and rapidly changing environment, as numerous factors can affect final decision-making, such as decision-makers, interest conflicts, decision-making importance, and the numerous criteria involved in the issue, among other things (Bhushan and Rai 2004). In the spatial context, decision-making is often dynamic and involves detailed knowledge from different sources that a wide variety of decision-makers interpret in relation to varying requirements, priorities, and/or alternatives (Sugumaran and Degroote 2010). Decision-making is complicated and difficult in the health sector as it requires numerous variables, choices, incomplete knowledge, and a range of desires (Marsh et al. 2017). Spatial decision support systems, which can identify priority areas in the geographical region through epidemiological steps and detailed knowledge of a disease, to avoid epidemiological outbreaks, are especially involved in epidemiological surveillance. Various strategies combined with new technology are being used to better target restricted monitoring and prevention at research levels like mapping, GeoComputation, decision support system (DSS), and spatial and temporal modelling. The ability of GeoComputation-based DSS to manage vast data sets and integrate satellite images improves the feasibility of researching malaria transmission determinants and promptly provides data for monitoring and policy formulation.

A collection of academic and government modellers has produced valuable studies in many of the recent communicable disease's public health crises, but only a few have been directly assigned and funded with organized functions. In public health crises, the reactive deployment (rather than prospective) of modellers and epidemiologists is no surprise. Models helped decision-makers to consider the likely outbreak course, the danger of international spread, and the effect of the action (Rivers et al. 2019). Through quantitative and statistical modelling, the use of vaccinations (Bellan et al. 2015), antivirals (Gani et al. 2005), and other countermeasures (e.g. school closures, social dissociation) may also be a valuable tool through tailored response needs of outbreaks. Nonetheless, little effort has been planned to enhance modelling optimization and further epidemic data collection during public health emergencies due to emerging infectious diseases, contrary to research and development of medical countermeasures. Compartmental modelling is the most common epidemiological modelling method. Compartmental models can be determinist or stochastic, and individuals may be categorized by illness or symptom. A complex compartmental malaria model that illustrates features such as human hosts' superinfector is taken into account by Alonso et al. (2019), and a pathogen transmission model using mosquitoes is used to discuss the effect of hurricanes on disease dynamics (Chowell et al. 2019).

In the past few decades, the availability of complex and accurate knowledge has improved disease modelling. Modern model maker like the susceptible-infected-removed (SIR) model and its comparison to the Indian plague have been developed for the purpose of capturing epidemiologic human pathogens by using the numbers or death figures observed during each individual event and have been focused on simplistic conventions about geographies, for example, host mixing within populations. Epidemiological data like seroprevalence surveys have guided many researches (Farrington 1990). In addition, a variety of large-scale data sets are redesigned to understand the transmission interactions. Mobile phone data, for example, has been used recently to identify mode of travel in order to inform epidemiological models (Wesolowski et al. 2014; Klepac et al. 2018). We have examples in this issue of different forms of data used for the guiding of model growth, likewise the practice of census information to forecast the appearance of imported pathogens, data on comprehensive transport to recognize the possibility of a major pandemic in current times, and data on animal migration to prevent pathogen propagation.

7.3 Multi-criteria Decision Approach (MCDA)

The decision-making process is characterized as a collection of procedures for decision-makers to consider possible options based on multiple criteria and establish an order of preference for alternatives. Multiple-criteria decision-making process (MCDA), a tool which is widely used in environmental, industrial, and business management, is focused on the field of organizational analysis. MCDA algorithms are available to evaluate different types of decision-making problem, including the positioning and comparison of alternatives based on several parameters, using both quantitative and qualitative indicators (Behzadian et al. 2010). Such approaches are intended to facilitate decision-making, to reduce the final decision-maker's obligations, and to ensure that a compromise is achieved according to the parameters concerned (Youngkong et al. 2012). Diaby and Goeree (2014) proposed that three phases (Fig. 7.1) were involved in MCDA:

  (i) Define the limits of the problem: A well-defined and explicit study should serve as a guide to the research itself if a central issue is to be predefined.
 (ii) Define criteria for assessment: The next step is the selection of criteria for test review – i.e. which criteria should be used to determine the study question.
(iii) Select a multi-criteria model: This model must be selected in view of the problem identified in stage I after the application methodology has been completed.

MCDA approaches are, therefore, entirely suitable to tackle dynamic, cross-disciplinary, and multisectoral decisions such as the management of zoonotic diseases (Mourits et al. 2010). Aenishaenslin et al. (2013) present the findings of a review of an MCDA approach to Lyme disease control in Quebec, Canada. The decision model, developed in Quebec under a new epidemiological framework, adapts and analyses Aenishaenslin et al.'s findings (2015), where Lyme disease has been prevalent for the past 30 years. The decision model is based in Switzerland. The adaptation model took place using a participatory approach with a group of Swiss stakeholders. For multi-criteria analysis, the PROMETHEE approach was used. The MCDA model defines and compares the main elements with findings of the Quebec model. Frazão et al. (2018) review and synthesize literary papers that concern MCDA in the healthcare field, assess general problems and methodologies, and organize them into one piece of research. Finally, the study of the problem showed that the MCDA has comprehensive application in the health sector to solve the problem of the rating.

7.3.1 Case Study: Risk-Zone Identification of Visceral Leishmaniasis Based on MCDA Approach

The transmission of kala-azar is hooked on environmental micro (physiographic and climate) and socio-economic variables just like other infectious or vector-borne diseases. Physiographical factors mainly contribute to the abundance of the vectors, while climate factors directly and through vector survival influence the extrinsic

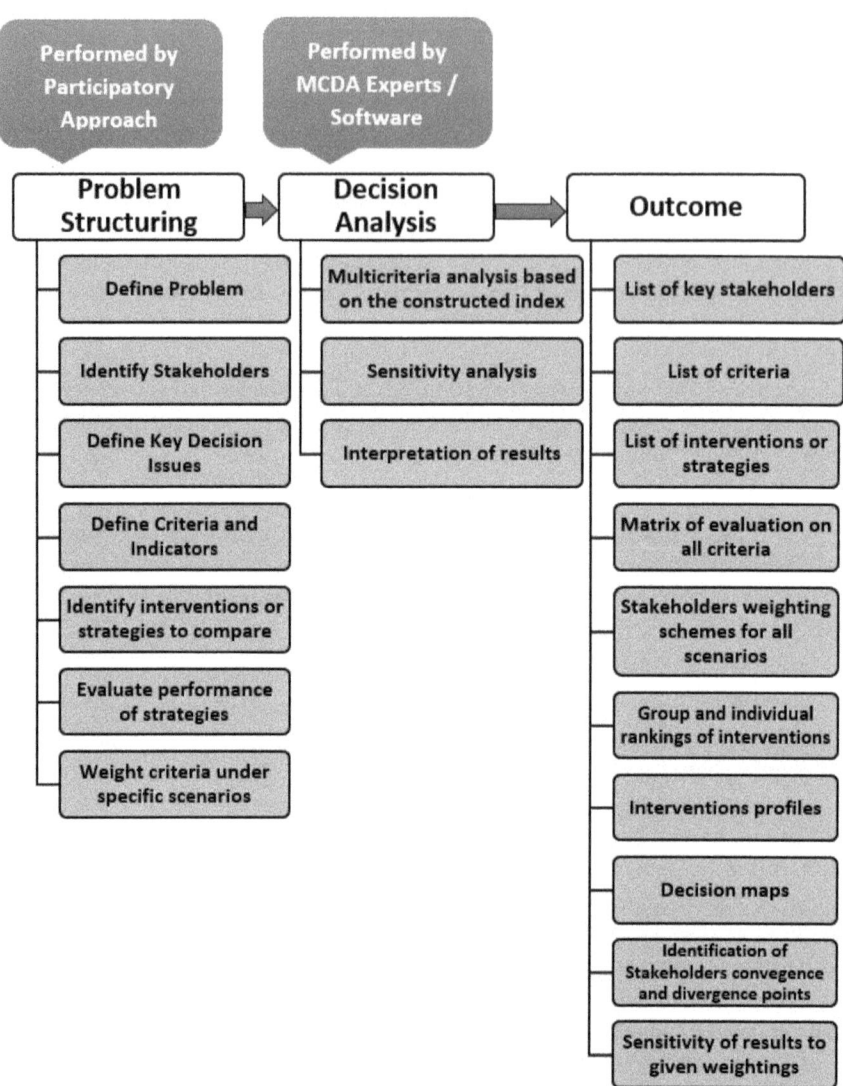

Fig. 7.1 Steps of multi-criteria decision-making approach (MCDA). (Modified after Aenishaenslin et al. 2013)

incubation (of parasites). Once environmental conditions are conducive to the occurrence of kala-azar, the chief determinants become the human (demographic) factors (Bhunia et al. 2012). The model adapted environmental and demographic factors that encompassed the series of dissimilar variables related to kala-azar transmission. The incidence of kala-azar is considered to be correlated with variables, for example, vegetation, soil moisture, land use/land cover humidity, temperature, and illiterate population and low living people either independently or in a group. All the nine variable datasets were used to compute a composite index, using weighted overlay function with QGIS software to produce the risk model kala-azar. The risk score was calculated using an integrated epidemiologic data analysis with the related factors which include the ranges of indoor room temperature (23–32 ° C), indoor room humidity (55–83%), NDVI (−0.023 to 0.54), wetness index (WI) (−70.56 to 31.04), vector density (2.35–6.96), and point prevalence rate per thousand population (0–10.24), in addition to the kala-azar risk index for illiterate and non-working population forms. In order to create an integrated analysis, this technique is typically used to put on a mutual measurement scale of values for different and individual inputs. To achieve this, scores (in the range/characteristics provided above) were given for each value of these risk variables to generate patterns over different geographic areas. The risk map (Fig. 7.3) was classified into various spatial entities, from 'very high risk' ('red') to 'very low risk' ('blue') units, according to risk values. The measured risk index values ranged between 23 per cent and 57 per cent according to region, with no kala-azar registered, case detection and Kala-azar control only in areas where the kala-azar risk index is greater than 37 per cent, the theoretical limit is that it is the maximum kala-azar risk index map values.

The kala-azar risk map generated (by using environmental, climate, entomological, epidemiological, demographic, etc.) shows a high promise of identifying areas at high risk for kala-azar outbreaks. However, it is not intended to estimate the prevalence of incidents, but simply to classify areas with a high-risk potential of kala-azar transmission at the micro level. The findings now show that areas with 'non-risk' kala-azar transmission can also be found at micro level using this risk model. The kala-azar risk map can be valuable to health authorities and policymakers who can use it to establish and plan the needs of kala-azar monitoring and the required mitigation programmes in the identified areas (Fig. 7.2). In order to continually track the condition of the kala-azar outbreak, evidence from the kala-azar risk map can also be worn as an administration controller.

7.4   Weighted Linear Combination (WLC) for Decision-Making

The WLC has two elements: the weight and value functions for each criterion. The weight is the value of each criterion, which is important for the decision-maker (expert). There must be one equal amount of weight. The WLC method has been consistently employed in the MCDM with GeoComputation, which is a simple, easy to understand process. It can be described as a technique used for the production of decision maps via spatial data and decision-maker preferences (Malczewski 2000). The weight is the value of each criteria in the decision-maker's understanding. The number of weights must suit one (Malczewski and Rinner 2015). The value

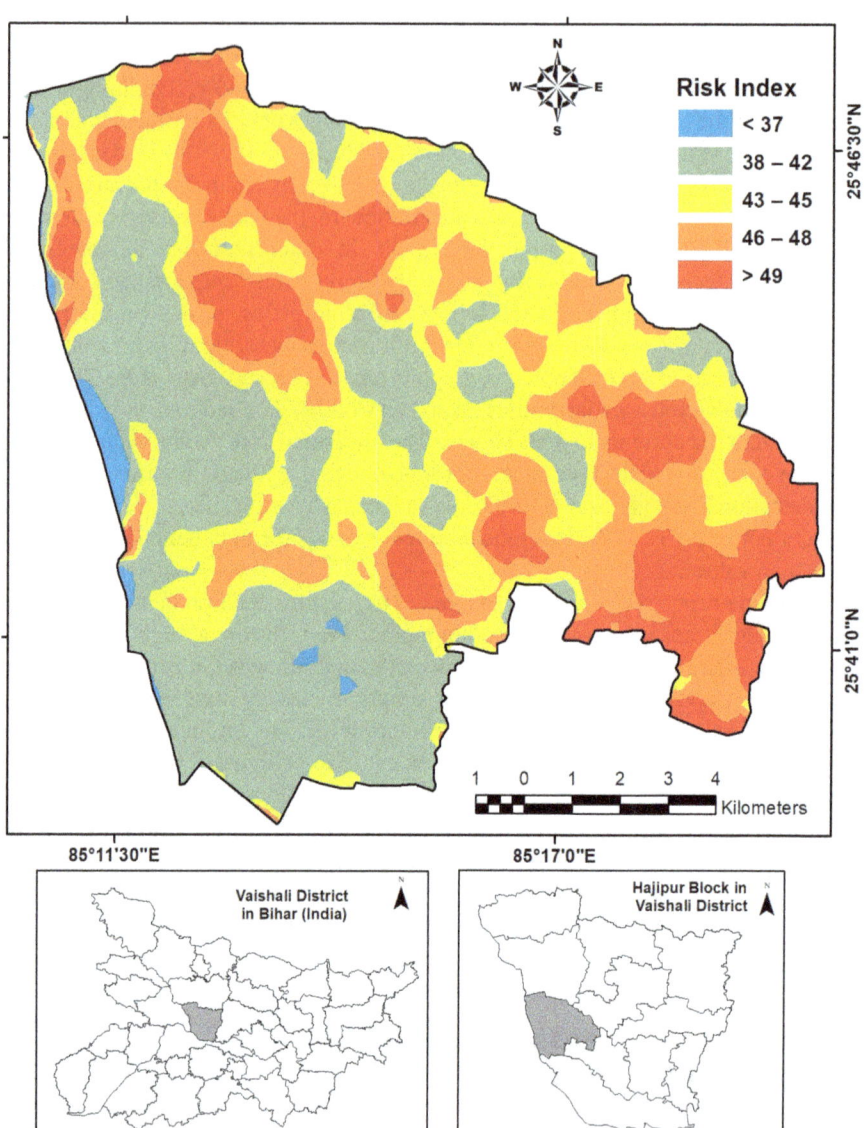

Fig. 7.2 Risk zone analysis of visceral leishmaniasis using multi-criteria decision-making approach

function transforms a criterion on a comparable scale at the different levels (Malczewski 2011). For each geo-object, the WLC method to select the best alternative can be expressed in the equation:

$$P*(a) = \max_i \left\{ \sum_{j=0}^{K} z_{ij}(a) w_j \right\}, \forall i = 1, 2, \ldots, m;$$

$$j = 1, 2, \ldots K \text{ and } a = 1, 2, \ldots, n.$$

where $P*(a)$ is the best score among the m substitutes for each geo-object a, z_{ij} is the value event of the ith alternative in terms of the jth criterion for each geo-object, and w_j is the weight attributed to each of the K criteria.

7.4.1 Case Study: Visceral Leishmaniasis Risk Analysis Based on Population Characteristics Using WLC Methods

In light of the demographic properties such as population density, family size, unemployed people, illiteracy rate, and agricultural and nutritional density of the area, the weighted linear combination model has been applied either individually or in combination with visceral leishmaniasis. A standard weighted linear blend approach is used to generate an integrated analysis for the common measurement scale of values for specific and unequal inputs (Bhunia et al. 2016; Malczewski 2000). In order to estimate their effect on VL transmission in this particular area, a relationship between population property and the average annual village-speaking VL incidence was calculated using the Spearman's rank correlation (rho). With <0.05 per cent, the study was done. Easy weightings and scores have been calculated for all input variables according to the relative value of each kala-azar transmission parameters. We used rating systems based on values from 1 to 5, where '5' means extremely suitable, '4' extremely suitable, '3' moderately suitable, '2' less suitable, and '1' much less/unsuited. Finally, a per cent influence is weighted on each input raster based on its importance to the rho value model. The total effect is equivalent to 100% for all rasters. Then, the cell values of each input raster are multiplied by the weight of the raster. In this weighted linear combination model, population density per km^2 has an influence of 10 per cent, family size has an influence of 30 per cent, illiteracy has an influence of 25 per cent, the unemployed population has an influence of 15 per cent, agrarian density has an influence of 10 per cent, and nutritional density has an influence of 10 per cent.

It is observed that with the rise in population density in the study area, the rate of occurrence is increasing. A quantitative analysis of VL distribution with respect to population density shows a positive association (rho = 0.37, p < 0.005). In this area, the family size ranges from 2.02 to 10.41, with a mean average size of 5.46. It is concluded, therefore, that the occurrence of VL increases with an increase in family size. In addition, the level of occurrence of VL in the study area shows a favourable and important relationship (rho = 0.48, p < 0.05) to the rate of illiteracy rate, suggesting that these two variables continue to develop together. The overlay analysis indicated that the rate of occurrence of VL is also increasing when the percentage of

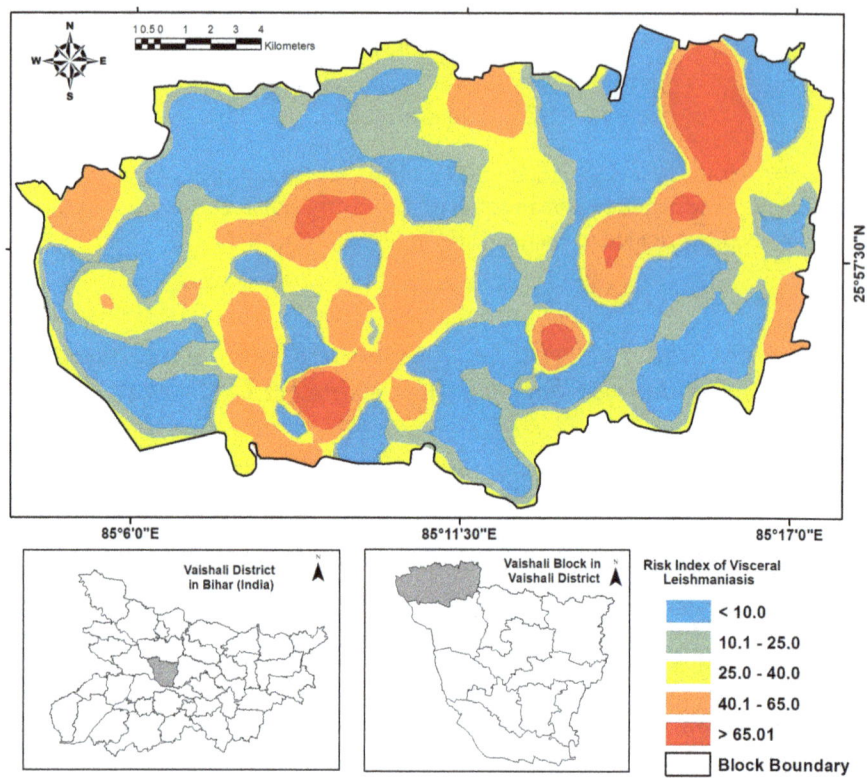

Fig. 7.3 Risk index analysis of visceral leishmaniasis based on population characteristics of Vaishali Block in Vaishali District of Bihar (India)

the unemployed population is rising. The strong positive relationship between these two variables was verified by a statistical study of the Spearman's rank correlation coefficient approach. A correlation coefficient (rho) of 0.47 was found with $p < 0.001$, indicating that the two variables tend to grow together. In addition, a simple linear relationship was formed between the villages with agricultural density and the annual incidence rate of VL cases and showed a significant relationship (rho = 0.30; $p < 0.003$). The risk index value has ranged from 10.01 to 65.0. Based on the geometric interval, the risk index value was categorized into (i) 'very high'-risk areas (risk index value >65.10), (ii) 'high'-risk areas (risk index values 40.10–65.0), (iii) 'medium-risk' areas (risk index values 25.01–40.10), (iv) 'low'-risk areas (risk index value 10.1–25.0), and (v) 'very low-risk' area (risk index value <10.1) (Fig. 7.3). The 'high-risk' areas on the map are shown in 'dark red' colour, while 'deep blue' displays the 'low-risk' areas. However, the high-risk areas were located mainly in the central and north-western part of the block and some small patches of the north-east portion. On the other hand, the block's southern, west, and northern portions showed a very low risk of transmission of VL.

7.5 Analytical Hierarchy Process (AHP)

AHP approach has been introduced into the GIS-based appropriate protocol (Saaty 1977; Marinoni 2004), which is a proven method for multi-criteria approaches. This approach can integrate both quantitative and qualitative knowledge on policy issues. In order to select alternatives that depend on their relative performance on over one value criteria, AHP is a versatile, quantitative approach (Boroushaki and Malczewski 2008). A comparison with the AHP approach enables the ratios of parameters:

Step 1 – define possible solutions: Define the full set of choices.

Step 2 – coordinate guidelines: Formulate policy requirements and ensure that the stakeholders recognize and appreciate the guidelines.

Step 3 – build surveys: Express sentiments in pairs and attain each criterion's weight.

Step 4 – gather input: Applicants complete surveys and ask participants who is a domain expert.

Step 5 – consistency test: Measure the consistency ratio, and verify that the allowances are included. The consistency ratio (CR) is always less than 0.1 in the AHP system.

Step 6 – identify group values: Following the similarities, discover shared values.

Step 7 – criterion weights: The values are extracted from a comparison of pairs, which may be the preferences of the experts involved in the decision.

Step 8 – identify alternatives, obtain access to the end results, and then allocate them according to measure.

Step 9 – the final results (Fig. 7.4).

There are the same number of rows and columns in the AHP comparison matrix where scores are noted on one side of the diagonal and standards of 1 are recorded on the matrix diagonal (Gorsevski et al. 2012). To make the assessment, Saaty (2008) indexes the scale of numbers that shows how many times one item dominates the other. The scale is shown in Table 7.1.

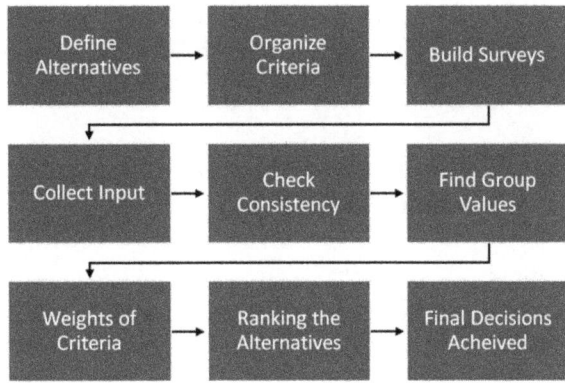

Fig. 7.4 AHP process for disease mapping

Table 7.1 Saaty (2008) index

Intensity of importance	Description	Explanation
1	Equal importance	Two actions make an equal contribution to the item
2	Weak to slight	
3	Moderate importance	Experience and decision support one operation marginally over another
4	Moderate plus	
5	Strong importance	Experience and decision support one behaviour strongly over another
6	Strong plus	
7	Very strong or demonstrated importance	An action is strongly favoured over another; its predominance proved in practice
8	Very strong	
9	Extreme importance	Evidence promoting one operation over another is of the highest order of affirmation possible
Reciprocal	Values for inverse comparison	A reasonable assumption

If the matrix of comparison is not totally consistent or contradictory, then various situations tend to be resolved; one acceptable calculation defined by Saaty (1977) is measured accordingly:

$$CI = \frac{\lambda_{max} - n}{n - 1}$$

where λ_{max} is the largest value of its own and n is the sum of alternatives equated. Sometimes the judge is not entirely steady; in order to quantify this inconsistency, Saaty established the consistency (CR) ratio:

$$CR = \frac{CI}{RI}$$

This value can be derived from a lookup table where RI is random index. Saaty had needed acceptability of only the CR value below 0.1.

7.6 Technique for Order of Preference by Similarity to Ideal Solution (TOPSIS)

TOPSIS is a key MCDA tool that is commonly used in various decision-making issues, originally designed by Hwang and Yoon (1981). TOPSIS's basic definition includes the thorough geometric aloofness chosen between the substitute solution and the positive ideal solution (PIS) and the longest geometric distance between the ideal alternative and negative ideal explanation (NIS). TOPSIS is essentially a utility-based approach that explicitly compares each alternative. It depends on the data in the matrices and weights of the assessment (Cheng et al. 2002). TOPSIS consists of defined decision frame, decision matrix standardization, calculation of space, and integrating operations. The procedures of TOPSIS are as follows:

Step 1: The performance matrix is formed; we have a matrix $(X_{ij})_m$ portion of each alternative and the criterion given as X_{ij} to form an appraisal matrix containing of m substitutes and n criteria. The matrix structure can be expressed accordingly:

$$D = \begin{array}{c} \\ A_1 \\ A_2 \\ \vdots \\ A_m \end{array} \begin{array}{cccc} C_1 & C_2 & \cdots & C_n \\ \left[\begin{array}{cccc} x_{11} & x_{12} & \cdots & x_{1n} \\ x_{21} & x_{22} & \cdots & x_{2n} \\ \vdots & \vdots & \ddots & \vdots \\ x_{m1} & x_{m2} & \cdots & x_{mn} \end{array} \right] \end{array}$$

Step 2: Regularize the decision matrix, using the normalization scheme:

$$r_{ij} = \frac{x_{ij}}{\sqrt{\sum_{i=1}^{m} x_{ij}^2}}, i = 1, 2, \ldots, m, j = 1, 2, \ldots, n$$

Step 3: Allocate a weight vector to the group feature, and create the normalized weighted decision matrix:

$$v_{ij} = w_j r_{ij}, i = 1, 2, \ldots, m, j = 1, 2, \ldots, n$$

Step 4: Prevent the perfect positive and perfect negative solutions. Ideal approach for positive effects:

$$A^* = \left\{ V_1^*, \ldots, V_n^* \right\}, V^* = \left\{ \max \left(V_{ij} \right), j \in J; \ \min \left(V_{ij} \right), j \in J' \right\}$$

Negative ideal solutions:

$$A^- = \left\{ V_1', \ldots, V_n' \right\}, V' = \left\{ \max \left(V_{ij} \right), j \in J; \ \min \left(V_{ij} \right), j \in J' \right\}$$

Step 5: Determine the measure of separation for each alternate distance (measure the distance):

$$S_i^* = \sqrt{\sum_{j=1}^{n} V_{ij} - V_j^*)^2}, i = 1, \ldots, m$$

Separating the negative ideal alternative is:

$$S_i^- = \sqrt{\sum_{j=1}^{n} V_{ij} - V_j')^2}, i = 1,.....,m$$

Step 6: Measure relative proximity to the ideal C_i^* solution:

$$C_i^* = S_i' \left(S_i^* + S_i^- \right), 0 < C_i^* < 1$$

Step 7: The preference order ranks, the higher the index value, the better the substitute, the higher a C_i^* value below 1.

7.7 Ordered Weighted Averaging (OWA)

The ordered weighted averaging (OWA) is a class of Yager (1988) multi-criteria accumulation or combined weights which involve two sets of weights: weighing criterion weights and ordered weights. W_j criterion weight is allocated for all sites to the jth criterion index, which indicates that the assessment criteria have a relative value in accordance with the decision-maker's penchants. The well-organized weights are linked to the location-to-location criteria values. They are applied to the attribute value of a given position in a decreasing order without considering the map value of the criterion. For the OWA mix operators, the ordered weights are essential. As part of the GIS-IDRISI decision support module, Eastman (1997) and Jiang introduced OWA operations.

The OWA appeal is to reorder and adjust criteria parameters so that different maps and scenarios can be created. In comparison to Boolean overlay AND operator, which is the lowest risk, OWA approach involves a broad range of risk scenes between the ANDs and ORs (Gorsevski et al. 2012), whereas OR operators are highest-risk decision-making. Yager (1988) suggested OWA as a family of amalgamation machinists that was parameterized. The OWA function defined as follows for a set of n criterion maps formally an OWA operator of dimension n is a mapping F: Rn as R which has an allied array of methodical weight W = [w_1, w_2, ..., w_n] deceitful in the unit recess and summing to 1, given the set of feature value a_1, a_2, ...,a_n and the OWA calculation is as follows:

$$F\left(a_1, a_1, ..., a_n\right) = \sum_{j=1}^{n} w_j b_j$$

where b_j is the a_i's jth largest. For example, the first-order weight, w_1, is assigned to the highest suitability attribute value with a particular location of the attribute values; w_2 is allotted to the second main fittingness attribute value, while w_n is assigned to the lowest aptness characteristic. Yager (1988) introduced two characterizing features relating to OWA operator's weighting vector w. The first of these is the attitudinal character (ORness), defined as:

$$A - C(W) = \frac{1}{n-1} \sum_{j=1}^{n} (n - j) w_j, A - C(W) \in [0,1]$$

Additionally, A–C(max) is 1, A–C(med) is 0.5, and A–C(min) is 0. Therefore, the A–C shifts from 1 to 0 as we switch from an average of Max to Min. The attitudinal character exemplifies the resemblance between accumulation and OR operation, defined by OR as the Max. The second characteristic is the diffusion. This would be demarcated as:

$$H(W) = -\sum_{j=1}^{n} w_j \ln(w_j)$$

This importance characterizes the degree of the used arguments.

This can be done by changing the ordered weights, and a continuous aggregation process will produce what is in return. The weights ordered are $W = [w_1, w_2, ..., w_n]$, where w_n epitomizes the rank ordered. When $w_{min} = [1, 0, ..., 0]$ is AND (and ness) operator. It is OR Operator (ORness) if $w_{max} = [0, 0, ..., 1]$. It has the similar explanation as the WLC (weighted linear combination) when $w_{mean} = [1/n, 1/n, ..., 1/n]$. The degree of weight dispersion gives TRADE-OFF point, which reflects the calculation of compensation. The deciding trade-off was:

$$TRADE - OFF = 1 - \sqrt{\frac{n \sum_r \left(w_r - \frac{1}{n}\right)^2}{n-1}}$$

where n is the number of criteria, r is the order of criteria, and w_r is the weight of the criteria for the rth order. The OWA combination operator in GIS-OWA model is associated with the location of jth and a set of order weights $V = (v_1, v_2, ..., v_3)$, where $v_j \in [0, 1]$ for $j = 1, 2, ..., n$ and $\sum_{j=1}^{n} v_j = 1$. The GIS-OWA method had been described by Malczewski (2006) as:

$$OWA_i = \sum_{j=1}^{n} \left(\frac{u_j v_j}{\sum_{j=1}^{n} u_j v_j} \right) z_{ij}$$

where z_{ij} obtained decreasingly by rearranging the criterion weights and u_j is the rearranged weight of the jth criterion. In this analysis, the OWA order weights are determined using equation based on Yager (1996):

$$W_j = \left(\frac{j}{n}\right)^\alpha - \left(\frac{j-1}{n}\right)^\alpha$$

Table 7.2 The decision strategies corresponding to specific ORness value (Jelokhani-Niaraki and Malczewski 2015)

α	ORness	Aggregation strategy	Decision strategy
0	1	Logic OR(MAX)	Extremely optimistic
0.1	0.9		Very optimistic
0.4	0.7		Optimistic
1	0.5	WLC	Neutral
2	0.3		Pessimistic
10	0.1		Very pessimistic
∞	0	Logic AND(min)	Extremely pessimistic

where α is the degree of optimism which specifies the policy for decision. α is ORness connected as the following calculation:

$$\text{ORness} = \frac{1}{1+\alpha}, \alpha \geq 0$$

The decision strategies corresponding to specific ORness value is shown in Table 7.2.

7.8   Spatial Modelling in Hotspot Assessment

There has been growing awareness of the importance of spatial clusters or 'hotspots' in epidemiology of infectious diseases, and targeting hotspots also is seen as an essential component of disease control strategies. Hotspots have been identified in many ways, including high incidences and/or prevalence, increased transmission or risk, or increased possibility of disease. Hotspot may also be a region at high risk for the emergence or reappearance of infectious diseases (Preston 1995). In books such as *The Hot Zone*, this growing usage caught the imagination of people. The roots are biological, which defines a high biodiversity area hotspot (Myers et al. 2000).

The public health researchers and policymakers have been interested in spatial clusters or 'hotspots' of disease and activity in conjunction with health. A disease hotspot is a geographical structure, which is characterized as an infrequent number of cases in populations, locations, and times, and can be detected, viewed, and examined through GIS and spatial analysis methods (Table 7.3). Hotspots can give researchers clues to the aetiology of disease and risk behaviour, signifying local ecological or social physiognomies which endorse risks. An effective public health action approach is likely to be tailored specifically for policymakers and planners. Over these last 20 years, the approaches for perceiving hotspots of disease have progressed suggestively to integrate more vigorous statistical and geographical

Table 7.3 Overview of definitions for different types of 'hotspots'

Suggested and/or alternative term of hotspot	Definition
Transmission hotspot, transmission foci	An area of elevated transmission efficiency(i.e. elevated reproductive number, R)
Emergence hotspot	An area with a high frequency of emergence or re-emergence of diseases or drug-resistant strains
Burden hotspot, high burden, hyperendemic region, case cluster	An area of elevated disease incidence or prevalence or a geographic cluster of cases

Source: Lessler et al. 2017

exploration formulations needed to accurately detect hotspots; the principles that are needed for aetiology and disease transmission are, however, given less attention.

In the 1970s and 1980s, interest in studying and identifying local space clusters of illness arose in reaction to two major interrelated social movements: increasing consciousness of the health effects of environmental pollution and swelling public concern regarding communities of disease. Officials and scientists of public health required accurate mappings resources in order to assess if risk in particular areas was higher (Neutra et al. 1992). At the same time, the advances in the GeoComputation techniques offered opportunities to perform large-scale spatial analytical research and for data on complex social, demographic, and environmental patterns to layer spatial information on the occurrence of disease. Such cross-linked socio-political and technological developments drive an active, multicultural area of research on the detection of space clusters that continues today.

Usually, two interrelated components are methods for detecting local spatial clusters: first, a geography search tool used to classify the local disease concentrations to be clustered. The search strategies comprise field-based, site-side, and case-side approaches that periodically scan the research area to look for local clusters and artifacts by creating clusters around cases of disease or units comprehending case rates (Cromley and McLafferty 2011). Geographical search also includes decisions about the extent and the search window/filter settings for local clusters. A variety of methods use basic geometrical procedures such as circles or ellipses, while other methods use an established set of weights to detect the area in clusters; hotspot calculation methods are based on a mathematical model to assess whether or not the concentration of the local disease is exceptional. Frequentist and Bayesian methods for cluster testing have been developed. The test element may also contain spatial and GIS models that have specific features and/or local population and environment that can influence the development of the clusters. Throughout the process of spatial search as well as statistical hotspot identification, variables such as unequal population distribution, transport networks, and topography may be integrated.

Geographers and statistics were able to determine early methods of the identification clusters of local diseases, such as GAM (Openshaw et al. 1988), Besag and Newell (Besag and Newell 1991), disease mapping and analysis (DMAP) system, Rushton and Lolonis (1996), spacious scan (Kulldorff 1997), and Fotheringham and Zhan methods (Fotheringham and Zhan 1996), and local indicators of spatial autocorrelation, such as local Moran's I and $G*$ (Anselin 1995). The technique is usage in clinical research, most commonly used in cancer survival (Wan et al. 2015), medication-based prescribing (King and Essick 2013). To classify hotspots of tuberculosis (TB) and better understand their relationship with social and economic influences, Gehlen et al. (2019) used an adaptive quadratic radial kernel algorithm. TerraView 5.3.2 (National Space Research Institute – Instituto Nacional de Pesquías Espaciais (INPE)) was used as the framework for this project. The adaptive quadratic kernel takes into consideration the distance variation and the field, according to the developers of the programme. Therefore, a distance threshold or a number of neighbouring should not be predefined, as this is the algorithm itself. In addition, the Department of Public Health and current disease control programs have implemented these approaches. The large usage in research and policy fields of these approaches calls for replication on current methodological advances and upcoming instructions.

Over the previous couple of decades, the application of hotspot study both in public health and epidemiological investigation and in other fields (for instance, a large part of hotspot analysis literature comes from disease or environmental mapping and study), primarily because of the implementation of GeoComputation techniques, has developed significantly. The hotspot analysis consists of the following data type:

- Points – Positions of objects/events in a sample (e.g. violent, crime, earthquakes, avian flu, etc.)

Depending on the query, the hotspot analysis can include the succeeding information type:

- Attributes – Unconditional or incessant variable describing objects/events in more details
- Period – Periods or times of occurrences
- Other covariates – Other explanatory variables

7.9 Hotspot Analysis Techniques

Spatial pattern and hotspot analysis, including spatial autocorrelation and cluster analysis, can be analysed in different ways. Specific mapping techniques are used to display hotspots, including lines, space ellipses, thematic and quadrats, interpolation, and estimate of kernel density (Chainey et al. 2008). The spread of isolated spatial (i.e. entity or event) anomalies is shown in point mapping using similar points. The hotspot is a dot at a given address in this case. Identifying clusters as points in a field can be difficult by witnessing single events since it depends on the observer's visual perception. Spatial ellipses show hotspots on the map with standard deviational ellipses around dot clusters. These are described as below.

7.9.1 *Nearest Neighbour Index (NNI)*

The NNI calculates the average distance to the nearest neighbour from each point. The index is the relation between the measured distance and the projected distance (a hypothetical allocation of the same number of features, casing a similar total area is assumed to be used). We may contemplate the points grouped if the NNI is below 1. When the NNI is greater than 1, we assume that the pattern of the spot distribution is uniform (NIJ 2005).

Figure 7.5 represented the calculation of the nearest neighbour index based on the average distance of HIV cases from each district to its nearest neighbouring district in 2019. The observed mean distance is calculated as 39895.24 m, and the expected mean distance is calculated as 25067.02 m and is calculated based on the Euclidean distance method. The calculated nearest neighbour ratio is 1.59 with z-score of 6.97 and *P-value* < 0.0000.

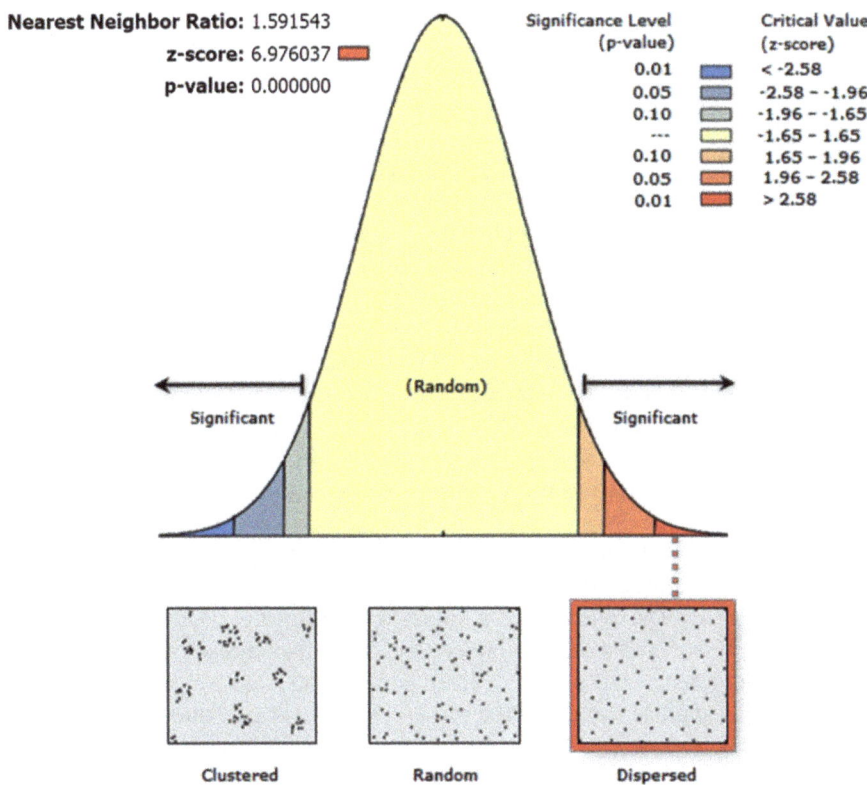

Fig. 7.5 Average nearest neighbour analysis of HIV cases of Bihar. (Data collected from Bihar State Health Society, 2019). Results also shows given the z-score of 6.97603681016, there is a less than 1% likelihood that this dispersed pattern could be the result of random chance

7.9.2 Spatial Autocorrelation

Global *Moran's I* and *Geary's I* are two figures estimating the total spatial self-relation degree. In other words, they calculate the tendency of clustering events or the degree to which near points have average similar values to those that are more distant. These two figures calculate local spatial self-relationships, with versions. They vary in that they refer to each spatial analysis unit from the global statistics. The points must be aggregated by placing a structure on databases (grid or geographical unit) that limits the number of neighbours to be considered before the Moran's can be determined. This is done to measure a matrix of weight. This matrix may be a measure of the cell contiguity or a distance-based weight.

Moran's values ranges from -1.0 to $+1.0$, in which a value of zero is viewed as an RSO. A greater than zero value is considered a positive spatial self-relation, and a less than zero value are considered a negative spatial autocorrelation (NIJ 2005). *Geary's I* is identical to *Moran's I* except for the product word in the numerator, which implies that the intensity value difference at each point is compounded over the intensities (e.g. the intensity difference with each other at each measurement location). The values of this formula range from 0 to 2, where a value of 1 is assumed if any position has no relation to other places. A value below 1 implies positive spatial self-correlation, while a value below 1 is a negative spatial autocorrelation (NIJ 2005).

Figure 7.6 represents the spatial cluster and outlier type (COType) analysis of tuberculosis in Bihar State (India) based on Anselin Local Moran's Index statistics during the period between 2017 and 2019. The test was conducted on the basis of contiguity boundaries, where the only computations of the target polygon feature are neighbouring polygon features that share a border or overlay. The study actually decides whether the apparent similitude (a high- or low-value spatial cluster) or dissimilarity (a spatial outer surface). The COType analysis would show whether the function has a high value and is surrounded by low-value characteristics (HL) or if it is small in value and is surrounded by high-value disease incidence (LH). The z-scores and p-values are statistically significant measurements. An elevated positive z-score for a feature indicates similar values (both high and low) in the corresponding functions.

The statistics for the high HIV district incidence (hotspots) and the low HIV (coldspot) spatial clusters using *Getis-Ord* G_i^* statistics (Getis and Ord 2010) are shown in Fig. 7.7. The z-score and p-values are statistically valid indicators that differentiate between districts, to reject the zero hypothesis or not. For the corresponding z-notes and p-values, districts for high or low values are spatially clustered. This approach works in the sense of surrounding features by looking at each area. A high-value district is important, but may not be statistically relevant. A function is high in value and is also surrounded by other districts with high values to be a statistically important hotspot.

Distance measurements are used in this study of district boundary centroids. In this analysis, the nearest neighbour distance is used in which neighbouring

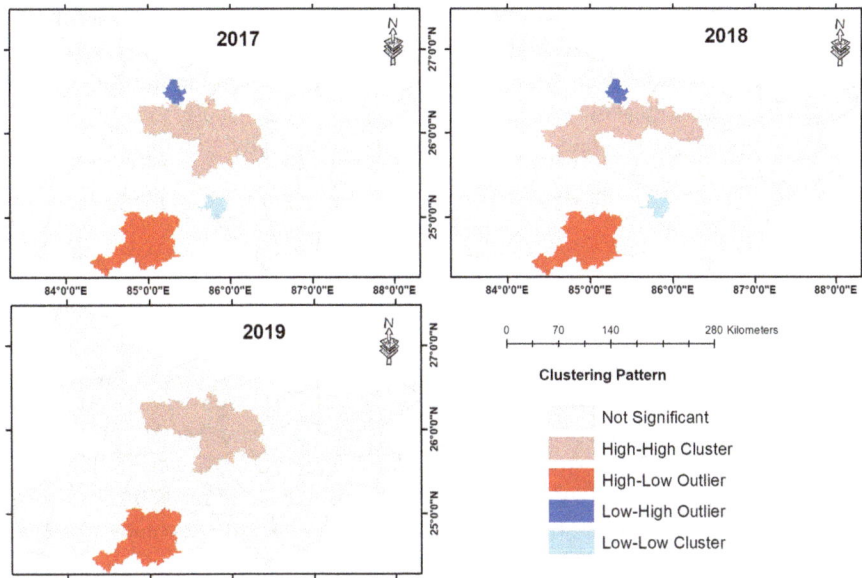

Fig. 7.6 Spatial cluster and outlier analysis of tuberculosis (TB) incidences (2017–2019) of Bihar using Moran's I index. (Data source: Bihar State Health Society, Bihar, India)

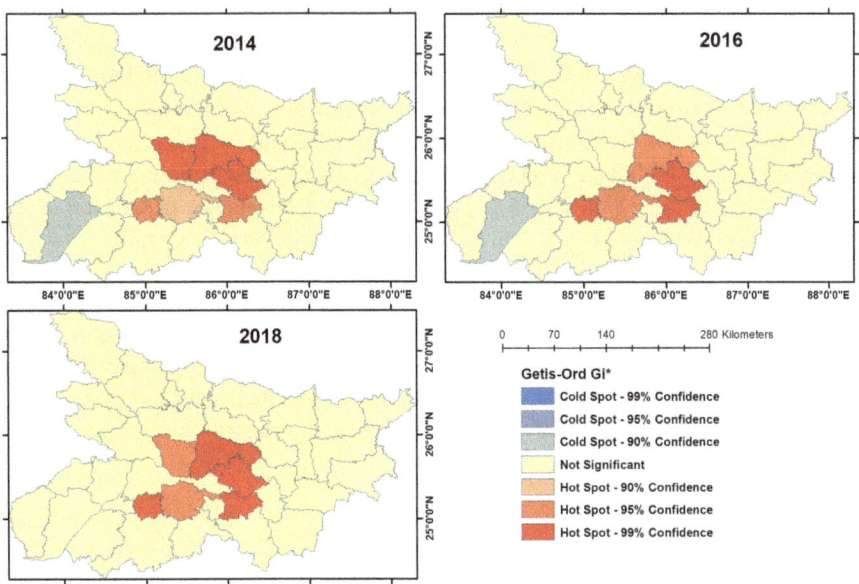

Fig. 7.7 Hotspot analysis of HIV incidences during the period between 2014 and 2018. (Data Source: Bihar State Health Society, Bihar, India)

characteristics have a greater effect than features far away on calculations for a destination. For the increasing function of the dataset, the given G_i^* statistic is a z-point. The larger the z-score, the more intense the clustering of high values (hotspot) is for statistically significant positive z-score tests. The lower the z-score, the more severe the clustering of low values (coldspot) is for statistically significant negative z-scores.

7.9.3   Kernel Density Estimation (KDE)

KDE produces a smooth, continuous surface map, displaying variation gradients across the study areas without limiting them to themes. The result of non-parametric intensity estimations in cell grids is a weighting function centred on the surface area that is generated by a persistent bandwidth or search radio (Waller and Gotway 2004; Chainey et al. 2008). Although KDE is a more versatile method for the display of hotspots on a globe, it still has the same drawbacks as thematic mapping since statistically significant hotspots and coldspots cannot be defined.

7.9.4   MapReduce-Based Framework

The MapReduce-based framework has been developed by researchers at the Center for Visual and Decision Informatics (CVDI). They are focused on modern big data architecture. Conventional methods of hotspot detection use interpolation which appears to include non-hotspot areas. That can lead to difficulties. The groundbreaking computational CVDI hotspot tool can detect hotspots with a substantial reduction in false positives in a spatial/temporal sense – a major advantage (Raghavan et al. 2016). CVDI researchers broadened the algorithmic approach by creating a variant of MapReduce-based distributed polygon propagation. This method is based on the MapReduction implementation of a polygon propagation approach for the identification of compact, regionally adapted hotspots. Propagation of polygons is computer-costly. MapReduce is a model for the simultaneous treatment of large data sets on a group of commodity machines. The MapReduce algorithm will reduce the runtime by 90% compared to serial implementations during analytical evaluations.

7.9.5   Scan Statistics-Based Techniques

The scan statistics has become one of the most common tools for detecting clusters of diseases and is commonly used by many departments and researchers of public health (Kulldorff 2001). In order for the spatial clusters to be evaluated and the position of the estimated clusters to be identified (Kulldorff 1997), the number of cases, for example, accidents, can be expected to be distributed between Poisson and Bernoulli. Spatial scan statistics that rely on the availability of data for the collection or for particular cases of precision geographic coordinates are given because each area is at risk for only one person. Algorithmically, a circular window on the studied map is placed by the spatial scan method and allows the centre of the circle to travel across the region such that the window contains various sets of adjacent cases in different places. The spatial scan statistic S is defined as the highest probability ratio over all possible z circles, based on the total number of cases observed:

$$S = \frac{\max\limits_{z} L(Z)}{L_0}$$

where L(Z) is the likelihood to circle Z (the probability of observed data is implied, providing the differential event rates both inside and outside the zone) and where L_o is a null hypothesis probability function. Because this likelihood ratio is maximized in all cycles, the circle that is most likely clustered is defined. Prospective space-time models for online surveillance depend on mathematical principles and techniques (Sonesson and Bock 2003). Statistical surveillance means electronic monitoring of a stochastic process to identify a 'major' process changes as quickly and accurately as possible at an unknown time point. A broad variety of approaches for surveillance, including multiple composite forms (Jacquez et al. 1996) and different statistical scans (Kulldorff 2001), has been proposed for public health surveillance in several different disease circumstances.

7.9.5.1 Case Study: Hotspot Cluster Detection Using Scan Statistics

Spatial scan statistics were used for the identification in various areas, including astronomy, bio-surveillance, natural catastrophes, and forestry, of hotspot clusters (or coldspot clusters). This method is based on the concept of defining a related geographical subset that maximizes probability in the entire field of analysis (Ishioka et al. 2019). We only consider this cluster of all patterns of linked regional subsets to identify a hotspot cluster which actively agglomerates high-risk regions so as to ensure maximum probability. Recently, the spatial scan statistics (Kulldorff 1997) and the publicly available SaTScan™ applications (Kulldorff 2018) were commonly used for cluster identification.

Suppose a research area is broken down into m areas. This approach is designed to find a linked regional subset, Fenster Z, which can be a space cluster. The cluster is based on an LR of zero probabilities with an alternative probability model. Here, we use for the Poisson model-based spatial scan statistics consisting of the number of cases detected and predicted. The hypotheses for the presence of a hotspot cluster that is more pronounced than expected are described as:

$$H_0 : \text{Expectation of } O(Z) = E(Z)$$

$$H_1 : \text{Expectation of } O(Z) > E(Z)$$

where if the random cases and the estimated number of cases are defined by $O()$ and $E()$ within the specified window Z, respectively. (The sign of uniformity of H_1 can be reversed for a coldspot detection.) Alternatively, the number of the observed region i can be indicated under null hypothesis of no clusters in the research field, expressed as:

$$O_i \sim \text{Poisson}(E_i), i = 1, 2, \ldots, m$$

where O_i and O_j are independent ($i \neq j$). The LR statistics are calculated by the location and size of each scanning window, expressed as:

$$LR(Z) = \left(\frac{o(Z)}{E(Z)} \right)^{o(z)} \left(\frac{o(Z^c)}{E(Z^c)} \right)^{o(Z^c)} I(o(Z) > E(Z)),$$

where $o()$ refers to the number of cases observed in the Z window specified and Z_c refers the entire region outside Z. $I()$ is the function indicator. The LR logarithm (LLR) is commonly used instead of the ratio itself for statistical simplicity. The hotspot cluster is described in this article as a Z window with maximum LLR.

Mosha et al. (2014) identify the hotspots of *Plasmodium falciparum* malaria infections to allow interventions and to reduce malaria burden more rapidly in the population. A wide radius cluster of 2.88 km covering 141 families and a small radius of 0.1 km covering 5 households was identified using SaTScan analysis for the identification of nPCR hotspots (Fig. 7.8). SaTScan research found that in the first-year people living in a nPCR cluster had a risk of testing nPCR in the second year for malaria positive four times than people living in NPCR colds.

As a second scan process, echelon scan is introduced for the determination of a hotspot by moving the scan window. Echelon (Kurihara 2004) analyses divide the field of study into structural cells that support bases or foundations. For example, suppose ten regions from (A) to (J) were tagged as illustrated in Fig. 7.9 [left], which holds its own value in each area.

7.9.6   Risk-Adjusted Nearest Neighbour Hierarchical (RNNH) Clustering

RNNH is based on the popular approach of the nearest neighbour hierarchy (NNH) (Jain et al. 1999) combining the power of a cluster with interpolating kernel density techniques. RNNH shall mainly be designed to identify baseline data point clusters.

Fig. 7.8 Location of the study site within Tanzania (inset map) and clustering of malaria infection using SaTScan (coldspot significantly lower infection, hotspot significantly greater infection). (Source: Mosha et al. 2014)

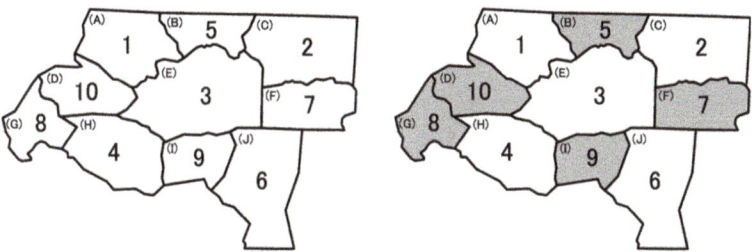

Fig. 7.9 Evaluation of hotspot cluster detection using spatial scan statistic based on exact counting. (Source: Ishioka et al. 2019)

This adapts the threshold distance algorithmically to a calculation of the density of the baseline factor in a variable manner (e.g. in places where the population is high, the threshold will be lower). Such density measurements are based on the distances between the site and any or all other data points by means of the density of the kernel. The main steps of the RNNH method are summarized below.

Define a grid over an area of interest; calculate the density of the kernel of the basis points for each cell of the grid. Measure the threshold distances between data points for hierarchical grouping purposes using the following formula:

$$0.5\sqrt{\frac{A_i}{N_i}} \pm t\frac{0.26136}{\sqrt{\frac{N_i^2}{A_i}}}$$

where A_i indicates the range of grid cell i, N_i indicates the average number of data points in cell row i, and t indicates the random distribution on the basis of estimates of the kernel density of the null point distance.

7.9.7 Support Vector Machines (SVMs)

SVMs are the most common algorithms of a class using the node substitution principle (Bennett and Campbell 2000). Motivated by the principle of statistical analysis, SVMs are systematically designed to optimize and have no local minima for
complicating the learning process. Informally the following procedures are followed: Firstly, the input data are implicitly mapped to a high-dimensional kernel
function space (typically the Gaussian kernel $K\left(x_i, x_j\right) = e^{-q\|x_i - x_j\|^2}$ with width
parameter q). Secondly, in the function's region with the lowest range that contains
most of the data, these methods consider a hypersphere. To allow data sets outside
the sphere, slack variables can be added. The problem with finding this hypersphere
can be formulated as a quadratic or a linear system depending on the distance function used. Thirdly, the kernel function is built to estimate the support for the underlying data distribution, and parameters are learned in the second stage.

7.9.8 Spatial Contact for Risk Analysis

Spatial interaction usually happens between two people, whether in an open
environment or in an enclosed room, where they can touch each other directly or
indirectly quickly or easily. For example, if an infected person coughs or passes in
a bus, microorganism droplets will reach the body of another person, allowing a
disease to spread. This concept makes space interaction between humans one of the
most important factors in the transmission of most diseases (Perez and Dragicevic
2009); the integration of interaction patterns into epidemic modelling will contribute
to a deeper understanding of potential epidemic transmission trends within a
sensitive population (Mossong et al. 2008). A key role can be played by the spatial
aspect of the health data in explaining the heterogeneity in risk because of the
various types of health condition, environmental risks, population numbers, and
demographic and socio-economic characteristics (e.g. vulnerability and exposures)
that vary in space. This is an important contribution to the study and understanding
of connections between geography, climate, and human health in the sense of
epidemiology and public health. Geographical information science and
GeoComputation are now widely used for exposure assessments. In conjunction
with Tobler (1970), GIS may be used for simple spatial analysis, where all objects
are more connected than distant ones that are more related, or it can also be applied
to data algorithms for data analysis that allow spatial testing with more complex models.

The spatial clustering within sexual networks of similar strains of chlamydia has
shown the significance of people who connect distinct geographic and demographic
groups (Wylie et al. 2005). The study of the environments of social geography is
then a conceptual perspective for location and space. For instance, two people can
live in different locations on a permanent basis but can bring the two very close by
their relationship space – attended the same school for some time or visited the

same place while travelling. Similarly, the only noticeable physical connection between persons living very far away could be a bath house where people gather for an unprotected, anonymous sex. Thus, cartography links between people and places have shown that special locations, at both short and long distances, are related to risk interplay between social contacts (Rothenberg et al. 2005). Infections of sexual infection (STIs such as chlamydia, gonorrhoea, and infectious syphilis) and blood-borne pathogens (BBPs) are treatable diseases concentrated in smaller parts of our community's HIV specifically and hepatitis C (WHO 2012). The position where individuals participate in risk behaviour, as well as the individuals with whom they interact, may be part of an activity network. Cummins et al. (2007) examined the idea of space application in health solutions too much dependent upon the conventional Euclidean view of space in the past.

Logan et al. (2016) investigate the role of street-level connectivity in blood-borne and sexually transmitted pathogens selected by respondent-guided sampling. In Winnipeg, Canada, 600 people were recruited from January to December 2009 for a survey of 600 injection drug users, men who have sex with children, street young people, and homeless people. In 2009, social network data were collected from respondent-driven sampling: injection drug users (IDUs), street youth, people with guns, and homeless in Winnipeg, Manitoba. The questionnaire received geographic coordinates of the named intersections for the following: 'What is the closest intersection to the area you have most frequently lived in over the last 6 months?' 'Thinking of the locations above where you most often injected, what is the nearest intersection to that place?' 'Thinking about where you meet your customer partners most frequently, what is the closest intersection to that place?' 'Thinking of the place where you meet your sex worker partners most frequently, what is the closest intersection to that location?' 'OpenStreetMap' (www.openstreetmap.org) has been used to check each geographic location as a base layer and a point manually. The author has reconstructed participants' social network to explain how participants were interconnected and how big those components were. Relations were created by linking the residence of each person to their sites of risk and the residences of recruits. The relative significance of the positions has been determined by immediately counting the number of connections (degree) to neighbouring nodes. Betweenness centrality (Brandes 2001), represented as:

$$C_B(\vartheta) = \sum_{s \neq \vartheta \neq t \in V} \frac{\sigma_{st}(\vartheta)}{\sigma_{st}}$$

provides the relationship between the shorter paths between the s and t nodes at each node (υ). This helps in calculating the likelihood of any residence or hazard position over a short distance between two other locations. This provides opportunities to move people who may not otherwise be connected through the socio-spatial network.

The combination of social network analysis and spatial analysis showed the social topology among the participating networks. Kernel density estimation (KDE) was used here to deduce risk areas where spatial cluster correlates with high social

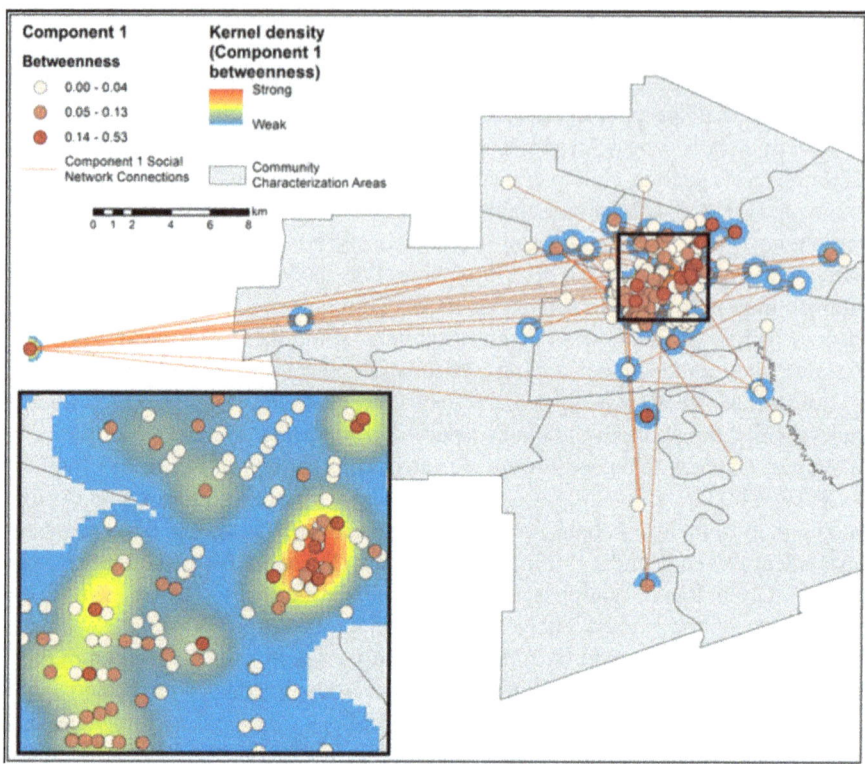

Fig. 7.10 Geographic extents of component. Points that indicate spatial distribution of homes and sites of risk activity are influenced by the key value ratings. The portion 1 KDE surface is determined by weighting the socio-spatial network weighting scores. (Logan et al. 2016)

networking findings from social network analysis. The KDE function fits a surface over each point in the dataset, using the points or point values within a chosen radius, here described as 500 meters to represent one large city block in Winnipeg (Fig. 7.10).

7.10   Ecological Risk Assessment and GeoComputation

An ecological risk assessment is used to determine the probability that exposure to one or more of the environmental stressors, including chemical contaminants, land changes, diseases, invasive organisms, and climate changes, may have an impact on the ecosystem. Ecological risk assessment is a method for determining the possibility of adverse environmental consequences or events arising from exposure to one or more stressors. In order to understand and avoid the relationship between stressors and ecological effects in a manner that is useful for the decision-taking of the community, this technique systematically analyses and organizes data, knowledge, assumptions, and uncertainty. An assessment may include chemical, physical, or biological stressors, and consideration may be given to one or more stressors. Several factors – those associated with poor health, injury, cancer, or death – affect health and wellbeing. On the one hand, the overestimation of climate change effects can lead to the ignorance of important non-climate altering or confounding factors (e.g. habitats and land use changes, agricultural and other economic activities, urbanization, human migration, health infrastructure and technologies, and host population and behaviour) (Tabachnick 2010; Lafferty 2009). According to the World Health Organization, global and regional health environmental risks include 'climate change, loss of stratospheric ozone, biodiversity-related ecosystem changes, changes in hydrological and freshwater sources, habitat degradation, urbanization and food systems stress' (WHO 2017).

7.11   Ecological Factor and Disease Risk Modelling

Increased risk of the rapid transmission of infectious diseases is highest due to travel and transport (Jaffry et al. 2009). In particular, ecological transformation is most important for tropics that cause new human, wildlife, and domestic animal health pathogens to emerge (Jones et al. 2008). The living distance of hosts and vectors, which influences the distribution of natural foci, is defined, to some degree, by climate factors such as temperature, lights, and rainfall (Wu et al. 2014). It is clear that rising anthropogenic emissions of greenhouse gasses have led to world warming and that weather conditions like temperatures and precipitation are more frequently extremely variable and have a wide impact on human health and on the ecosystem (Epstein 2001). Many infectious diseases are highly prone to climatic change due to the impact on pathogens, vectors, hosts, and their living conditions of weather changes (IPCC 2013). In the distribution of vectors and their hosts, topographer influences like elevation, slope, and aspect play a significant role by influencing the relocation of the hydrothermal combination. The vegetation will provide ticks and their vectors with an appropriate living environment. The vector density is in accordance with the forest form and structure. The current and possible natural focal points of the vectors should therefore be identified and relative environmental factors of high disease risk established (Li et al. 2017). In several developing countries with tropical forests (Lambin and Meyfroidt 2011), significant shifts in land use occur in particular. For example, population growth may have a structural as well cumulative impact (Rosa et al. 2004). Anthropogenic drivers of global environmental change can have different effects.

Wilcox and Colwell (2005) suggest that changing ecological systems produce feedback affecting the natural ecosystem and ultimate pests, animal host species, and humans as a result of numerous facet-based meetings with anthropogenic environmental changes, such as urbanization. These altered host pathogens allow pathogens into 'new' hosts possible, quickly pathogens adapt, a new pathogen variant more frequently generates new diseases and re-emerging diseases as well as spread and increase epidemic and frequency of the diseases existing. As a consequence, new pathogens and hypotheses (e.g. epidemiological models) are now understood as characteristically generating emerging phenomena, like 'social-ecological processes', and potentially unpredictable. Therefore, almost all new diseases bear a vector or zoonotic disease (i.e. sustained by humans in natural pathology cycles) (Woolhouse and Gowtage-Sequeria (2005). The host parasitic complexes or their immediate progenitors, which usually comprise non-linear, cross-scale action, exist as part of naturally co-evolved ecosystems (Horwitz and Wilcox 2005). There is no adequate theory or method to explain complex phenomena such as EID and the relationships between factors affecting illness. Integrated methods – bridging social sciences – are daunting and challenging but probably the most effective if thorough description comes and the analytical capacity for solving a complex problem is enhanced. Due to its significance and complexity, we conclude that the integration of the social sciences includes the theoretical rationale and a methodological methodology driven by a model. This method will produce combined results from many studies to explore various aspects of the model.

7.12 Case Study: Sandfly Density (Kala-Azar) Prediction Using Ecological Variables

The intrinsic and extrinsic factors are both determined by the geographical distribution of sand flies (*Phlebotomus argentipes*), their history, their hosts, and their ability to transmit infection. The factors that determine the sandfly's response to external conditions are biochemical and physiographical properties. Biotic and abiotic elements of the environment that influence sandfly biology are extrinsic influences. The sandfly is closely related to environmental variables (Bhunia 2014). All these layers were combined with the weighting index overlay system to distinguish sandfly abundance areas. In consideration of the effect on the abundance of sandflies, weights of different subjects were assigned on a scale of 1–5. The weights were allocated to various groups on each subject at 1–5 according to their relative influence on the abundance of sandfly. A scale was used to allow a standard assessment of many classes of a given theme: very low abundance (weight = <2.0), low abundance (weight = 2.1–4.0), moderate abundance = 4.1–6.0), high abundance (weight = 6.1–8.0), and very high abundance (weight = above 8.1). A pairwise comparison matrix for individual themes and features was then developed by Saaty's analytic hierarchy process (AHP) (Saaty 1980). Increasing environmental variable requires pairwise comparisons for the value of the hierarchy objects. In order to normalize, we evaluate the overall column and divide each cell value by the sum. This brings us the regular matrix. It gives us a clear average of the lines. The weighted linear combination (WLC) technique shall add one to the weight. The consistency ratio of the random matrix with the same dimension was measured as a

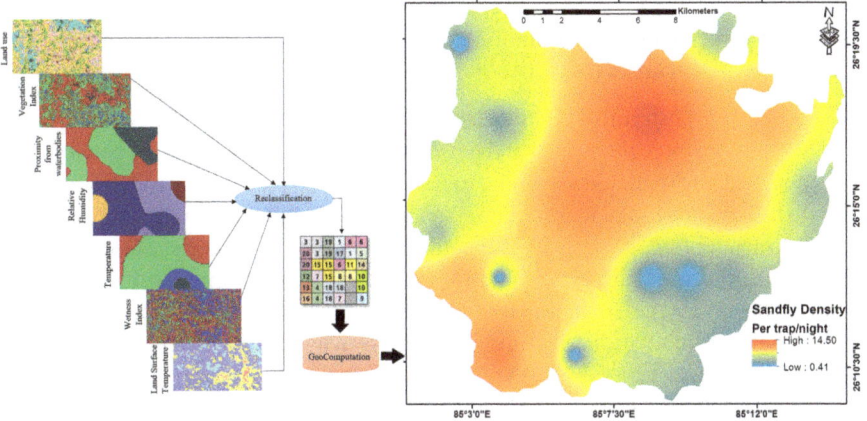

Fig. 7.11 Prediction of sandfly density based on ecological variables of Motihari Block of Muzaffarpur District (Bihar, India)

coherence index ratio in the initial matrix. It shows the possibility of randomly generating matrix scores. When the proportion accuracy reaches 0.10, a relation with pairs will be checked. The following is obtained from the combined weights of different polygons in the interconnected layer. Sandfly abundance map (SAM) is a dimensionless quantity that helps in indexing probable sandfly abundance potential zones in the area (Fig. 7.11). Overall, on the basis of the Saaty's AHP procedure, the proposed method is a comparatively straightforward method for representing predictors and compiling pair-sided association data, particularly in subjective cases. The system can also be used for collective decision-making – which is important for the analysis – in the hierarchical structure of this model, specific perspectives from decision-makers should be used.

7.13 Methodological Issues of Modelling Approach

An increased spatial epidemiological study evaluating the geographical distribution of possible health threats and associating them to environmental risk factors has contributed to an increase in available spatial climate, health, and population data coupled with improved statistical and spatial analysis methods. Their data is also increased. The capacity to identify disease clusters rapidly and determine the spatial distribution of disease risk and link environmental data with health outcomes provides a powerful tool to determine the spatial relationships between disease and environmental risks. During most spatial epidemiology studies, aggregate data at the region level are obtained because of data limitations. A variety of statistical methods and instruments are available to quantify uncertainty associated with the estimated risks in order to determine area-level risks and confidence intervals.

Standardization of the summary level permits the consideration and the type of indirect or direct standardization of identified risk factors such as age, gender, and socio-economic status:

- *Data* – Data constraints also include ecological measures of spatial epidemiology. Although ecological studies can be used for evaluating connections between the exposure and the incidence of the disease, there have been associated problems with the use of aggregate data. The standardized mortality/morbidity ratio (SMR) is the most common summary indicator for mapping disease risk. In small areas or in rare diseases, this form of indirect standardization compares the number of study cases reported to the number of cases predicted using age-specific statistics of a normal population. Sampling variability would dominate the estimates in these situations (Elliot et al. 2000).
- *Misclassification* – Incorrect circumstances and populations which directly influence the validity of a survey could be considered misclassified, and potentially erroneous time or space trends could be inserted into the risk (Oliver et al. 2005). The effect of geocoding errors can rely on space populations or sources of risk, but there are more errors in rural areas than in urban areas (Ward et al. 2005). The movement of populations would cause misclassification of access, which may lead to errors in risk patterns of time or space (Arnold 1999). This is especially problematic for findings with long delay times from the onset of an exposure to a disease, but duration, migration, and related exposure measurements in many cases are not well described.
- *Geographic Resolution* – The geographical resolution of the analysis may also affect the findings. The risk to health is mostly associated with fairly arbitrary administrative areas (i.e. the level of population and covariate data available), while risks can be vulnerable to adjustments of the production scale known as the 'modifiable area unit problem' (Openshaw 1984). Grouping data at the various spatial resolution levels (e.g. counties, census tracts, areas) or aggregating them to specific local structures invariably results that could influence the understanding of these effects. The effects of these results are specific.
- *Mapping* – In order to identify the areas at the highest risk in an epidemiological study, research into the spatial variations of disease result trends is also relevant. This typically is achieved by 'national' or 'focalized' study of clusters or by mapping diseases. Clustering considers the grouping of geographic disease cases in the sense of epidemiology in relation to non-casing events or the population at risk. Such residual variability, also called overdispersion, may be attributed either to the fact that the cases in one or more areas have been extra-aggregated or to the correlation between observations.
- *Modelling Approach* – In addition, disease modelling approaches tackle spatial risk distribution estimates. Small area studies can also indicate that data are sparse; the number of reported and predicted health results in small areas is often low, and so instable risk assessments are unreliable. Scarcity in disease mapping studies has been greatly tackled by Bayesian hierarchical models (Best et al.

2005). Those models reduce unstable risks to the local average risk by 'borrowing' data between areas.

- *Ecological Regression Model* – The relationship between risk factors and health outcomes at regional level is widely used in the assessment of ecological regression models. The above-noted overdispersion precludes the use of regular general linear models (GLMs) such as the Poisson and logistical regression (McCullagh and Nelder 1989). Unless spatial autocorrelation is ignored, narrower confidence intervals (Schabenberger and Gotway 2005) will lead to a skewed estimate of the regression coefficients and underestimation for their unsafety (i.e. false).
- *Uncertainty* – In terms of the existence and consistency of the data and reliability of the data, analysis and decision-taking of spatial data should be done (Longley et al. 2005). Information may undergo a number of transformations to generate derived data, from the acquisition of data from registered physical characteristics to geovisualization. In particular since spatial representation is typically the analyser's decision rather than an important data characteristic, data may be transformed between feature types (i.e. point, line, or area), interpolated, condensed, sampled, or quantified. Due to various geovisualization techniques (Cheung and Shi 2004), it can then be displayed with derived data.

7.14 GeoComputation: A Driver for Decision-Making in Public Health

Decisions including decision-making, strategy, and management should be considered by collective consensus and can be interactively applied taking human driving factors into account. Included in driving forces are population growth, health and prosperity, technology, politics, economics, and so forth, with human society setting priorities and targets to improve living standards. Richardson et al. (2013) offer a recent perspective on how GeoComputation advances help us better understand the disease trends, aetiology, transmission, and treatment. The new web-based applications in epidemiology have recently been cited by Delmelle et al. (2014), and an analytical module has been developed with the help of an expedited kernel density estimation (KDE) to track outbreaks of dengue fever in Colombia. Despite the dynamic generation of view-only maps and interactive maps (Gao et al. 2009), several web-based health applications permit users to study patient spatio-temporal distributions (Hammond et al. 2008). Web-based pattern detection software, in particular, a web-based framework (Mills and Curtis 2008), was nevertheless limited to some specific work.

Geospatial artificial intelligence (GeoAI) incorporates methods of space science (e.g. AI, data mining, and high-performance computing) to derive practical understanding from spatial large-scale data (VoPham et al. 2018). GeoAI is a focused area of health intelligence that includes location for information that can be used to improve human health (Kamel Boulos and Al-Shorbaji 2014). A popular topic in GeoAI applications at the population level as well as on the individual level is the use of new, large-scale sources such as social media, electronic records, satellite distance sensors, or personal sensors, to promote public health science (especially in the meaning of 'smart safe cities') and possibly precision medicine. The foundations of the GeoAI are machine learning, deep learning, and data extraction. Machine training especially involves AI methodologies and computer algorithms to acquire knowledge from patterns hidden in raw data by iteratively extracting and learning from it. Deep learning is commonly seen as a new form of machine learning, which allows computers to model brain functions so that they can more easily understand complex concepts in the real world (Kamel Boulos et al. 2019). In order to gain useful information and expertise from spatial big data for specific analytical needs, the GeoAI tools and framework use both of these approaches (Goodfellow et al. 2016).

The World Health Organization has developed HealthMapper as a method for tracking and disease mapping (Heymann and Rodier 2001). The focus will be on African countries where infectious diseases such as guinea worm disease, leprosy, tuberculosis, malaria, and HIV/AIDS are managed. It runs on a Windows operating system and is distributed freely or cheaply on the basis of the World Health Organization's institutional arrangements.

SIGEpi [Sistemas de Información Geográfica en salud (Martinez et al. 2001)] was established by the Pan American Health Organization to improve the analytical ability of epidemiology and public health in the Americas. The study covers a range of techniques for spatial data questioning and approaches for evaluating health outcomes and determinants, such as descriptive and exploratory approaches, disease mapping smoothing models, spatial clustering, and composite health index development. Benefit analysis may be performed to determine the relationship at both individual and aggregate levels between environmental factors and health outcomes. The software is distributed on a Windows operating system and is subject to a low-cost licence.

Applying a connection between the ESRI GIS (version 3.x, ESRI, Redlands, CA, USA, etc.) statistical software and Environmental Systems Research Institute (C++) that operates under any Microsoft Windows compatible framework, GeoDa was originally developed in the Spatial Analysis Lab at Illinois University (https://www.geoda.uiuc.edu/) but has been developed since then as a stand-alone program. In the form of a raw occurrence estimate (by ratio of the event count to a base population at risk), relative risk, and probability (by ratio of observed events to expectations), various spatial analysis and modelling tools are given. Concentrations can also be smoothened by using the empirical Bayes, the medium spatial window (with the total number of windows), and the spatial Bays which use the average window (rather than the overall average) for comparative purposes.

In the UK, the Rapid Inquiry Facility (RIF) was established. In order to evaluate the health hazards associated with the environmental exposure and to generate disease maps with and without statistical smoothing, Small Area Health Statistics Unit incorporates scientific statistics, statistical analysis, and quantitative epidemiological techniques. The RIF was originally designed to facilitate a risk assessment for a given population situation in the specified regions around a point source in a central reference area of the UK (Aylin et al. 1999). In addition to the EU-wide RIF project for the Exposure and Disease Mapping and Risk Assessment (EUROHEIS) European Health and Environment Information Network develop methods for integrating and analysing information on environmental exposure and human health (EUROHEIS 2003). Under the USA, the Environmental Public Health Tracking (EPHT) software of the Centers for Disease Control and Prevention (http://www.cdc.gov/nceh/tracking/) is a visual basic-developed RIF and functions as an ESRI ArcGIS-based application. The functionality of the RIF's disease mapping enables a consumer to map directly standardized rates and indirectly standardized risks. The relative risks can also be smoothed empirical Bayesian estimates. Currently, the RIF is designed to imagine and incorporate spatial uncertainty with other applications such as WinBUGS and SaTScan. It operates on a Windows framework and aims at free distribution of the software.

7.15   Application of GeoAI and Disease Modelling

For health and education, artificial intelligence (AI), like machine learning methods, has become more commonly used, predominantly with the growth of high concert and cloud computing (Topol 2019). The basic application of AI and the techniques and tools for data science to give reliable, effective, and successful insights into healthcare and medicine are referred to in health intelligence (Shaban-Nejad et al. 2018). Social media analytics for syndrome monitoring (Jane et al. 2018), prognostic modelling to recognize populations at a high risk of disease (Rajkomar et al. 2018), mobile health for healthcare provision (Istepanian and Al-Anzi 2018), and interpretation for medical imaging (Bi et al. 2019) were included in health intelligence applications. GeoAI incorporates methods for spatial science (e.g. geographic or GIS), AI, data mining, and high-performance computing in order to derive useful information from spatial big data (VoPham et al. 2018). GeoAI is a public intelligence system that integrates locations that provide valuable knowledge to enhance human health. Applications of new space-related data sources such as social media, electronic health records, remote satellite sensing, and personal sensors, to promote public safety sciences (particularly within the context of 'intelligent health cities') and possibly precision medicine are popular topics for population and individual GeoAI applications, providing new opportunities to gain insights into the field.

7.16   Artificial Neural Networks (ANN) and Spatial Modelling for Epidemiology

The ANNs are independent from the statistics distribution of data and do not need prior knowledge of data for derivative patterns as opposed to conventional statistical models. ANNs are simplistic math models that map the relationship between input and output layers with many examples. Artificial neural networks are one of the most common machine learning techniques (MLTs) used in environmental studies during the last years, inspired by human neural processing (Fig. 7.12). ANNs have several extremely integrated neurons working together to solve particular problems (Sordo 2002). In the spatial modelling of infectious diseases worldwide, few published ANN architectures have been developed. Aburas et al. (2010) used the Singapore National Environment Agency's data to predict the frequency of reported dengue cases using an ANN model with a context propagation algorithm. Global environmental TB incidence studies are insufficient for GIS and ANN integration in the USA, in particular (Mollalo et al. 2019) for epidemiological inferences. A dynamic model N-week forecast of the 2015–2016 Zika outbreak in the Americas has been implemented by Akhtar et al. (2019). The model used in this project is based upon multidimensional data from a time series of different countries, specifically epidemiological data, air travel volumes of passengers, suitability of the vector environment for the *Ae. aegypti* which primarily spread vector-borne diseases, and socio-economic and population data.

7.17   Fuzzy Models in Disease Distribution

In epidemiological research, diagnosing diseases entails a range of degrees of imprecision and confusion. In different patients and with a different disease diagnosis, one disease may manifest itself very differently. Despite its potential to resolve uncertainties, very few works have been presented to date which use flimsy logical principles to resolve epidemiological issues. The methods help one to understand and model their evolution in respect of the system's underlying dynamics, uncertainties relating to variables, parameters, boundary conditions, and initial states (Ortega et al. 2000). Mathematics are important instruments to show the relationship between cause and effect, as well as to test evidence that infectious diseases are definitive (John and Innocent 2005). The nature of infectious disease dynamics can be investigated by computer-assisted procedures. Epidemiological trends in the occurrence of a disease require computer resources (Alawieh et al. 2015). Studies have tried to clarify the mechanism of decision-taking in the medical sense using the Boolean or binary form to tackle the complexity of decision-making (Ahmadi et al. 2018). A degree of membership between *0* and *1* shall be granted to fuzzy groups. In both prediction and development of expert systems, fuzzification is an important technique. The main aim of using fluorescent logic is to address the

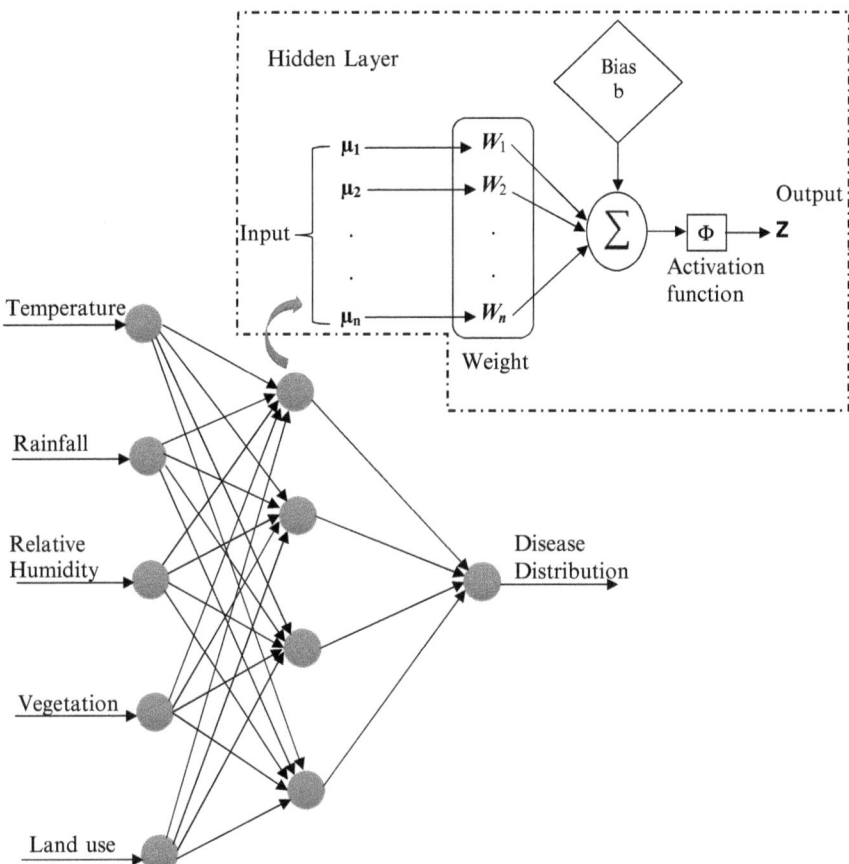

Fig. 7.12 Visualization of feedforward neural networks. In a recent analysis, an abundance of vector-borne disease expected using a simple back propagation method is identified. A neural network feed model with one hidden layer offers simulation time series that is commonly used in prediction. Input nodes, hidden nodes, and output nodes are present. In this regard, we set weight to decide the neuron, and the data transmit signals between neurons in order to analyse the competence of the ANN mechanism. A neural network feed-in model with a single hidden layer offers time-series modelling that is commonly used in prediction. Input nodes, hidden nodes, and output nodes are included. Here, we add weight to decide the corresponding neuron and data transmission between neurons in order to test the competence of the ANN method. Such signals are interpreted as an integrated feature, which incorporates signal and activation and then transmits the output

extreme limit problem. The quantitative data can be transformed to dynamic data by defining correct membership functions (Trapezoidal, Triangular, Gaussian, Bell). In medical practice, fuzzy logic is considered an important technique for modelling uncertainty (Zadeh 2008). Many scientific principles are flexible in the medical field (Tuncer et al. 2019). In an inconsistent, unpredictable, and incomplete setting, fuzzy logic makes a decision (Vieira et al. 2019). Fuzzy approaches deal with groups of vague and elusive boundaries. In 1965, the idea of fuzzy logic was introduced (Seising 2006).

The fundamental concept behind fluid logic is that the properties and geometry of spatial data have inaccuracies (Fig. 7.13). Fuzzy logic provides methods for addressing all types of error, but fluid logic focuses on errors in attribute data in relation to overlay analysis. In the class description and phenomenon calculation, the two key places needs to be focused where inaccuracies occur in the attribute data. Both these inaccuracies can cause imprecision in assigning cells to different classes, in particular in the class definitions. The fuzzy overlaying technologies available are fuzzy And, fuzzy Or, fuzzy product, fuzzy sum, and fuzzy gamma. Each of these techniques describes the connection of the cell with the input sets. The fuzzy and overlay sort, for example, produces an output raster where every cell value has the minimum value for every cell position set. Fuzzy overlays are based on set theory, while weighted overlays are based on linear method. The original values are changed by both techniques. The transition of a fuzzy overlay determines the probability of settlement, while the weighted overlay is at a quantitative scale of choice.

The key idea behind this approach is that a set of iteratively implemented rules can explain the dynamics of the system. Every rule is expected to have a fuzzy input and output. A panel of experts will produce a versatile membership function and variable and/or parameter as well as linguistic rules and a suitable inference system through their analytical expertise. The language model has the shape:

IF U is B_1 **AND** W_1 is A_{11} **AND**............**AND** W_n is A_{1n}
THEN $\bar{W}_i$ is $\hat{A}_{11}$ **AND**.....**AND** $\bar{W}_n$ is $\hat{A}_{1n}$ and V is D_1
ALSO
IF U is B_2 **AND** W_1 is A_{21} **AND** ......**AND** W_n is A_{2n}
THEN $\bar{W}_i$ is $\hat{A}_{21}$ **AND**.....**AND** $\bar{W}_n$ is $\hat{A}_{2n}$ and V is D_2

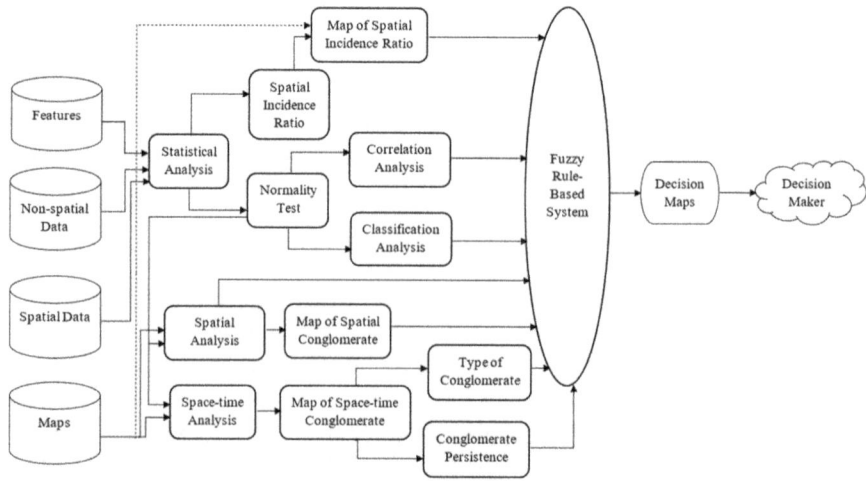

Fig. 7.13 Spatial decision approach based on fuzzy logic

ALSO

....
IF U is B_m **AND** W_1 is A_{m1} **AND**..............**AND** W_n is A_{mn}
THEN $\bar{W}_i$ is $\hat{A}_{m1}$ **AND**.....**AND** $\bar{W}_n$ is $\hat{A}_{mn}$ and V is D_m

where U is the input and W_i is the state variables of the systems; V and $\bar{W}_i$ are the output and state variables after each iteration, respectively; B_i and A_{ij} are the input fuzzy sets; and D_i and $\hat{A}_{ij}$ are the output fuzzy sets. Therefore, by using a model in each of the steps, we obtain the value of each state variable that is the input variable of the system in the following step, by choosing an adequate inference method and, if necessary, a defuzzification method, and so on. This is as follows:

$$U(1+1) = V(1)$$

and

$$Wi(l+1) = \bar{W}_i(l)$$

where $l + 1$ is the next step after l. This type of model is referred to as a MIMO (multiple input and multiple output) model.

There is another fuzzy dynamical system, and the preference for a specific model depends heavily on the type of information that the system provides. Part of the system conduct is often recognized, and rules may take the form:

$$\textbf{IF } U(l) \text{ is } B_m \textbf{ AND } W_1(l) \text{ is } A_{m1} \textbf{ AND} \ldots \text{..AND } W_{n(l)} \text{ is } A_m$$

$$\textbf{THEN } y(l+1) = f(U(l)), W_1(l), \ldots\ldots, W_1(l);$$

where y(l + 1) is some a priori function known from the system dynamics. Models of this kind are known as Takagi-Sugeno-Kang (TSK) models (Yager and Filev 1994).

7.18 Spatial Agent-Based Simulation Modelling for Disease Pattern

Agent-based models (ABM) are types of computer simulation consisting of agent and environment interactions. The agent may be anything from an individual to an organization or an institution, such as a nation state. In an ABM, 'agents' represent entities in a modelling framework in the real world. An ABM is also the world in which these entities 'reside'. Each of them has a state and has a specific behaviour. It was published in 1971 in Netlogo, Swarm, and Repast, and it was designed to allow the creation and understanding of ABMs by non-computer programmers (Gilbert 2008). Agencies are regulated by a variety of coded laws. In every phase, an agent determines what he should do. The actions can be just that simple to describe the directions in which the agent should take place on the basis of some artificial interpretation, or the actions can be more complex (MacNamee and Cunningham 2003). In addition, computer resource agent simulation has become a realistic tool for the study of epidemics by rising availability (Connell et al. 2009). Agent-based models are stochastic models of simulation, spatially transparent and discrete, where agents represent individuals that interact in space and time in accordance with specified laws (Balcan et al. 2010). Kelly et al. (2013) found that ABMs are suitable approaches when the model is structured to establish a desired device understanding. ABMs are strong methods, and various complex systems can be explored through ABM simulations, particularly in the field of process assumptions and interactions (Crooks and Hailegiorgis 2014). Such an approach enables the entire population to be plausibly modelled by providing a group model which behaves as though it lives in the city or in the country as a whole (Frias-Martinez et al. 2011).

In the epidemiology of contagious diseases, ABMs become common as the models can capture disease spread dynamics combined with the heterogeneous combination of agents with social networks (Bobashev et al. 2007). An ABM must model the characteristics of a disease (e.g. infection rates) and the characteristics of agents and the environment at an acceptable level of detail in order to model an outbreak in a reasonable manner (Hunter et al. 2017). ABMs and their capacity to generate emerging macro-effects from micro-regulatory processes have played a key role in developing various epidemiological methodological frameworks. Epidemiological applications based on ABM are designed to allow epidemiological researchers to perform a preliminary 'what if' study to determine the actions of the processes in different circumstances and to evaluate the alternative control policies to be taken with a view of combating epidemics.

Communicable diseases are diseases caused by a highly contagious infectious agent that can be spread by direct touch. Individuals who belong to a human society engage regularly in a series of activities. Some of the activities are fixed and some mobile. Stationary tasks in areas such as home, school, office, shopping, and retail areas take place at fixed locations. Individuals may engage in a group activity in these geographical locations. Mobile behaviours contribute to daily passenger travel through the public transport network. When a group disperses, person goes through time and space to another place, sometimes engaging with another group and joining them. The simulation of this diverse population is important to represent the direction of movement of individuals across space and time. The disease propagation model of this study shows this trajectory through the transport system and in the time as an hourly trajectory in space (movement from place to location).

Epidemiological applications based on ABM are designed to allow epidemiological researchers to perform a preliminary study "what if" to determine the actions of the processes in different circumstances and to evaluate the alternative control policies to be taken with a view to combating epidemics. In the earlier research, specific modelling has been developed to represent the spatial activity of various infectious diseases, including influenza (Bian and Liebner 2005), mumps

(Simoes 2005), the West Nile virus (Bouden et al. 2008), TB (Patlolla et al. 2004), Lassa virus (Dunham 2005), etc. Structured and functional urban networks models suggest spatial distribution and individuals' mobility, which allows the transmission of disease to model spatial heterogeneity. However, the lack of real landscape structures as well as integration with geospatial data and GeoComputing is one of its limitations to the continuous world in which the discreet person interacts.

7.19   Case Study: Contagious Spread of Disease Using Agent-Based Model

The best definition of ABMs is for their major components: climate, agents, and interactions. Figure 7.5 attempts to reflect the actions of the everyday path of individuals in an urban setting in a practical way and to depict the natural biotic process of the disease blowout among people. In the course of a distinct phase in time, the ABM operates through a geographic space in which a population of men, as agents, perform their daily business. It is supposed to be 10 h for an adult to follow a regular daily routine outside the house. Two hours of travel is spent by public transport, and the other 8 h are spent either in places of work or in schools (high schools, universities, community colleges, etc.) or in places of leisure such as shopping centres.

This model is implemented using georeferences in GIS data layers from urban areas, with the intention of geographically representing the normal urban landscape with typical people contacting. The model was developed to provide georeferenced information on population densities and various land use and transportation networks to take into account some of the variables that may influence an outbreak in urban areas. The geographical area closed where individuals communicate with each other and transmit disease is shown in Fig. 7.14. The movement of individuals by transport simulates the movement of people using a GeoComputation comprising all space sites on roads and topology stores. Therefore, if an officer is needed to get to a destination, he or she must check the list of the transport network to find out which roads he or she is expected to travel along: The first one (A) is the population of agents that reflect the actions of individuals living in urban areas. The second (B) refers to the density information for the population; this dataset allows an estimate of the number of people per square metre, thus giving people who do some stationary operation in this region greater risk of infection. The third (C) is the transport method used by officials to move inside the municipality from one location to the next. Finally, the fourth input (D) reflects a total of three different forms of land use: residential (houses, townhouses, and apartments) research and study areas (institutional and industrial buildings) (Fig. 7.15).

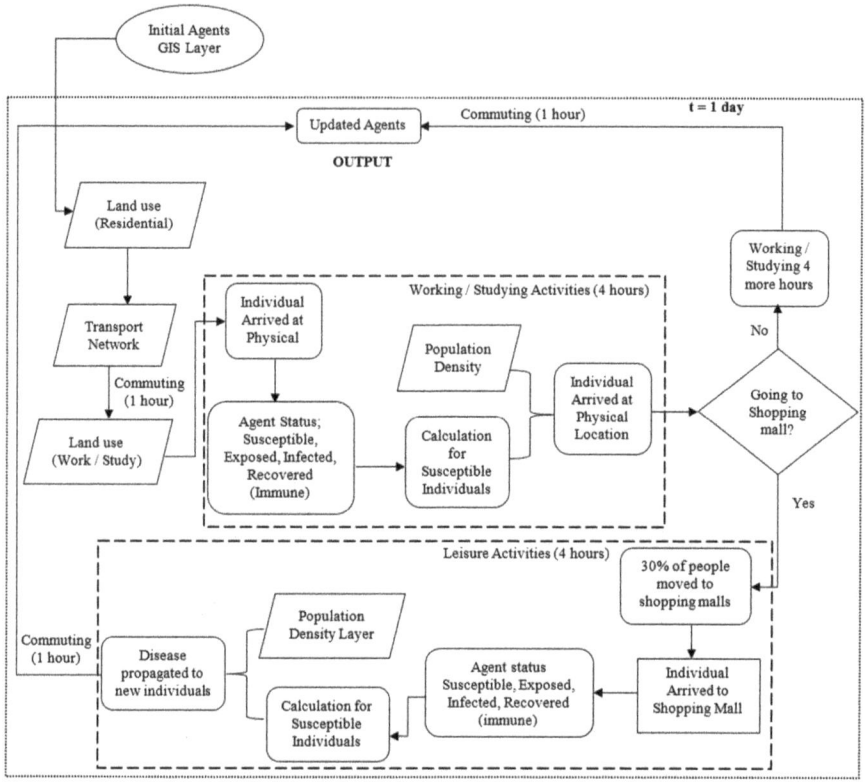

Fig. 7.14 Epidemic spreading mechanism AB model for a single time phase reflecting everyday operation of the behaviours of individuals and their interactions within an urban environment. (Source: Perez and Dragicevic 2009)

7.20   GeoComputation for Spatial Planning of Public Health Services

For health policymakers and planners, GeoComputational techniques have been made accessible to use also in low- and middle-income countries in political and implementation processes (Sugimoto et al. 2007); however, their degree of implementation differs from field to field as well as from goal to objective. These approaches are used to track the spread of infectious agents (e.g. human immunodeficiency virus (HIV)), as done by Coburn and Blower (2013), as well as sophisticated statistical methods (e.g. allocation of medical resources) and prediction of high-risk areas of the potential disease-related outbreaks (Kim et al. 2014).

There are many explanations, for example, the effective adoption of evidence-based approaches based on scientific methods is commonly documented as being faced with challenges that outweigh their benefits. GeoComputational techniques are not commonly used in public policy decisions or in the treatment of health

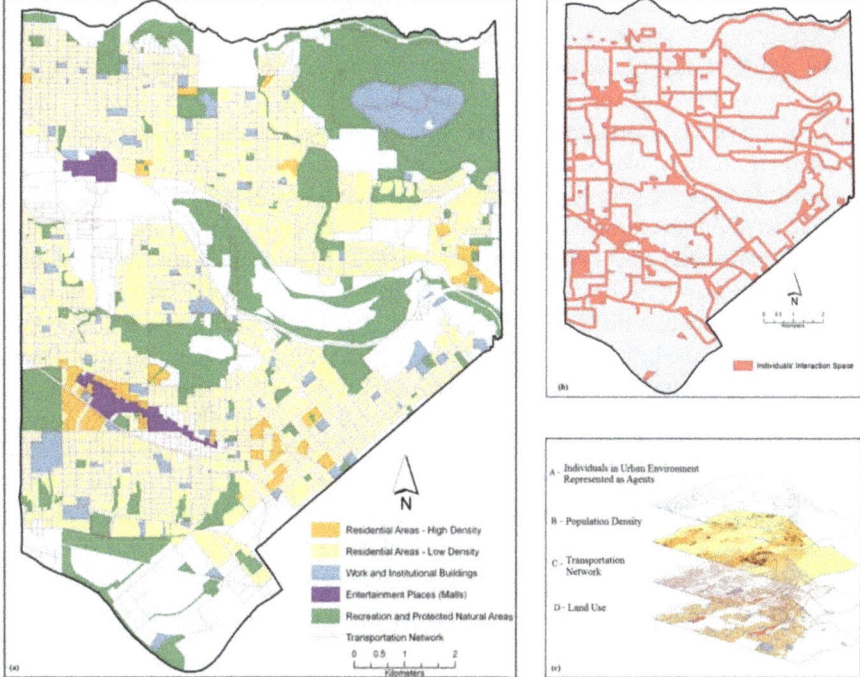

Fig. 7.15 Four specific georeferenced inputs are available in the agent-based model. (Source: Perez and Dragicevic 2009)

problems in most developed and underdeveloped countries. Although these methods are commonly used for health-related issues, they are primarily only used for visualization and explanation. The majority of the use of spatial methods for health data, according to literature, focuses on the basic use of charts. For example, disease maps, forecasting trends, and allocation of solutions to health resources in a health planning or policy framework are not commonly adopted or implemented. Despite the increasing numbers of publications using GeoComputational health analysis (Noor et al. 2009), this lack of implementation remains problematic. Furthermore, no attempts are being made to explore possible obstacles to the usage or application of geospatial instruments in public health policy processes:

• *Lack of professional skill and preparation* – Often this means a shortage of professional staff, practical expertise, and exposure to science and advanced technologies. The cost of GIS training for workers is typically high, both because of the steep curve associated with required software and statistical methods (Walshe 2017). Since GIS or geospatial modelling needs some technical knowledge, these issues are especially important. Personnel preparation should also be planned as a long-term, highly technological, and infrastructure-supported project.

- *Ignorance or lack of faith in the benefits* – The implementation of GeoComputational tools is impossible if policymakers do not fully understand the shortcomings of current practices and the corrective position of new tools. The use of these instruments is not justifiable without a strong understanding of the importance of an alternative solution. Although politicians agree to adjust planning strategy, they still ignore the effects of geospatial data because they feel too difficult or are simply unaware of the use of maps (Berger et al. 2001).
- *Restricted or inaccurate data uncertainty* – Records are small or inaccurate, one of the main sources of ambiguity. There are also significant drawbacks to public or government data bases, such as the sharing of resources and reliability of databases between different organizations. The fact that many health services either use, in the best case, old, obsolete computer systems or, in the worst cases, hard copy systems is historically of low quality. Extreme focus therefore needs to be extended to the use of health data (Ali et al. 2001). The lack of high-quality data and access to current data are main challenges when applying spatial strategies efficiently and broadly for decision-making support in the health field (Cockings et al. 2004). While some methods for avoiding disclosure of patient details have been developed, they typically are complicated and difficult to implement (Zimmerman and Pavlik 2008). Policymakers may be hesitant to follow spatial solutions to routine health policies and procedures without sufficient experience using suitable methods for purposes of reducing transparency.

7.21 Conclusion

GeoComputing offers spatial relationships that help spatial analysis techniques such as geocoding for the resolution of problems. Geocoding is a computational technique to turn textual knowledge definitions into addresses in numerical coherence at a spot on the surface of the Earth. Decision-makers want to consider how local neighbourhoods are impacted by change. There is good evidence to derive more value from existing data because of the hazards involved in decision-taking. The use of health GeoComputing is one of the cheapest ways of extracting intelligent information from existing data. GeoComputational GIS health calculations allow you to identify disease clusters, inappropriate cost, quality, and outcome variations in various policies and actions. The increase in data availability, methods, and technology is obviously important in the future of spatial epidemiology. Epidemiological studies require a diverse methodology and therefore increased demands on workers using techniques from a variety of different fields, such as statistics and geographical information sciences. Rather than simply implementing, multidisciplinary teams should be working together to ensure that accepted methods are fully understood and supplemented.

In surveillance systems and decision support systems for public health monitoring, convergence of diverse data sources was a common topic. To do so, collections from local health providers, federal or state agencies can be combined such that reportable disease information can be collected within the same system combining sources such as media, genetic history, socio-economic data, and other characteristics of the environment. In the recently developed SDSS review, however, incorporation of many data sources was not universal. The quality and use of social media, news, landscape, climate, and socio-economic data is mirrored here, as a variety of networks covered by these increasingly accessible information systems provide even more data to better understand what helps spread disease and outbreaks. In order to determine geographical regions which zoonosis spread and circulate, understanding the phylogenetics of a disease in the context of spatial distribution, environmental features, and possible hosts helps. Changes in human population, industrialization, world trade, and travel also contribute to the introduction or discovery of new pathogens worldwide, and phylogenetic analyses facilitate the monitoring of disease diffusion through human interactions.

In the development of SDSS to manage outbreaks of zoonosis, data integration and availability still present a number of challenges, including variability among available sources such as scale, completeness, and timeliness. Potential data sources span the microscopic and global scale, which allow data transformation and forecasting to be properly analysed and viewed. Emerging genetic data sets pose some difficulties, although georeferenced datasets documenting environmental and climatic phenomena are easier to use, as separation is typically manually collected from public records or from publications. Efforts have been made to extract geospatial metadata from GenBank information including location and host in order to visualize genetic sequence data and to allow for the automated linkage between sequence data related to spatial modelling diseases. In clinical decision support systems, machine learning algorithms were used for treating missing patient information in order to assist professionals in the decision-making and can also be useful to address this limitation in spatial decision support. The social media, emerging outlets, and different types of citizen science are becoming more widely used methods to monitoring disease reporting. One SDSS analysed here included a hotline or web report on public disease morbidity to increase monitoring. In the emergence of mobile telephone in an ever-connected digital age, a kind of citizen science has emerged that has been described by the words of Kamel Boulos as 'wikification of the GIS by the masses' (WGM). In this process, data quality and integrity must also be taken into account. Increased sophistication and integration between online tools and new SDSS in terms of public health would be needed in order to further enhance the incorporation of new data sources and to minimize cognitive overload. Such strategies should be based on epidemiological research, in order to best meet the purpose of the end consumer. To better address the needs of the customer, an emphasis on a coherent approach in SDSS creation is important. The relation between a number of available systems and the end user, as stated here, is not constrained in many of the recent designs and system evaluations.

References

Aburas HM, Cetiner BG, Sari M (2010) Dengue confirmed-cases prediction: a neural network model. Expert Syst Appl 37(6):4256–4260

Aenishaenslin C, Hongoh V, Cissé HD, Hoen AG, Samoura K, Michel P, Waaub JP, Bélanger D (2013) Multi-criteria decision analysis as an innovative approach to managing zoonoses: results from a study on Lyme disease in Canada. BMC Public Health 13(1):897

Aenishaenslin C, Michel P, Ravel A et al (2015) Acceptability of tick control interventions to prevent Lyme disease in Switzerland and Canada: a mixed-method study. BMC Public Health 16:12. https://doi.org/10.1186/s12889-015-2629-x

Ahmadi H, Gholamzadeh M, Shahmoradi L, Nilashi M, Rashvand P (2018) Diseases diagnosis using fuzzy logic methods: a systematic and meta-analysis review. Comput Methods Prog Biomed 161:145–172

Akhtar M, Kraemer MU, Gardner LM (2019) A dynamic neural network model for predicting risk of Zika in real time. BMC Med 17(1):171. https://doi.org/10.1186/s12916-019-1389-3

Alawieh A, Sabra Z, Bizri AR, Davies C, White R, Zaraket FA (2015) A computational model to monitor and predict trends in bacterial resistance. J Glob Antimicrobial Resist 3(3):174–183

Ali M, Emch M, Ashley C, Streatfield PK (2001) Implementation of a medical geographic information system: concepts and uses. J Health Popul Nutr:100–110

Alonso D, Dobson A, Pascual M (2019) Critical transitions in malaria transmission models are consistently generated by superinfection. Philos Trans R Soc B 374(1775):20180275

Anselin L (1995) Local indicators of spatial association—LISA. Geogr Anal 27(2):93–115. https://doi.org/10.1111/j.1538-4632.1995.tb00338.x.

Arnold, R. (1999). Small area health statistics unit procedures for estimating populations in small areas. Studies on medical and population subjects-office of population censuses and surveys, pp 10–23

Aylin P, Maheswaran R, Wakefield J, Cockings S, Jarup L, Arnold R, Wheeler G, Elliott P (1999) A national facility for small area disease mapping and rapid initial assessment of apparent disease clusters around a point source: the UK small area health statistics unit. J Public Health 21(3):289–298

Balcan D, Gonçalves B, Hu H, Ramasco JJ, Colizza V, Vespignani A (2010) Modeling the spatial spread of infectious diseases: the GLobal epidemic and mobility computational model. J Comput Sci 1(3):132–145

Beard R, Wentz E, Scotch M (2018) A systematic review of spatial decision support systems in public health informatics supporting the identification of high risk areas for zoonotic disease outbreaks. Int J Health Geogr 17(1):38. https://doi.org/10.1186/s12942-018-0157-5

Behzadian M, Kazemzadeh RB, Albadvi A, Aghdasi M (2010) PROMETHEE: A comprehensive literature review on methodologies and applications. Eur J Operat Res 200(1):198–215. https://doi.org/10.1016/j.ejor.2009.01.021

Bellan SE, Pulliam JR, Pearson CA, Champredon D, Fox SJ, Skrip L et al (2015) Statistical power and validity of Ebola vaccine trials in Sierra Leone: a simulation study of trial design and analysis. Lancet Infect Dis 15(6):703–710. https://doi.org/10.1016/S1473-3099(15)70139-8

Bennett KP, Campbell C (2000) Support vector machines: hype or hallelujah?. SIGKDD Explorations 2(2):1–13

Berger J, Suzuki T, Senti KA, Stubbs J, Schaffner G, Dickson BJ (2001 Dec) Genetic mapping with SNP markers in drosophila. Nat Genet 29(4):475–481. https://doi.org/10.1038/ng773. PMID: 11726933

Besag J, Newell J (1991) The detection of clusters in rare diseases. J R Stat Soc A Stat Soc 154(1):143–155. https://doi.org/10.2307/2982708

Best N, Richardson S, Thomson A (2005 Feb) A comparison of Bayesian spatial models for disease mapping. Stat Methods Med Res 14(1):35–59. https://doi.org/10.1191/0962280205sm38 8oa. PMID: 15690999

Bhushan N, Rai K (2004) Strategic decision making: applying the analytic hierarchy process, vol 9. Springer, Berlin, pp 11–21. http://www.springer.com/978-1-85233-756-8

Bhunia GS, Chatterjee N, Kumar V, Mandal R, Das P, Kesari S (2012, February) Remote sensing and GIS: tools for the prediction of epidemic for the intervention measure. In: Proceedings of the 14th annual international conference and exhibition on geospatial information technology and application. India Geospatial forum (PN-31), Gurgaon, India

Bhunia GS, Kesari S, Chatterjee N et al (2013) Spatial and temporal variation and hotspot detection of kala-azar disease in Vaishali district (Bihar), India. BMC Infect Dis 13:64. https://doi.org/10.1186/1471-2334-13-64

Bhunia GS, Siddiqui NA, Shit PK et al (2016) Spatial clustering of Plasmodium falciparum in Bihar (India) from 2007 to 2015. Spat Inf Res 24:639–648. https://doi.org/10.1007/s41324-016-0061-7

Bi WL, Hosny A, Schabath MB, Giger ML, Birkbak NJ, Mehrtash A et al (2019) Artificial intelligence in cancer imaging: clinical challenges and applications. CA Cancer J Clin 69(2):127–157

Bian L, Liebner D (2005) Simulating spatially explicit networks for dispersion of infectious diseases. GIS, Spatial Analysis and Modelling, pp 245–264

Bobashev GV, Goedecke DM, Yu F, Epstein JM (2007) A hybrid epidemic model: combining the advantages of agent-based and equation-based approaches. In 2007 Winter Simulation Conference, pp 1532–537

Boroushaki S, Malczewski J (2008) Implementing an extension of the analytical hierarchy process using ordered weighted averaging operators with fuzzy quantifiers in ArcGIS. Comput Geosci 34(4):399–410

Bouden M, Moulin B, Gosselin P (2008) The geosimulation of West Nile virus propagation: a multi-agent and climate sensitive tool for risk management in public health. Int J Health Geogr 7(1):35

Brandes U (2001) A faster algorithm for betweenness centrality. J Math Sociol 25(2):163–177

Breiman L, Friedman JH, Olshen RA, Stone CJ (1984) Classification and regression trees, vol 432. Wadsworth. International Group, Belmont, pp 151–166

Brownstein JS, Freifeld CC, Reis BY, Mandl KD (2008) Surveillance Sans Frontieres: internet-based emerging infectious disease intelligence and the HealthMap project. PLoS Med 5(7):e151

Chainey S, Tompson L, Uhlig S (2008) The utility of hotspot mapping for predicting spatial patterns of crime. Secur J 21(1–2):4–28

Cheng Y, Canuto VM, Howard AM (2002) An improved model for the turbulent PBL. J Atmos Sci 59(9):1550–1565

Cheung CK, Shi W (2004) Estimation of the positional uncertainty in line simplification in GIS. Cartogr J 41(1):37–45

Chowell G, Mizumoto K, Banda JM, Poccia S, Perrings C (2019) Assessing the potential impact of vector-borne disease transmission following heavy rainfall events: a mathematical framework. Philos Trans R Soc B 374(1775):20180272. https://doi.org/10.1098/rstb.2018.0272

Coburn BJ, Blower S (2013) Mapping HIV epidemics in sub-Saharan Africa with use of GPS data. Lancet Glob Health 1(5):251–253

Cockings S, Dunn CE, Bhopal RS, Walker DR (2004) Users' perspectives on epidemiological, GIS and point pattern approaches to analysing environment and health data. Health Place 10(2):169–182

Connell R, Dawson P, Skvortsov A (2009) Comparison of an agent-based model of disease propagation with the generalised SIR epidemic model (No. DSTO-TR-2342); Defense Science and Technology Organisation: Canberra, Australia, 2009; 1–22

Cromley EK, McLafferty SL (2011) GIS and public health. Guilford Press, New York

Crooks AT, Hailegiorgis AB (2014) An agent-based modeling approach applied to the spread of cholera. Environ Model Softw 62:164–177

Cummins S, Curtis S, Diez-Roux AV, Macintyre S (2007 Nov) Understanding and representing 'place' in health research: a relational approach. Soc Sci Med 65(9):1825–1838. https://doi.org/10.1016/j.socscimed.2007.05.036. Epub 2007 Aug 13. PMID: 17706331

Delmelle EM, Zhu H, Tang W, Casas I (2014) A web-based geospatial toolkit for the monitoring of dengue fever. Appl Geogr 52:144–152

Diaby V, Goeree R (2014) How to use multi-criteria decision analysis methods for reimbursement decision-making in healthcare: a step-by-step guide. Expert Rev Pharmacoecon Outcomes Res 14(1):81–99

Dobson MW (1983) A high resolution microcomputer based color system for examining the human factors aspects of cartographic displays in a real-time user environment. In: Presented at the 6th international symposium on computer assisted cartography, vol 1, pp 352–361

Dunham JB (2005) An agent-based spatially explicit epidemiological model in MASON. J Artif Soc Soc Simul 9(1)

Eastman JR (1997) IDRISI for windows, version 2.0: tutorial exercises graduate School of Geography. Clark University, Worcester, MA

Eisen L, Eisen RJ (2011) Using geographic information systems and decision support systems for the prediction, prevention, and control of vector-borne diseases. Annu Rev Entomol 56:41–61

Elliot P, Wakefield JC, Best NG, Briggs DJ (2000) Spatial epidemiology: methods and applications. Oxford University Press, Oxford

Epstein PR (2001) Climate change and emerging infectious diseases. Microbes Infect 3(9):747–754

Farrington CP (1990) Modelling forces of infection for measles, mumps and rubella. Stat Med 9(8):953–967

Fotheringham AS, Zhan FB (1996) A comparison of three exploratory methods for cluster detection in spatial point patterns. Geogr Anal 28(3):200–218

Fraccaro P, Plastiras P, Dentone C, Di Biagio A, Weller P (2015) Behind the screens: clinical decision support methodologies–a review. Health Policy Technol 4(1):29–38

Frazão T, Camilo D, Cabral E et al (2018) Multicriteria decision analysis (MCDA) in health care: a systematic review of the main characteristics and methodological steps. BMC Med Inform Decis Mak 18:90. https://doi.org/10.1186/s12911-018-0663-1

Frias-Martinez E, Williamson G, Frias-Martinez V (2011) An agent-based model of epidemic spread using human mobility and social network information. In 2011 IEEE third international conference on privacy, security, risk and trust and 2011 IEEE third international conference on social computing. IEEE, pp 57–64

Gani R, Hughes H, Fleming D, Griffin T, Medlock J, Leach S (2005) Potential impact of antiviral drug use during influenza pandemic. Emerg Infect Dis 11(9):1355

Gao S, Mioc D, Yi X, Anton F, Oldfield E, Coleman DJ (2009) Towards web-based representation and processing of health information. Int J Health Geogr 8(1):1–14

Getis A, Ord JK (2010) The analysis of spatial association by use of distance statistics. In: Perspectives on spatial data analysis. Springer, Berlin/Heidelberg, pp 127–145

Gehlen M, Nicola MRC, Costa ERD, Cabral VK, de Quadros ELL, Chaves CO, Lahm RA, Nicolella ADR, Rossetti MLR, Silva DR (2019) Geospatial intelligence and health analitycs: its application and utility in a city with high tuberculosis incidence in Brazil. J Infect Public Health 12(5):681–689

Gilbert N (2008) Agent-based models (Vol. 153). Sage Publications, Incorporated

Goodfellow I, Bengio Y, Courville A, Bengio Y (2016) Deep learning, vol 1. MIT press, Cambridge

Gorsevski PV, Donevska KR, Mitrovski CD, Frizado JP (2012) Integrating multi-criteria evaluation techniques with geographic information systems for landfill site selection: a case study using ordered weighted average. Waste Manag 32(2):287–296

Hammond D, Barzyk TCK, Zartarian V, Schultz B (2008) Application of GIS mapping tools to prioritize community air pollution issues. Epidemiology 19(6):S173

Hastie TJ, Tibshirani RJ (1990) Generalized additive models, vol 43. CRC press, New York

Heymann DL, Rodier GR (2001) Hot spots in a wired world: WHO surveillance of emerging and re-emerging infectious diseases. Lancet Infect Dis 1(5):345–353

Horwitz P, Wilcox BA (2005) Parasites, ecosystems and sustainability: an ecological and complex systems perspective. Int J Parasitol 35:725–732

Huang Z, Das A, Qiu Y, Tatem AJ (2012) Web-based GIS: the vector-borne disease airline importation risk (VBD-AIR) tool. Int J Health Geogr 11(1):1–14

Hunter E, Mac Namee B, Kelleher JD (2017) A taxonomy for agent-based models in human infectious disease epidemiology. J Artif Soc Soc Simul 20(3)

Hwang CL, Yoon K (1981) Multiple attribute decision making: methods and applications. Springer-Verlag, New York. https://doi.org/10.1007/978-3-642-48318-9

IPCC (2013) Intergovernmental panel on climate change (IPCC, 2013) Climate change 2013: The physical science basis. www.ipcc.ch. Accessed 16 June 2015

Ishioka F, Kawahara J, Mizuta M, Minato SI, Kurihara K (2019) Evaluation of hotspot cluster detection using spatial scan statistic based on exact counting. Jpn J Stat Data Sci 2(1):241–262

Istepanian RS, Al-Anzi T (2018) M-health 2.0: new perspectives on mobile health, machine learning and big data analytics. Methods 151:34–40

Jacquez GM, Waller LA, Grimson R, Wartenberg D (1996) The analysis of disease clusters, part I: state of the art. Infect Control Hosp Epidemiol 17(5):319–327

Jaffry KT, Ali S, Rasool A, Raza A, Gill ZJ (2009) Zoonoses. Int J Agric Biol 11(2):217–220

Jain A, Murty M, Flynn P (1999) DataClustering: a review. ACM Comput Surv 31(3):264–323

Jane M, Hagger M, Foster J et al (2018) Social media for health promotion and weight management: a critical debate. BMC Public Health 18:932. https://doi.org/10.1186/s12889-018-5837-3

Jelokhani-Niaraki M, Malczewski J (2015) A group multicriteria spatial decision support system for parking site selection problem: a case study. Land Use Policy 42:492–508

John RI, Innocent PR (2005) Modeling uncertainty in clinical diagnosis using fuzzy logic. IEEE Trans Syst Man Cybern B Cybern 35(6):1340–1350

Jones KE, Patel NG, Levy MA, Storeygard A, Balk D, Gittleman JL, Daszak P (2008) Global trends in emerging infectious diseases. Nature 451(7181):990–993

Joyce K (2009) "To me it's just another tool to help understand the evidence": public health decision-makers' perceptions of the value of geographical information systems (GIS). Health Place 15(3):801–810

Kamel Boulos MN, Al-Shorbaji NM (2014) On the internet of things, smart cities and the WHO Healthy Cities.13–10

Kamel Boulos MN, Peng G, VoPham T (2019). An overview of GeoAI applications in health and healthcare. https://doi.org/10.1186/s12942-019-0171-2

Kelly RA, Jakeman AJ, Barreteau O, Borsuk ME, ElSawah S, Hamilton SH et al (2013) Selecting among five common modelling approaches for integrated environmental assessment and management. Environ Model Softw 47:159–181

Kim D, Lauria DT, Poulos C, Dong B, Whittington D (2014) Effect of travel distance on household demand for typhoid vaccines: implications for planning. Int J Health Plann Manag 29(3):e261–e276

Klepac P, Kissler S, Gog J (2018) Contagion! the bbc four pandemic–the model behind the documentary. Epidemics 24:49–59. https://doi.org/10.1016/j.epidem.2018.03.003

Kulldorff M (1997) A spatial scan statistic. Communications in Statistics - Theory and Methods 26(6):1481–1496. https://doi.org/10.1080/03610929708831995

Kulldorff M (2001) Prospective time periodic geographical disease surveillance using a scan statistic. J R Stat Soc Ser A 164:61–72

Kulldorff M (2018) Information management services, inc. SaTScan™ v9. 3: Software for the spatial and space-time scan statistics. 2014

Kurihara K (2004) Classification of geospatial lattice data and their graphical representation. In: Classification, clustering, and data mining applications. Springer, Berlin/Heidelberg, pp 251–258

Lafferty KD (2009) The ecology of climate change and infectious diseases. Ecology 90(4):888–900. https://doi.org/10.1890/08-0079.1

Lambin EF, Meyfroidt P (2011) Global land use change, economic globalization, and the looming land scarcity. Proc Natl Acad Sci 108:3465–3472. https://doi.org/10.1073/pnas.1100480108

Lessler J, Azman AS, McKay HS, Moore SM (2017) What is a hotspot anyway? Am J Tropical Med Hygiene 96(6):1270–1273

Li Y, Wang J, Gao M, Fang L, Liu C, Lyu X, Bai Y, Zhao Q, Li H, Yu H, Cao W (2017) Geographical environment factors and risk assessment of tick-borne encephalitis in Hulunbuir, northeastern China. Int J Environ Res Public Health 14(6):569. https://doi.org/10.3390/ijerph14060569

Logan JJ, Jolly AM, Blanford JI (2016) The sociospatial network: risk and the role of place in the transmission of infectious diseases. PLoS One 11(2):e0146915. https://doi.org/10.1371/journal.pone.0146915

Longley PA, Goodchild MF, Maguire DJ, Rhind DW (2005) Geographic information systems and science. Wiley, Hoboken

MacNamee B, Cunningham P (2003) Creating socially interactive no-player characters: The µ-siv system. Int J Intell Games Simulation 2(1):28–35,186–221

Madoff LC, Woodall JP (2005) The internet and the global monitoring of emerging diseases: lessons from the first 10 years of ProMED-mail. Arch Med Res 36(6):724–730

Malczewski J (2000) On the use of weighted linear combination method in GIS: common and best practice approaches. Trans GIS 4(1):5–22

Malczewski J (2006) GIS-based multicriteria decision analysis: a survey of the literature. Int J Geogr Inf Sci 20(7):703–726

Malczewski J (2011) Local weighted linear combination. Trans GIS 15(4):439–455

Malczewski J, Rinner C (2015) Multicriteria decision analysis in geographic information science. Springer, New York, p 331

Mantey PE, Carlson ED (1980) Integrated geographic data bases: the GADS experience. In: Blaser A (ed) Data Base techniques for pictorial applications, Lecture notes in computer science, vol 81. Springer, Berlin, Heidelberg. https://doi.org/10.1007/3-540-09763-5_9

Marinoni O (2004) Implementation of the analytical hierarchy process with VBA in ArcGIS. Comput Geosci 30(6):637–646

Marsh K, Goetghebeur M, Thokala P, Baltussen R (eds) (2017) Multi-criteria decision analysis to support healthcare decisions, vol 10. Springer, Berlin, pp 978–973

Martinez R, Vidaurre M, Najera P, Loyola E, Castillo-Salgado C, Eisner C (2001) SIGEpi: geographic information system in epidemiology and public health. Epidemiol Bull 22(3):4–5

McCullagh P, Nelder JA (1989) Generalized linear models. CRC Monographs on Statistics & Applied Probability, Springer Verlag, New York

Mills JW, Curtis A (2008) Geospatial approaches for disease risk communication in marginalized communities. Prog Community Health Partnersh 2(1):61–72

Mollalo, A., Mao, L., Rashidi, P., & Glass, G. E. (2019). A GIS-based artificial neural network model for spatial distribution of tuberculosis across the continental United States. Int J Environ Res Pubic Health, 16(1), 157.https://doi.org/10.3390/ijerph16010157

Morens DM, Fauci AS (2013) Emerging infectious diseases: threats to human health and global stability. PLoS Pathog 9(7):e1003467

Mosha JF, Sturrock HJ, Greenwood B, Sutherland CJ, Gadalla NB, Atwal S, Hemelaar S, Brown JM, Drakeley C, Kibiki G, Bousema T (2014) Hot spot or not: a comparison of spatial statistical methods to predict prospective malaria infections. Malar J 13(1):53

Mossong J, Hens N, Jit M, Beutels P, Auranen K, Mikolajczyk R, Massari M, Salmaso S, Tomba GS, Wallinga J, Heijne J (2008) Social contacts and mixing patterns relevant to the spread of infectious diseases. PLoS Med 5(3):e74

Mourits MCM, Van Asseldonk MAPM, Huirne RBM (2010) Multi criteria decision making to evaluate control strategies of contagious animal diseases. Prev Vet Med 96(3–4):201–210

Myers N, Mittermeier RA, Mittermeier CG, Da Fonseca GAB, Kent J (2000) Biodiversity hotspots for conservation priorities. Nature 403:853–858

Neutra R, Swan S, Mack T (1992) Clusters galore: insights about environmental clusters from probability theory. Sci Total Environ 127(1–2):187–200. https://doi.org/10.1016/0048-9697(92)90477-A

Noor AM, Alegana VA, Gething PW, Snow RW (2009) A spatial national health facility database for public health sector planning in Kenya in 2008. Int J Health Geogr 8(1):13

Oliver MN, Matthews KA, Siadaty M, Hauck FR, Pickle LW (2005) Geographic bias related to geocoding in epidemiologic studies. Int J Health Geogr 4(1):29

Openshaw S (1984) Ecological fallacies and the analysis of areal census data. Environ Plan A 16(1):17–31

Openshaw S, Charlton M, Craft A (1988) Searching for leukaemia clusters using a geographical analysis machine. Pap Reg Sci 64(1):95–106

Ortega NRS, Sallum PC, Massad E (2000) Fuzzy dynamical systems in epidemic modelling. Kybernetes

Owens DK (2002) Analytic tools for public health decision making. Med Decis Mak 22(1_suppl):3–10

Patlolla P, Gunupudi V, Mikler AR, Jacob RT (2004) Agent-based simulation tools in computational epidemiology. In: International workshop on innovative internet community systems. Springer, Berlin/Heidelberg, pp 212–223

Perez L, Dragicevic S (2009) An agent-based approach for modeling dynamics of contagious disease spread. Int J Health Geogr 8(1):50

Phillips, S. J., Dudík, M., & Schapire, R. E. (2004, July). A maximum entropy approach to species distribution modeling. In Proceedings of the twenty-first international conference on Machine learning. pp 83. https://doi.org/10.1016/j.ecolmodel.2005.03.026

Preston R (1995) The hot zone: the terrifying true story of the origins of the Ebola virus. Anchor, Harpswell, ME

Raghavan VV, Hu X, Gabbouj M. (2016). Visual and decision informatics (CVDI). Compendium of Industry-Nominated NSF I/UCRC Technological Breakthroughs. Available at: http://www.iucrc.org/sites/default/files/breakthroughs/pdf/CVDI-2016.pdf

Rajkomar A, Oren E, Chen K, Dai AM, Hajaj N, Hardt M et al (2018) Scalable and accurate deep learning with electronic health records. NPJ Digital Med 1(1):18

Richardson EA, Pearce J, Mitchell R, Kingham S (2013 Apr) Role of physical activity in the relationship between urban green space and health. Public Health 127(4):318–324. https://doi.org/10.1016/j.puhe.2013.01.004. Epub 2013 Apr 12. PMID: 23587672

Rivers C, Chretien JP, Riley S, Pavlin JA, Woodward A, Brett-Major D, Berry IM, Morton L, Jarman RG, Biggerstaff M, Johansson MA (2019) Using "outbreak science" to strengthen the use of models during epidemics. Nat Commun 10(1):1–3

Robertson C, Long JA, Nathoo FS, Nelson TA, Plouffe CC (2014) Assessing quality of spatial models using the structural similarity index and posterior predictive checks. Geogr Anal 46(1):53–74. https://doi.org/10.1111/gean.12028

Rosa H, Kandel S, Dimas L (2004) Compensation for environmental services and rural communities: lessons from the Americas. Int For Rev 6(2):187–194

Rothenberg R, Muth SQ, Malone S, Potterat JJ, Woodhouse DE (2005) Social and geographic distance in HIV risk. Sex Transm Dis 32(8):506–512

Rushton G, Lolonis P (1996) Exploratory spatial analysis of birth defect rates in an urban population. Stat Med 15(7–9):717–726

Saaty TL (1977) A scaling method for priorities in hierarchical structures. J Math Psychol 15(3):234–281

Saaty TL (2008) Decision making with the analytic hierarchy process. International Journal of Services Sciences 1(1):83–98. https://doi.org/10.1504/IJSSCI.2008.017590

Schabenberger O, Gotway CA (2005) Statistical methods for spatial data analysis.. Chapman Hall/CRC

Seising R (2006) From vagueness in medical thought to the foundations of fuzzy reasoning in medical diagnosis. Artif Intell Med 38(3):237–256

Shaban-Nejad A, Michalowski M, Buckeridge DL (2018) Health intelligence: how artificial intelligence transforms population and personalized health

Simoes J (2005) Modelling the spreading of infectious diseases using social mobility networks. Centre for Advanced Spatial Analysis, UCL

Sonesson C, Bock D (2003) A review and discussion of prospective statistical surveillance in public health. J R Stat Soc A Stat Soc 166:5–21. https://doi.org/10.1111/1467-985X.00256

Sordo M (2002) Introduction to neural networks in healthcare. Open Clinical knowledge management for medical care

Sugimoto JD, Labrique AB, Salahuddin A, Rashid M, Klemm RD, Christian P, West KP Jr (2007) Development and management of a geographic information system for health research in a developing-country setting: a case study from Bangladesh. J Health Popul Nutr 25(4):436

Sugumaran R, Degroote J (2010) Spatial decision support systems: principles and practices. Crc Press

Tabachnick WJ (2010) Challenges in predicting climate and environmental effects on vector-borne disease episystems in a changing world. J Exp Biol 213(6):946–954

Tobler WR (1970) A computer movie simulating urban growth in the Detroit region. Econ Geogr 46(sup1):234–240

Topol EJ (2019) High-performance medicine: the convergence of human and artificial intelligence. Nat Med 25(1):44–56

Tuncer T, Dogan S, Abdar M, Ehsan Basiri M, Pławiak P (2019) Face recognition with triangular fuzzy set-based local cross patterns in wavelet domain. Symmetry 11(6):787

Vieira LCPFDS, Rizol PMDSR, Nascimento LFC (2019) Fuzzy logic and hospital admission due to respiratory diseases using estimated values by mathematical model. Ciencia & Saude Coletiva 24:1083–1090

VoPham T, Hart JE, Laden F, Chiang YY (2018) Emerging trends in geospatial artificial intelligence (geoAI): potential applications for environmental epidemiology. Environ Health 17(1):40

Waller LA, Gotway CA (2004) Applied spatial statistics for public health data (Vol. 368). John Wiley & Sons

Walshe N (2017) Developing trainee teacher practice with geographical information systems (GIS). J Geogr High Educ 41(4):1–21. https://doi.org/10.1080/03098265.2017.1331209

Wan LB, Levitch CF, Perez AM, Brallier JW, Iosifescu DV, Chang LC, Foulkes A, Mathew SJ, Charney DS, Murrough JW (2015 Mar) Ketamine safety and tolerability in clinical trials for treatment-resistant depression. J Clin Psychiatry 76(3):247–252. https://doi.org/10.4088/JCP.13m08852. PMID: 25271445

Ward MH, Nuckols JR, Giglierano J, Bonner MR, Wolter C, Airola M et al (2005) Positional accuracy of two methods of geocoding. Epidemiology 16(4):542–547

Wesolowski A, Stresman G, Eagle N, Stevenson J, Owaga C, Marube E, Bousema T, Drakeley C, Cox J, Buckee CO (2014) Quantifying travel behavior for infectious disease research: a comparison of data from surveys and mobile phones. Sci Rep 4:5678

Wilcox BA, Colwell RR (2005) Emerging and reemerging infectious diseases: biocomplexity as an interdisciplinary paradigm. EcoHealth 2(4):244

Woolhouse ME, Gowtage-Sequeria S (2005) Host range and emerging and reemerging pathogens. Emerg Infect Dis 11(12):1842

World Health Organization (2005) International health regulations. World Health Organization, Geneva. http://whqlibdoc.who.int/publications/2008/9789241580410_eng.pdf

World Health Organization (2012) Global incidence and prevalence of selected curable sexually transmitted infections-2008.

World Health Organization (2017) Global environmental change. Available online (http://www-whoint/globalchange/environment/en/), Accessed on 4 July 2017

Wu X, Tian H, Zhou S, Chen L, Xu B (2014) Impact of global change on transmission of human infectious diseases. Sci China Earth Sci 57(2):189–203

Wylie JL, Cabral T, Jolly AM (2005) Identification of networks of sexually transmitted infection: a molecular, geographic, and social network analysis. J Infect Dis 191(6):899–906

Yager RR (1988) On ordered weighted averaging aggregation operators in multicriteria decisionmaking. IEEE Trans Syst Man Cybern 18(1):183–190

Yager RR (1996) Quantifier guided aggregation using OWA operators. Int J Intell Syst 11(1):49–73

Yager RR, Filev DP (1994) Essentials of fuzzy modeling and control. New York, 388: 22–23

Youngkong S, Baltussen R, Tantivess S, Mohara A, Teerawattananon Y (2012) Multicriteria decision analysis for including health interventions in the universal health coverage benefit package
in Thailand. Value Health 15(6):961–970

Zadeh LA (2008) Is there a need for fuzzy logic? Inf Sci 178(13):2751–2779

Zimmerman DL, Pavlik C (2008) Quantifying the effects of mask metadata disclosure and multiple
releases on the confidentiality of geographically masked health data. Geogr Anal 40(1):52–76

Chapter 8
GeoComputation and Participatory Epidemiology

Abstract Participatory epidemiology is a new field that is focused on the use of participatory methods for gathering of data, veterinary information, and conventional oral history. Epidemiology involves group findings. Epidemiology is being used to bridge the gap between definition and practice through participatory research. Participatory GeoComputation approach is a kind of community-based spatial study in archaeology for spatial planning, spatial information, and communication management. The present chapter illustrates various of types of participatory GeoComputational techniques for emerging infectious disease and epidemiological control strategy. This chapter described about current participatory GeoComputational scenario, like crowdsourcing, crowdcasting, Web 2.0, Geotagging, Geo-Wiki, GeoBlog, GeoRSS feeds, and GeoCitizen, in disease control programme. This chapter also described about the social network, social distancing, social inequality, social relationships, and social capital and integration with the geospatial data for epidemiological application.

Keywords Epidemiology · GIS · Social network · Social distance ·
Crowdsourcing · Crowd casting · Geotagging, Geo-Wiki, GeoBlog, GeoRSS feeds
· GeoCitizen · Disease control programme

8.1 Introduction

Across several ways, epidemiology has helped to recognize specific disease risk factors and to improve public health. Epidemiology has developed models for genetic, population, and environmental data over the past few decades in order to establish information applicable to policymakers and public health practice (Trinidad et al. 2015). However, there are still some aspects of epidemiology in progress. In 1980, in two headings, Stallones widened the view on disease causation: (i) biomedical individualism or individualization of risk and (ii) the connection between risk exposure and outcome in terms of integrating individual-level with social- and environmental-level risk factors. Factors arising from interactions

G. S. Bhunia, P. K. Shit, *GeoComputation and Public Health*,
Springer Geography, https://doi.org/10.1007/978-3-030-71198-6_8

between individuals (e.g. community-based initiatives) and social or political institutions (i.e. public health programmes) are often difficult to define. Subsequently, while many macrosocial influences in the health and societal factors are accounted in the discipline, there are still challenges with illustrating the 'complexity of disease at different interacting rates'. The basis of participatory research are large social movements that aim to create a more democratic and equitable society. There is a general understanding that science is more than conformity to epistemological or methodological guidelines; it is basically a resource for knowledge development to enhance the lives of people. Various scholars have been arguing for further epidemiological studies for the past 20 years. Different terms, including traditional epidemiology (Brown 1993), lay epidemiology (Olsen and Banwell 2013), community-based research (Wing et al. 1996), and participatory health research (Freifeld et al. 2010), are used for participatory types of epidemiology.

Participatory approaches to epidemiology may help identify research targets that are important for researchers as well as for public health practitioners as they frequently concentrate on health antecedents and therefore function as an empirical foundation for health equity through social and policy intervention. For example, government promoted large-scale behaviour change interventions since the COVID-19 pandemic, such as keeping physical distance, handwashing, cough etiquette, and enforcing the lockout guidelines to curb the spread of infection. However, such initiatives may not be successful until the society is concerned to the extent that these strategies are actively adopted (Fig. 8.1). By using a participatory approach, epidemiologists may define the research issues that are important to those impacted specifically by the issue and provide more specific information for public health programme growth (Krieger 2010). Participatory epidemiological work uses technologies that transcend the person level and may thus extend to other models.

Fig. 8.1 Community participation in disease-spreading activities. Community engagement can increase awareness about ways to recognize signs of COVID-19 and the value of taking safety measures such as wearing facial masks, keeping personal distance, and regular washing of hands. Public consultation, in partnership with local communities, could play a vital role in coordinating local action to acknowledge disadvantaged individuals, provide help for the elderly and those in quarantine, establish improved communication mechanisms, and assist in interaction monitoring

Through collaboration among epidemiology and anthropology, for instance, awareness of health and disease and local and tribal attitudes and awareness of globalization and emigration-related factors have been improved, and so detailed standards have been established for defining the population being studied (Weiss 2001). For large-scale surveys to better target other groups, participatory methods can also be implemented somehow. This involved many events of public awareness, 'by way of a variety of public meetings [and] key informant interviews […] with various state and local public health departments, healthcare organisations, research community and advocacy groups' (Brown et al. 2005). In using participatory methods, epidemiologists will take a 'realistic' approach to the generation of data and thus follow a more embedded approach.

Heterogeneous data from multiple sources representing various levels (e.g. individual, communities, networks) are used for participatory epidemiology. A conceptual model was built based on a participatory group appraisal that included physical, social, and individual variables. The participatory mapping of sites for the creation of malaria breeding sites was modified to classify (Dongus et al. 2007) landfill disposal sites (McMahan and Burke 2007) and high sexual risk zones (Power et al. 2007) and to explain trends of healthcare uses (Rutta et al. 2005). In the 1990s, participatory mapping approaches with exposure to spatial data technology, such as geographical data systems and global positioning systems, changed, although there remains a central method of translating local information into 'expert' data for understanding the physical and social organization of locations. The convergence of community growth with the geospatial technologies has become known as participatory/public participation geographic information system (PGIS). GIS has supported the study of health facility accessibility and disease risk in various communities, along with participatory methodologies (Brabyn and Barnett 2004) and health and social care planning.

8.2 Participatory GeoComputation

In order to facilitate the development of awareness by local and non-governmental organizations, a public participation geographic information system (PPGIS) is intended to introduce the scientific practices of GIS and mapping to the local level (Schuurman 2008). The concept was coined by the National Center for Geographic Information and Analysis (NCGIA) in 1996. Participatory GeoComputation (PGC) is a consultative methodology in the area of spatial planning, information management, and communication. PGC incorporates a number of methods and techniques for geospatial information management, including sketch maps, participatory 3D (P3DM), aerial photography, satellite imagery, and global positioning system (GPS) data, which are used as space education interactive vehicles in two- to three-dimensional (virtually or physically) maps, which depict spatial awareness in different ways (Rambaldi et al. 2006). A successful PGC activity is integrated into long-term spatial decision-making strategies, is scalable and tailored to different

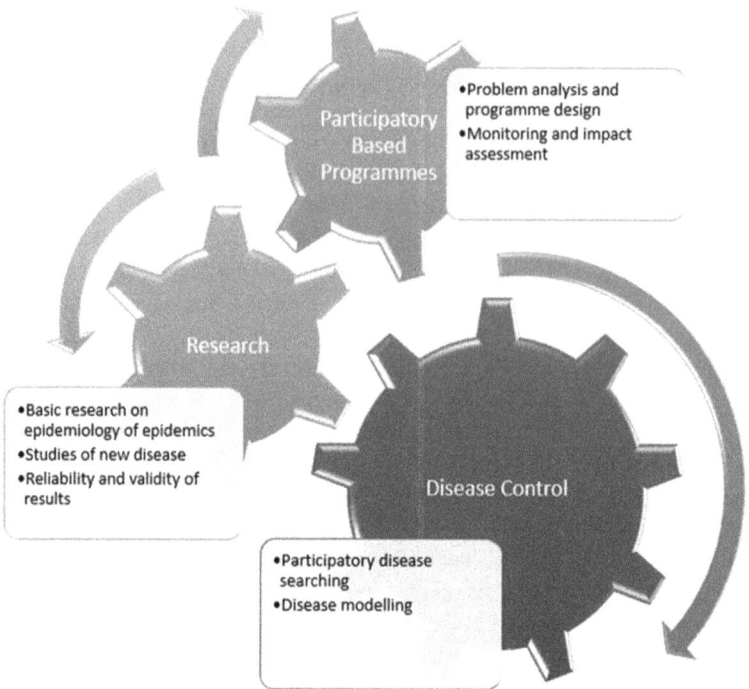

Fig. 8.2 The role of geospatial technology in participatory mapping

social and cultural and bio-physical settings, and relies on multidisciplinary dissemination and competence (Fig. 8.2). The technique includes various tools and methods while relying often on the combination of 'specialist' expertise and socially distinct local knowledge. It encourages collaborative participation of stakeholders in space knowledge generation and management and employs landscape knowledge to encourage specific decision-making frameworks that enable efficient collaboration and community advocacy.

8.2.1   Volunteered Geographic Information

Volunteered Geographic Information (VGI) is the use of geographic data resources freely given to individuals to obtain, compile, and disseminate them (Goodchild 2007). VGI can be seen as an extension of GeoComputation's essential and participatory approaches. VGI is a unique case of the biggest process known as user-generated content and enables people to become more active in epidemic prediction and mapping activities. Wikimapia, OpenStreetMap, and Yandex are examples of this phenomenon. The site contains information about general basic maps and allows users to construct their own contents by marking locations in which specific events or features have happened, but not already depicted on the base map. VGI can be regarded as georeferenced data generated inside services like Tripadvisor, Flickr, Twitter, Instagram, and Panoramio.

8.2.2 Collaborative Mapping

Collaborative mapping is the compilation from a community of individuals or organizations of Web mapping and user-generated content and can take a variety of different forms (Goodchild 2007). Since the expanding of technology to store and distribute maps, collaboration maps in OpenStreetMap or components such as Google Maker and Yandex Map Editor have become competitors in commercial services. It is a special case for a growing crowdsourcing trend, which allows people to engage in a collaborative effort to achieve a goal. Especially when data, information, expertise, and aggregation of data are available in a specific group, collaborative mapping may help people or community activities with an electronic planning platform. Extensions of strategic and participatory methods to information management integrate technological methods with collaborative collaboration to achieve the group goal. In addition, the accumulated data can be used for position service like the geolocation where a mobile device is actually used, for instance, with the available disease location (GPS-sensor) (Fig. 8.2). The importance of fuzzy logic or a fuzzy architectural analysis can be portrayed.

8.2.3 Counter-Mapping

Counter-mapping is the mapping method in which groups use state-structured mapping techniques and create their own maps as an alternative to government-used mapping techniques. The word 'counter-mapping' was coined in 1995 by Nancy Peluso, professor of forest policy, who investigated the implementation in Kalimantan of two forest mapping strategies (Peluso 1995). It refers to the use of maps and methods for mitigating or condemning inequality. The importance of counter-mapping is focused on specific ideologies, advocacy, and activism. Geographical concerns and conflicts regarding landscape ecology and conservation are most often related to counter-mapping. The mechanisms of neoliberalism and technical democracy have enabled these counter-mapping efforts. Examples of counter-mapping include attempts to demarcate and protect traditional lands, community mappings, publicly accessible geographically oriented participatory information systems, and mapping to combat resource claims of a developed area with a relatively developing or underdeveloped zone. The use, scope, and geographical scale of counter-mapping have shifted in the present round of competitions over extractives and indigenous sovereignty; the maps are being used in wider transmedia campaigns of indigenous sovereignty. The counter-mapping as data justice during both periods involved the convergence of redistributive, transformative, and restorative justice in large scale.

8.3   Community Participation for Disease Control

Community means different things in different ways, with different individuals. A community provides a range of possible intervention goals and is also viewed as an integrated, proximal, and holistic system that offers incentives and services that form the lifestyle of people (Macintyre and Ellaway 2000). Participation in the community refers to the participation of those involved in changing certain circumstances that impact society wellbeing. Community-based operation includes members of the society involved in preparing, designing, executing, and reviewing the services and approaches (Cargo and Mercer 2008). A network also provides the opportunity to pool capital and cooperate among network-based organizations, some of which are state and national affiliates that can funnel money to promote local projects and test their developments (Kreuter et al. 2000).

Community-based preventive initiatives focus on public wellbeing and may also tackle improvements in the social and physical environment, incorporate intersectoral action, promote group engagement and mobilization, demonstrate the history, or use a program approach. Changes in environmental social and physical characteristics are important findings for community-based interventions as the representation of risk factors, health conditions, and fitness metrics in a population is primarily influenced by social and physical environments. Evidence has also demonstrated that intersectoral cooperation (involving and integrating actors from a number of related sectors in the preparation, delivery, and management of interventions) is an essential component of public health initiatives (Gibson et al. 2007). Since much of the social and environmental determinants of public health remain outside of the health sector's sphere of control, these intersectoral alliances are important mechanisms by which improvements in the major health determinants can occur.

The Community Organization Model (CRM) is a participatory method of the policy taking that empowers neighbourhoods to improve their health (Fig. 8.3). It stresses the active community engagement in recognizing an important health issues and approaches to resolve them. Communities are building on their resources and mobilizing together to build strategies that meet health goals. The Model Community Organization characteristics include:

- Comprehending the context and underlying causes of health issues
- Shared decision-making and problem-solving
- Focusing on practical issues
- Active participation by numerous individuals and organisations within the neighbourhood

Fig. 8.3 Intersectoral
community models for
public health controls

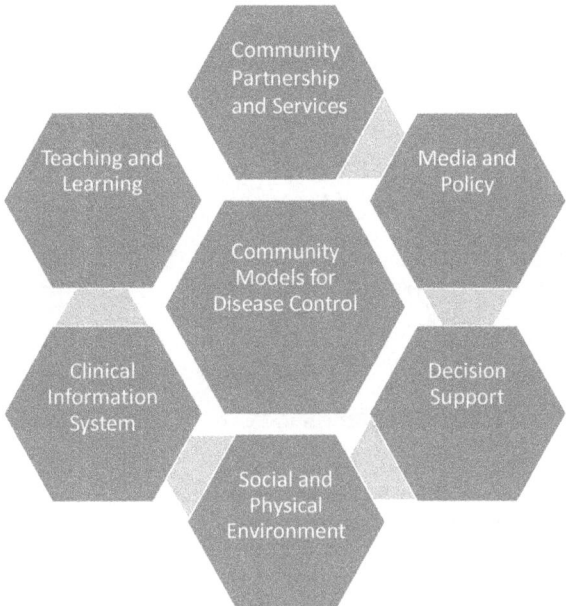

- Developing and retaining the resources and the ability to bring about positive change
- Providing collective reviews

8.4   GeoComputation and Community Engagement

Geography should act as a starting point for developing a plan for civic participation. So GeoComputational analysis should include the resources needed to do this. Maps and spatial analytics create a common interpreter that helps individuals who live and work in communities to establish solid, long-lasting relationships with others simply by exchanging geographic information.

8.4.1   Crowdsourcing

Crowdsourcing is the mechanism by which people gather volunteered knowledge (Hudson-Smith et al. 2008). One aspect of crowdsourcing is surveys. They can be used to build inventories, gather knowledge about the views of people, or take stock of what a group feels on such topics (Fig. 8.4). Combined with geography, surveys may reveal how important a location or the local community is to the problems at issue. Crowdsourced processing of spatial data is made possible by tools such as

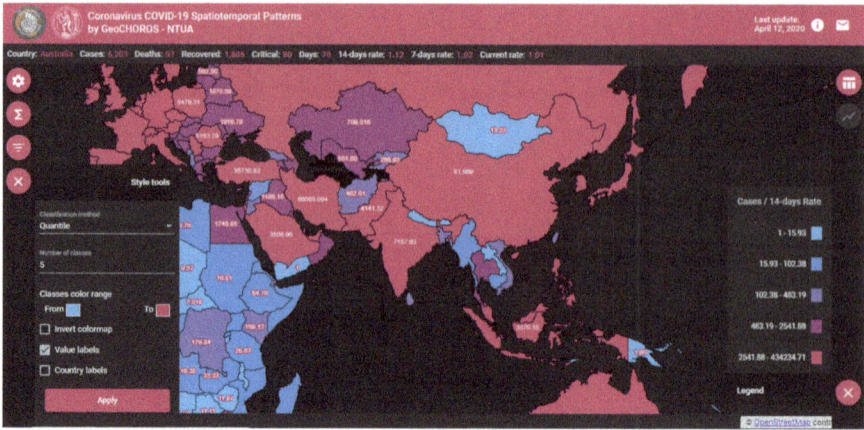

Fig. 8.4 An online platform for the monitoring, analysis, and mapping of the geographical distribution and evolution of the coronavirus pandemic (Web-GIS) has been created by the GeoCHOROS research team. (Source: http://geochoros.survey.ntua.gr/coronavirus/). This offers the forum for presenting cases in terms of population, deaths in terms of cases, and cases in terms of population age or particular categories, such as elderly. Around the same time, it gives you the ability to analyse how this pandemic progresses by region

portable GPS devices, digital notebooks, smart apps (equipped with open-source applications, images, video, voice recorders). Facebook and Flickr are possible crowdsourcing sources. GeoForm makes surveys easy and clear to administer and makes them feel better of urgency which a group can appreciate. All can download the source code at GitHub. This allows developers to customize intuitive apps that work for smartphones, laptops, and web browsers to optimize user interaction.

Enabling people to submit feedback is one aspect; focusing on the real data is another possible option. A crucial method for managing a crowdsourcing platform is to collect, interpret, and then act on the crowdsourced results. It provides data in both map and table formats and contains resources for pattern identification, information analysis, status monitoring, and assigning of responsibilities.

8.4.2 Crowdcasting

Crowdcasting is media group. The push factor makes potential users aware of the authority's urge for them to engage in crowdcasting, often through various websites and associated digital media such as email. The reasons to build processes in this manner obviously rely on the interest the crowd finds in engaging in this manner. Through providing bonuses, mostly in monetary interest, some crowdcasting programmes also provide opportunities for participants to rely on their gathered or own results.

8.4.3   Web 2.0

Web interfaces serve as portals that can be delivered as a central infrastructure or that can be designed in a collaborative manner from the grassroots with a distinct function for architects and consumers. These are only one type of cloud-forming technologies and services. Web users can collaboratively and independently generate content on the Internet, allowing for a customized online experience through wikis, tweeting, photo-sharing, and other technology. There are different frameworks for engaging in the latest mode of geospatial technology and cartography: Geotagging, GeoBlogs, Geo-Wiki, GeoRSS updates, Google MyMaps, and Yahoo Pipes. Three approaches of improved Web 2.0 location recognition are primarily used for content georeferencing.

8.4.4   Geotagging

If a user executes location-based searches, the information (such as photo, video, or website) to a set of spatial coordinates can then be identified. This is good for smartphones and vehicle navigation systems. The key problems are GeoSpam and material quality depending on the user.

8.4.5   Geo-Wiki and GeoBlog

A Geo-Wiki is a wiki based on a map that lets users define features on a map and report on them. For example, https://wikimapia.org uses Google Maps as the context map and facilitates location development and user editing. By inserting links to local landmarks/features/malaria, users annotate charts. Geo-Wiki demonstrates how crowdsourcing is used to verify data and boost datasets. Geo-Wiki Project is a collaborative network of people who wish to help improve the accuracy of maps of collaborative ground cover. As there are significant variations in different maps of the global ground cover, actual environment, and ground use, critical reliable data are missing. Volunteers are asked to study global land cover hotspot maps to decide if the land cover maps are accurate or inaccurate, depending on what they currently see in Google Earth to their local awareness. Datasets of land cover are crowdsourced using a Web Map Service (WMS). Their feedback is registered in a database, along with submitted images, and will be used in the future to construct a modern and updated global hybrid land cover map.

A blog is a diary-like web page allowing a person to upload text, images, photos, and links to other websites. Users will read the blog posts and comment. A GeoBlog helps users to annotate maps by inserting connections to nearby landmarks, markets, institutions, and other places.

8.4.6 *GeoRSS Feeds*

An RSS Feed delivers information including local weather warnings and news stories to a website. RSS Feeds are XML/HTML files with coordinate details that can
be accessed either by connecting to current map sites or through a special view-
finder. A GeoRSS feed provides external interface parameters for location information to help tagging content. (a) GeoRSS Simple, simplistic format restricted to
WGS84 latitude/longitude; (b) GeoRSS GML, richer version containing more functions and coordinate systems; and (c) W3C GeoRSS, partly obsolete version, are
still commonly used.

8.4.7 *GeoCitizen Approach*

The GeoCitizen system helps residents and groups to collaboratively communicate
findings, share proposals, and address and track urban planning challenges, as well
as the delivery of public infrastructure and other topics at local level using a geospatial web platform (Fig. 8.5). This framework – the GeoCitizen platform – fuses
GeoWeb technology and social media into a common, detailed, and collaborative
resource for participatory strategic planning, community participation, and citizen
engagement. Because people are strong listeners of issues happening in their neighbourhoods, this forum helps them to interconnect their experiences and opinions of
their living situation to other residents of the same city or group on a local scale.
This also offers an important platform for communicating with other residents,
organizations, and municipal projects outside of their immediate areas or groups
facing the same type of challenges and sharing common thoughts or suggestions
and potentially exchangeable approaches for solutions.

The GeoCitizen platform offers a connectivity system that for people and their
projects comprises a social network focused on georeferenced experiences about
what is impacting their quality of life. Through addressing these concerns in the
World Wide Web's public room, ineluctable power ties between people, public policy, and other players in (spatial) decision-making processes might become more
accountable, and the flow of information should be improved. The emerging
GeoCitizen framework provides an open-access point for each citizen to define a
problem geographically and link it to a problem-solving mechanism that is unique
to thematic analysis. It helps users to locate friends in their community and to be
aware about the geographical context of the problem and the problems associated
with it. The GeoCitizen platform is built on personal computers and handheld
devices such as smartphones and tablets as a cross-platform for web browsers. This
uses the mobile web browser jQuery (a JavaScript library), the Google Map API,
and a PostGres/GIS database as a platform for web apps.

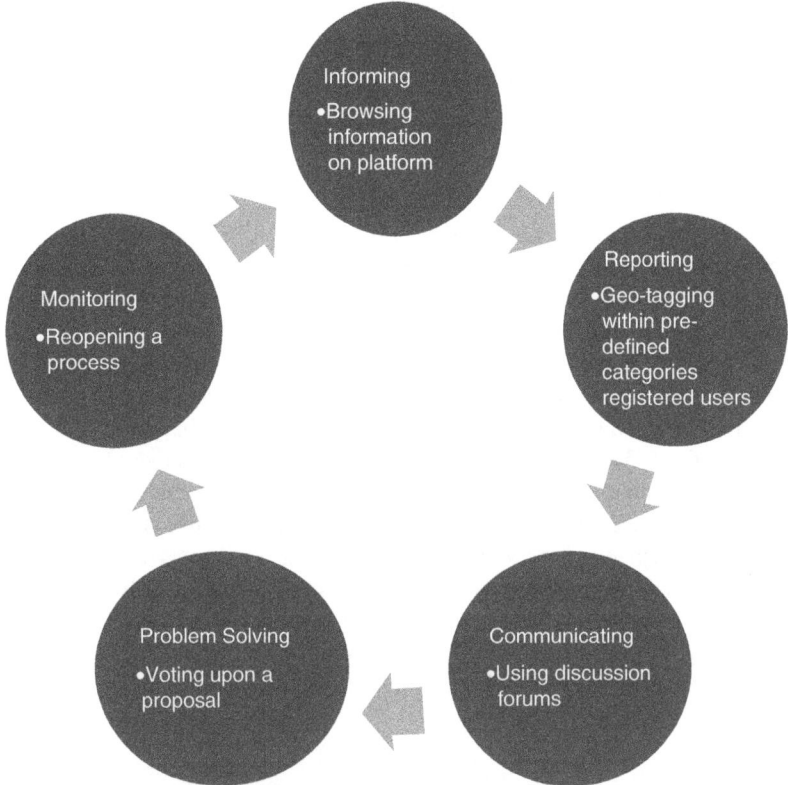

Fig. 8.5 Framework for GeoCitizen process. (Source: Atzmanstorfer et al. 2014)

8.4.8 *Online Mapping Services*

Additionally, some sites require users to create their own maps using the mapping services online. Google MyMaps is an example of mapping system online. Google Maps API helps developers to quickly merge maps with data from GeoRSS streams or other websites. The resulting mashup will be posted on the user's page instead of the Google platform. Living Science, a Google Map mashup found at www.living-science.ethz.ch, helps users to dynamically scan a research paper index stored on the arxiv.org free archive. This helps users to determine how many papers in similar categories or in total were published at various locations (countries, cities) over each time span, and subcategories are displayed in cartographic ways.

Most of the above listed map mashups use Google Maps software, a free online mapping tool offered by Google (https://maps.google.com/), to embed a Google Map into a web page, as well as direct requests from a locally labelled database. Many free map providers include Yahoo Maps (https://maps.yahoo.com/Microsoft Virtual Earth (https://maps.live.com/), MapQuest (https://www.mapquest.com/), and OpenLayers (https://openlayers.org/), an alternate open source. Both of these map providers provide their own JavaScript application programming interface (APIs) to provide different services. Many of them have 'no coding' GUI that lets users use XML files to make charts.

8.4.9 Mobile-Based Platform

This development motivates GIS for a range of geospatial information systems to be implemented on new common hardware/software framework such as 'mobile phones with optimized advanced navigation tools.' The prospect of introducing 'location features' to mobile web browsers facilitates both active and passive crowdsourcing (e.g. geolocation for Twitter). Crowdsourcing of data via cell phones can produce real-time data and Geotagging of certain mobile applications such as short message service (SMS). Efforts are being made to build open standard for geoSMS, allowing for Geotagging of SMS messages. They have immense promise in emergency and crisis recovery systems. There will be plenty of live data in the future, referred to as human sensor streams, waiting to be incorporated into geospatial applications. Real-/near-time data produced by crowdsourcing/crowdcasting, tagging data, deriving trends, and connections in interlinked network environments with a strong repository of cloud-based resources would be the most possible resource for decision-makers and policymakers (Fig. 8.6).

8.4.10 Crowdsourcing and e-Governance

Crowdsourcing is a paradigm change in generating and editing the results. While the crowdsourcing implementation is now ongoing, it will take a bit more time to enter a mature phase with the requisite format, metadata, and authorization process (Fig. 8.7). A geospatial website would offer user-friendly, interactive data sets and map-based resources from federal, state, national, and local agencies as well as nongovernmental partners as part of the Obama Administration's Open Government Initiative. The website is connected to Data.gov, the national policy data depository and information repository; a large amount of the data currently accessible on the web is derived from current geographic data catalogues. The quality of all Data.gov databases and resources has been checked by the department to be compliant with the federal privacy, national security, and quality of information policies. The data will be remaining with the organizations that generated it, where it is most reliable, and transmitted across the web. Another example is the US Department of Defense developing a catastrophe-aware forum through the 'Pacific Disaster Center' programme. It tracks current meteorological and geological information flows and provides subscribers with real-time information and warnings. Users can post analyses

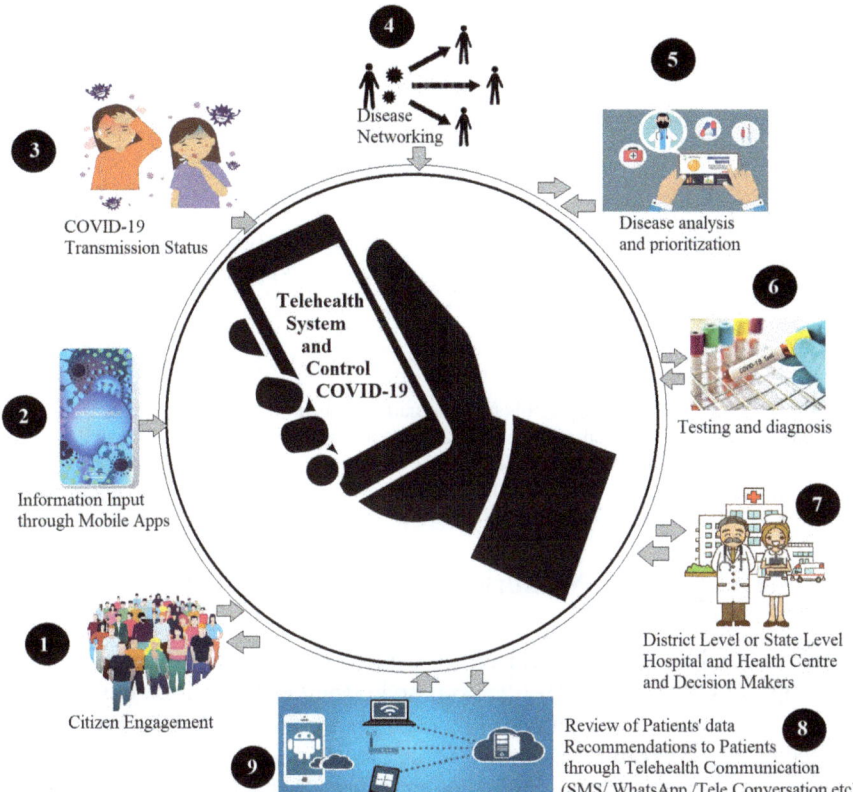

Fig. 8.6 Telehealth system and control of COVID-19. This framework enabled experts in the distance to communicate with patients by using just a computer equipped with a webcam, a microphone, and speakers through the asymmetric digital subscriber line between the expert site and local site of the patient. The method is simple. During the patient call, the telemedical assistant, after identifying the patient, asks about the patient's symptoms and medical problems and then registers it for physicians to evaluate and take further actions. Remote treatment eliminates the usage of services in health facilities, enhances access to care, and eliminates the chance of infectious disease spread from person to person. Aside from keeping people healthy, such as the general population, patients, and health professionals, another significant benefit is having widespread access to caregivers.

and situation updates and can access DisasterAWARE's underlying databases, which are available on Twitter. The device is ported to iPhones and iPads. Backend modules are based on ArcGIS platform and software for Java. Free and real-time geographic data can be gathered in emergency response scenarios that include crowding together with data integration to incorporate information from a wide range of sources.

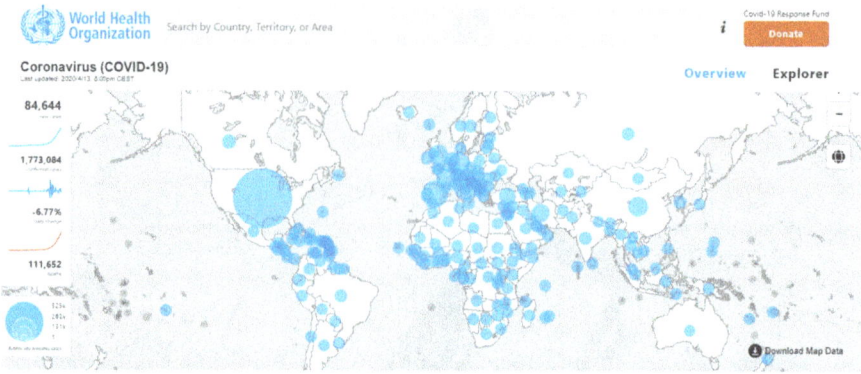

Fig. 8.7 The WHO directs and coordinates international health, combating communicable diseases through surveillance, preparedness, and response and applying GIS technology to this work. On April 13, 2020, the WHO unveiled its ArcGIS Operations Dashboard for COVID-19, which also maps and lists coronavirus cases and total number of deaths by country, with informational panels about the map and its data resources. (Source: https://who.sprinklr.com/)

8.5   Social Support Networks and Disease Control

Healthcare analysis has been criticized for frequently using a theoretical strategy which appear to consist in a specific study style input/output or black box. In addition, it was reported that a better theoretical basis would allow work on healthcare to become more insightful and effective (Christakis and Fowler 2009). Indeed, social networks impact health through a number of channels, including (1) social networking, (2) social power (e.g. expectations, social control), (3) social participation, (4) individual-to-person interactions (e.g. pathogen exposure), and (5) resource access (e.g. income, employment, information).

8.5.1   Social Network Elements and Attributes

Social networks are composed of two features: individuals (nodes) and their social relations (Fig. 8.8). If all the nodes and links are identified, one can draw representations of the network and infer the position of each person within it, putting each entity in comparable spatial space in social space. We can think about the 'distance' between two people within a network (also defined as the 'geodesic distance' or 'degree of separation'), which is the fastest route in the network from one person to another. Obviously, social network relations are not limited to acquaintances, so one may be linked to the friend of one's spouse's wife, or the sister of one's co-worker's husband, and so on. It is useful to appeal to 'egos' when addressing network effects or the persons under research, and their 'alters', or the people they are linked to.

Fig. 8.8 Elements of
social network

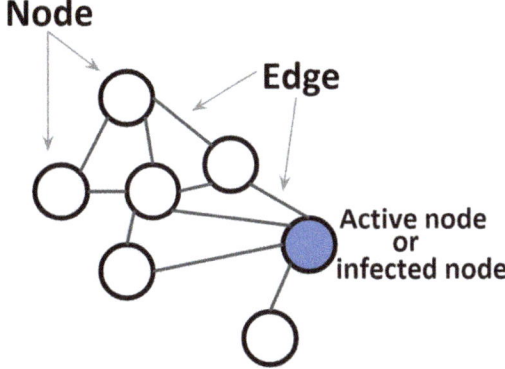

Social connections can be represented as edges (undirected node-to-node alliances) or arcs (targeted node-to-node relations). Examples of inchoate partnerships involve married partners and mothers. A direct friendship is one in which A considers B as a friend, but B does not show affection. Edges and arcs are mostly calculated on a linear (presence/absence) scale but can also be assessed (e.g. how well two individuals recognize each other or how well they get along very well). Real social networks also include arrays of sub-networks or 'components'. A node is a part of a network of which everybody in the same portion is linked to each other by at least one direction. Intuitively this implies that anyone in the first component should be associated with anybody in the second element with two separate components. When the biggest component comprises most nodes in a network, it is referred to as a 'giant component'. Graph analysis also contributes to the discovery of 'communities' or 'cliques' of subgroups of nodes that are closer to each other than they are in the rest of the network.

Network research in its basic form relies on interactions between homogeneous nodes, but more significantly important network data sets include details about node features, such as gender, employment, or health habits. Other features which differentiate nodes from each other may be dependent on the network itself being examined. For examples, the extent of a node in a network of unconstrained links is the number of other nodes to which it is directly linked. For a focused map composed of arcs, the number of arcs pointing to a node is *in-degree* of that node; it's *out-degree* is the number of arcs extending from a node. Higher-degree nodes are usually viewed as being more powerful and dominant within the network.

Centrality measurements in networks represent the degree to which a node interacts or sits between other nodes and thus its propensity to place itself near to the middle of its local network (Fig. 8.9). The shortest measure of centrality is the number of friends (known as 'degree' centrality) described above. Those with more mates seem to be more geographically positioned. But this calculation doesn't take into account variations in one's friends' centrality. Individuals connecting to multiple well-connected peers are more clustered than those linked to an equal number of inadequately connected peers. In other words, those who have friends with famous

Fig. 8.9 Illustration of a bidirectional transitive effect for two directed links

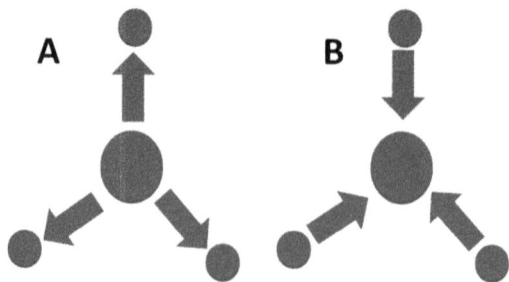

people would be more important than those who have friends with controversial persons. The centrality of the Eigenvector of a particular subject is a growing property of the sum of all the centralities of all the objects to which it is associated. The centrality values of the own vector are essentially relative: an entity associated with each other in the network would have the highest possible value, and a person not related to someone else would have a value of '0'. Certain centrality measurements capture the degree by which a specific node lays between other nodes, i.e. the degree to which that node lays along geodesic paths connecting other nodes and must thus be crossed if anything (e.g. knowledge, germs, money) is to pass between other nodes in the network.

8.5.2 Social Networks and Epidemiological Principle

Environmental epidemiology aims to define and explain the dynamics of the environmental factors that influence the pattern of illness and health distribution in a population. Several important social epidemiological principles are social distancing, socioeconomic inequality, social interactions, social capital, and career anxiety (Honjo 2004).

8.5.2.1 Social Distancing

Social distancing, or physical distancing, is a series of non-pharmaceutical procedures or steps taken to discourage an infectious illness from spreading by keeping a physical gap between individuals and minimizing the number of times people come into direct contact with each other (Valdez et al. 2012). The Centers for Disease Control and Prevention (CDC) has described social distancing as a series of 'methods to reduce the frequency and closeness of people-to-people interaction to reduce the risk of disease transmission'. This means standing 6 feet or 2 meters away from others, and not meeting in large groups. The World Health Organization (WHO) suggested the reference to 'physical' as an alternative to 'social' during the 2019–2020 coronavirus pandemic, in line with the notion that it is a physical

Fig. 8.10 Social distancing reduces the rate of disease transmission and can stop an outbreak

distance which prevents transmission; people can remain socially connected through technology.

The methods of social distancing date back to at least the fifth century BCE. Leviticus' ancient text provides one of the first recorded attempts to produce presumably as a solution to leprosy. A number of recent epidemics have effectively introduced social distancing strategies in contemporary times. Soon after the first influenza outbreaks were detected in the region during the flu pandemic of 1918, officials in St. Louis instituted school closures, bans on public meetings, and other social distancing measures (Fig. 8.10). Social distancing steps are most successful as the infectious disease spreads through droplet contact (coughing or sneezing); direct physical interaction, including sexual intercourse; indirect physical contact (e.g. contacting a polluted surface); or airborne dissemination (if the microorganism may persist in the air for long periods). The interventions are less successful in spreading an infection mainly through infected water or food or by predators such as mosquitoes or other insects (Maharaj and Kleczkowski 2012).

8.5.2.2   Social Inequality

Indicators of social disparity such as employment or income have an effect on access to healthcare and the use and quality of healthcare. The unequal distribution of goods, services, and opportunity in a community or society can be described as social inequality. Social epidemiologists also use indices of 'life resources' including schooling, jobs, and income to assess social disparity based on Max Weber's definition of social class (1958). The presumption is that these are the most significant factors for health relevant to aspects of distribution – the skills, knowledge, and resources of people who make up the main links between social stratification and the wellbeing of individuals. Research in social epidemiology, therefore, concentrate on the relationship of these socio-economic (education, occupation, and income) and health indicators. Instead, if health differences are due primarily to beyond the influence of the environment and circumstances, such disparities are perceived to be avoidable and unequal and contribute to health disparities. The lack of opportunities for good health due to the lack of social structures is important for health inequity, according to the capacity approach of Amartya Sen (2002).

8.5.2.3 Social Relationships

Social interactions impact compliance with medical care, help behaviour, the utilization of health facilities and outcomes. Social relationships contain three key components (Holt et al. 2010) which are defined and measured in various ways: (a) the assimilation grade in social media, (b) those which are to be supportive of interactions between people (i.e. received social support), and (c) beliefs and attitudes of the individual's accessibility of assistance (i.e. perceived social support). In addition, social support is usually divided into subtypes including emotional (e.g. awareness, appreciation) and functional support (e.g. practical assistance, financial assistance). Social cohesion and social engagement have been shown to improve wellbeing.

8.5.2.4 Social Capital

Social capital is a significant element for delivering highly organized treatment in healthcare organizations. Social capital can generally be described as facets which appear as resources for entities and encourage government action in social structures, including level of interpersonal trust and pre-emption norms, civic engagement, and mutual support (Kawachi and Berkman 2000). The macro level (i.e. historical and social, political, and cultural characteristics of societies), the meso level (i.e. organization and community characteristics), individual comportments (i.e. social participation) and individual norms (e.g. confidence and reciprocity), work in different areas of social capital according to Macinko and Starfield's reports (2001). The overall-level (i.e. ecological) and person-level, as well as multilevel, studies have repeatedly found positive ties between social capital and health (Gilbert et al. 2013).

8.5.2.5 Work Stress

Employment and engagement are of vital health value. The concept of demand control and the concept of effort-reward imbalance (ERI) are two common scientific methods aimed at recognizing demanding working environments that are likely to negatively impact the health of workers. The demand-control model postulates that task pressure stems from a mixture of high (quantitative) job demands and weak job power subdivided into discretionary expertise and decision-making authority (Karasek and Theorell 1990). The ERI model reflects on the historical lack of social reciprocity (Backé et al. 2012). Bad feelings and negative tension arise from a disparity between high expectations and poor incentives in terms of confidence, pay, work advancement, or workplace stability. Both models presume job stress contributions to reduced health (Nieuwenhuijsen et al. 2010).

8.6 Social Movement and GeoComputational Analysis

A social flow is a link that is generated by a decision by an individual to travel, interact, or state interaction between two locations. Social flow connects the systems of social network (SN) and GeoComputation: it is served as an edge between nodes in the traditional SN graph and can be positioned in geographical space by linking geolocated nodes as part of a spatial flow network. There are three major forms of social movements: (i) Transport and human movement statistics are derived from migrant traffic reports (Phithakkitnukoon et al. 2011), passenger flows (Limtanakool et al. 2009), and traffic volumes across different modes including traces of GPS (Liu et al. 2010). (ii) Evidence on telecommunications or information and communications technology (ICT) often reveals social interactions such as e-mail (Tyler et al. 2005), SMS, blogs, or instant messaging (IM) systems (Leskovec and Horvitz 2008); geotagged photo accesses (Crandall et al. 2010) authentication server in pages (Long et al. 2012), postal mail (Milgram 1967), and landline point-to-point or cell phone measurement. (iii) The reported associations are based on community databases, polls, Internet forums, self-reports, interviews, and collected evidence from populations, households, and organizations such as companies (Tyler et al. 2005), schools (Moody 2001), clubs (Zachary 1977), networks of political or public figures (Andris et al. 2015), or Internet connections (Crandall et al. 2010).

8.6.1 Geosocial Networking

Geosocial networking is a social networking form that utilizes Geotagging, geofencing, and geocoding resources to build additional social dynamics. The local data or geolocation methods provided by the user will allow social networks to associate and organize users with local individuals or activities that meet their interests. Web-based social media platform can be IP-based or trilaterated with hotspot technology (Fig. 8.11). Text location information or cell telephone monitoring may improve social networking with cell social networks.

We are collectively generated in a geographic and social network (SN) context, i.e. a system of connections that are formed and sustained by individuals (Fischer 1982). Our families, acquaintances, and professional connections are part of our SN. Throughout our lifespan, we allow SN and the geological ecosystem to evolve and build intertwined inextricable networks. However, GeoComputation is hardly a part of the debate if the human social activity is modelled. Geo-computational theorists, on the opposite, define the same dynamics of obesity by predicting spatial

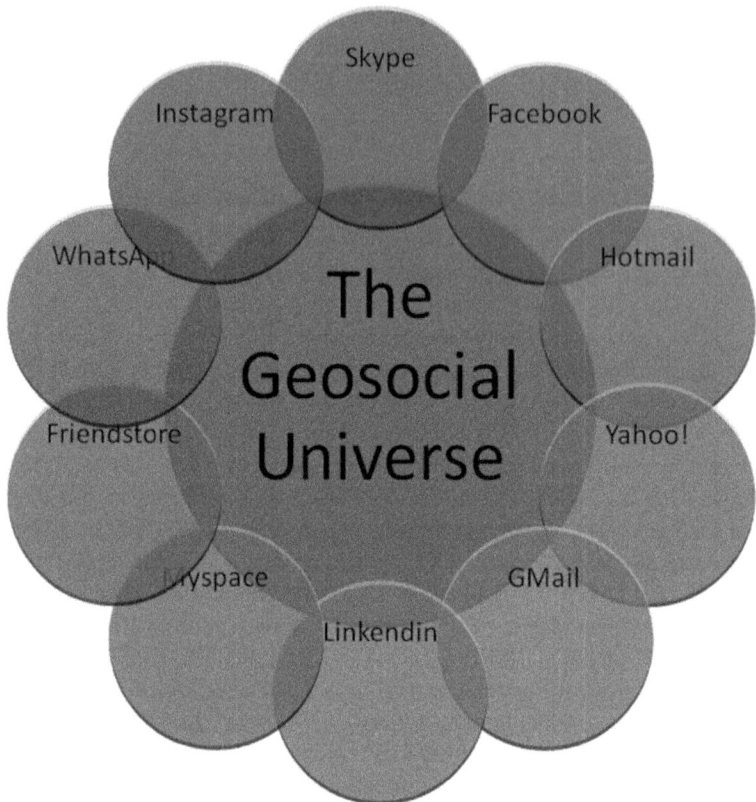

Fig. 8.11 An infographic illustrating and comparing the popularity of different geosocial networking services

disease exposure. Such kinds of analyses may be enhanced if merged in order to describe any social/spatial component. SNs will be configured within the GeoComputation, in order to understand this concept.

However, our GIS tools are poorly suited to model SN variables. We use them for people modelling. Although demographic data can effectively be modelled within geolocated census divisions by the Geo-Computational Science community, the framework has only begun to take benefit of incorporating social interpersonal link data, i.e. geolocated SNs or social network location datasets (Zheng 2012).

8.6.1.1 Relationship Between Social Network and GeoComputation

Social and interpersonal interactions with SNs and the study of social networks (SN) are depicted, albeit crudely, whereas spatial space is based on a GeoComputational framework (Fig. 8.12). A node is usually an individual or, less commonly, a community of people in standard NS. A boundary is a connection

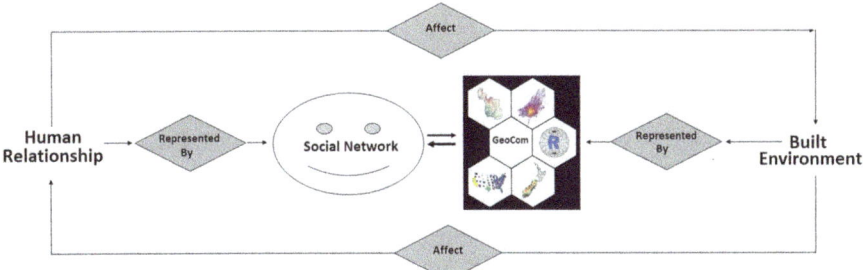

Fig. 8.12 Study of human relationship between SN and GeoComputation. (Modified after Andris 2016)

between nodes that indicate a relationship between individuals or groups. Nodes, such as positions for illness in an area or a person at the healthcare facility, must be meaningfully geolocated with the application of a GeoComputational study.

We have problems both with SNs and GeoComputation in one setting as they develop respectively in different fields of sociology and geography (Andris 2016). A variety of gaps pose problems in their communication: (a) SNs are represented as nodes and edges of a graphic structure, and GeoComputational device models are planar layers of structure; (b) SN models have edges between two entities, while geographical space has curvy and tortuous paths between entities; (c) geographic space has limitations, such as structural limits and geography (e.g. a train station may only have too many tracks added to it), while geography does not limit SNs; (d) SNs are represented in a function space, while GeoComputational models use a consistent Cartesian space with a structured spatial data setup; (e) the elements of an SN can be atomically reassembled at any size to edges and nodes, whereas the spatial space has various atomic amounts based on the geographical scale used; and (f) geographical infrastructure is technological, measurable, and functional, while SN infrastructure is mainly hypothetical.

8.7 Conclusion

Epidemiological models are powerful instruments for assessing strategies and envisioning disease propagation patterns, specifically in comparisons with intervention strategy. Participatory modelling (PM) attempts to redesign the power relationship between models and parties involved in the creation and implementation of the model using model experience. Essentially, modelling is a development-driven top-down exercise with minimal possibilities for participants to direct modelling, structure simulations, or use counterfactual analysis. In effect, health professionals have long considered traditional mapping as essential tools for tracking and combating contagion and recently GeoComputation. 'Collaborative' approaches to planning, such as participatory space planning (PSP), gain credence among policymakers and programmers and with community groups and civil society. In general, PGC tools allow more direct interaction with and within populations as 'users' of geographical information, partly by improving these multiple realities significantly. However, with health alert, all participatory mapping and PGC instruments can only mask structural inequalities, unless the pilot behind participatory space planning activities is to encourage engagement and good governance. Their use by GIS practitioners and communities deserves approval. Any technology is introduced, the type of social environment in which the technology operates will decide whether or not the technology can empower or early GIS architecture does not anticipate these inclusive GeoComputational tool styles, and only 'official' spatial databases were supposed to be spatially accurate. It was hard to predict the integration of different kinds of information (in contrast to data). It's important to improve device functionality and better produce, manage, and represent local spatial information, expectations, and preferences for hardware and software developers. The full complexity of local spatial information, human interaction with climate, or power relationship is never feasible for GeoComputational tools. Large technical advancements and fundamental conceptual improvements must be made before external experts can fully find spatial information and local geographic awareness to be complementary or incorporated.

References

Andris C (2016) Integrating social network data into GISystems. Int J Geogr Inf Sci 30:10, 2009–2031. https://doi.org/10.1080/13658816.2016.1153103

Andris C, Lee D, Hamilton MJ, Martino M, Gunning CE, Selden JA (2015) The rise of partisanship and super-cooperators in the US house of representatives. PLoS One 10(4):e0123507

Atzmanstorfer K, Resl R, Eitzinger A, Izurieta X (2014) The GeoCitizen-approach: community-based spatial planning–an Ecuadorian case study. Cartogr Geogr Inf Sci 41(3):248–259

Backé EM, Seidler A, Latza U, Rossnagel K, Schumann B (2012) The role of psychosocial stress at work for the development of cardiovascular diseases: a systematic review. Int Arch Occup Environ Health 85:67–79. https://doi.org/10.1007/s00420-011-0643-6

Brabyn L, Barnett AR (2004) Population need and geographical access to general practitioners in rural New Zealand

Brown P (1993) When the public knows better: popular epidemiology challenges the system. Environ Sci Policy Sustain Dev 35(8):16–41

Brown ER, Holtby S, Zahnd E, Abbott GB (2005) Community-based participatory research in the California health interview survey. Prev Chronic Dis 2(4)

Cargo M, Mercer SL (2008) The value and challenges of participatory research: strengthening its practice. Annu Rev Public Health 29:325–350

Christakis NA, Fowler JH (2009) Social network visualization in epidemiology. Norsk epidemiologi= Norwegian J Epidemiol 19(1):5

Crandall DJ, Backstrom L, Cosley D, Suri S, Huttenlocher D, Kleinberg J (2010) Inferring social ties from geographic coincidences. Proc Natl Acad Sci 107(52):22436–22441

Dongus S, Nyika D, Kannady K, Mtasiwa D, Mshinda H, Fillinger U et al (2007) Participatory mapping of target areas to enable operational larval source management to suppress malaria vector mosquitoes in Dar es Salaam, Tanzania. Int J Health Geogr 6(1):37

Fischer CS (1982) To dwell among friends: personal networks in town and city. University of Chicago Press, Chicago

Freifeld CC, Chunara R, Mekaru SR, Chan EH, Kass-Hout T, Iacucci AA, Brownstein JS (2010) Participatory epidemiology: use of mobile phones for community-based health reporting. PLoS Med 7(12):e1000376

Gibson L, Doherty J, Loewnson R, Francis V (2007) WHO health systems knowledge network. Challenging inequity through health systems: final report of the health systems knowledge network. WHO Commission on the Social Determinants of Health, Geneva

Gilbert KL, Quinn SC, Goodman RM, Butler J, Wallace J (2013) A meta-analysis of social capital and health: a case for needed research. J Health Psychol 18(11):1385–1399. https://doi.org/10.1177/1359105311435983

Goodchild MF (2007) Citizens as sensors: the world of volunteered geography. GeoJournal 69(4):211–221

Holt-Lunstad J, Smith TB, Layton JB (2010) Social relationships and mortality risk: a meta-analytic review. PLoS Med 7(7):e1000316. https://doi.org/10.1371/journal.pmed.1000316

Honjo K (2004) Social epidemiology: definition, history, and research examples. Environ Health Prev Med 9(5):193–199

Hudson-Smith A, Crooks MBM, Milton R (2008) Mapping for the masses: accessing web 2.0 through crowdsourcing. Ucl working papers series- (Paper 143) ISSN 1467-1298

Karasek R, Theorell T (1990) Healthy work. Stress, productivity, and the reconstruction of working life. Basic Books, New York

Kawachi I, Berkman LF (2000) Social cohesion, social capital, and health. In: Berkman LF, Kawachi I (eds) Social epidemiology. Oxford University Press, New York, pp 174–190

Kreuter MW, Lezin NA, Young LA (2000) Evaluating community-based collaborative mechanisms: implications for practitioners. Health Promot Pract 1(1):49–63

Krieger N (2010) Social inequalities in health. In: Olsen J, Saracci R, Trichopoulos D (eds) Teaching epidemiology. Oxford University Press, Oxford

Leskovec J, Horvitz E (2008, April) Planetary-scale views on a large instant-messaging network. In: Proceedings of the 17th international conference on World Wide Web, pp 915–924

Limtanakool N, Schwanen T, Dijst M (2009) Developments in the Dutch urban system on the basis of flows. Reg Stud 43(2):179–196

Liu L, Andris C, Ratti C (2010) Uncovering cabdrivers' behavior patterns from their digital traces. Comput Environ Urban Syst 34(6):541–548

Long X, Jin L, Joshi J (2012, September) Exploring trajectory-driven local geographic topics in foursquare. In: Proceedings of the 2012 ACM conference on ubiquitous computing, pp 927–934

Macinko J, Starfield B (2001) The utility of social capital in research on health determinants. Milbank Q 79(3):387–427, IV. https://doi.org/10.1111/1468-0009.00213. PMID: 11565162; PMCID: PMC2751199

Macintyre S, Ellaway A (2000) Ecological approaches: rediscovering the role of the physical and social environment. Soc Epidemiol 9(5):332–348

Maharaj S, Kleczkowski A (2012) Controlling epidemic spread by social distancing: do it well or not at all. BMC Public Health 12(1):679

McMahan B, Burke B (2007) Participatory mapping for community health assessment on the US-Mexico border. Pract Anthropol 29(4):34–38

Milgram S (1967) The small world problem. Psychol Today 2(1):60–67

Moody J (2001) Race, school integration, and friendship segregation in America. Am J Sociol 107(3):679–716

Nieuwenhuijsen K, Bruinvels D, Frings-Dresen M (2010) Psychosocial work environment and stress-related disorders, a systematic review. Occup Med 60(4):277–286

Olsen A, Banwell C (2013) Context and environment: the value of considering lay epidemiology. In: When culture impacts health. Academic, pp 85–93

Peluso NL (1995) Whose woods are these? Counter-mapping forest territories in Kalimantan, Indonesia. Antipode 27(4):383–406

Phithakkitnukoon S, Calabrese F, Smoreda Z, Ratti C (2011, October) Out of sight out of mind--How our mobile social network changes during migration. In: 2011 IEEE third international conference on privacy, security, risk and trust and 2011 IEEE third international conference on social computing. IEEE, pp 515–520

Power R, Langhaug L, Cowan F (2007) "But there are no snakes in the wood": risk mapping as an outcome measure in evaluating complex interventions. Sex Transm Infect 83(3):232–236

Rambaldi G, Kyem PAK, McCall M, Weiner D (2006) Participatory spatial information management and communication in developing countries. Electron J Inf Syst Dev Countries 25(1):1–9

Rutta E, Williams H, Mwansasu A, Mung'ong'o F, Burke H, Gongo R, Veneranda R, Qassim M (2005) Refugee perceptions of the quality of healthcare: findings from a participatory assessment in Ngara, Tanzania. Disasters 29(4):291–309

Schuurman N (2008) GIS: a short introduction. USA, UK, Australia. Blackwell Publishing, p 11. ISBN 978-0-631-23533-0

Sen A (2002) Why health equity? Health Econ 11:659–666

Trinidad SB, Ludman EJ, Hopkins S, James RD, Hoeft TJ, Kinegak A, Lupie H, Kinegak R, Boyer BB, Burke W (2015) Community dissemination and genetic research: moving beyond results reporting. Am J Med Genet A 167(7):1542–1550

Tyler JR, Wilkinson DM, Huberman BA (2005) E-mail as spectroscopy: automated discovery of community structure within organizations. Inf Soc 21(2):143–153

Valdez LD, Macri PA, Braunstein LA (2012) Intermittent social distancing strategy for epidemic control. Phys Rev E 85(3):036108

Weber M (1958) From Max Weber: essays in sociology (edited and translated by Gerth H and Mills CW)

Weiss MG (2001) Cultural epidemiology: an introduction and overview. Anthropol Med. 8:5–29

Wing S, Grant G, Green M, Stewart C (1996) Community based collaboration for environmental justice: South-East Halifax environmental reawakening. Environ Urban 8(2):129–140

Zachary WW (1977) An information flow model for conflict and fission in small groups. J Anthropol Res 33(4):452–473

Zheng Y (2012, April) Tutorial on location-based social networks. In: Proceedings of the 21st international conference on World wide web, WWW (vol 12, no 5)